Jeff Sachs's admirable book explains what sustainable development means, why it matters, and our history in relation to our environment. It shows how the pattern of growth we have followed has placed intense strains on our planet and generated severe risks to our lives and livelihoods. Above all, it explains clearly where we can go and how to get there. It describes a potential future of great attraction in terms of wellbeing, community, and prosperity, properly understood. It should, and I hope it will, strengthen both understanding of the great choices we face and the resolve to act. It is a splendid achievement.

▸ *Lord Nicholas Stern*, chair, Grantham Research Institute on Climate Change and Environment

There are very few authors that can so effectively rally the masses around an important cause—one that in this case concerns our collective future. There are still too few leaders with knowledge about global sustainable development challenges to affect the scalable changes that are needed.

Professor Sachs's book is a comprehensive summary of complex issues, but it breaks them down and makes them digestible and understandable while making the imperative clear. It will clearly contribute significantly in closing the gap between science and policy and hopefully serve as a catalyst for action.

▸ *Hans Vestberg*, CEO, Ericsson

As the concept of sustainable development comes of age, Jeffrey Sachs takes us on a journey from the intellectual and normative origins to the science of complexity that defines today's development choices. A compelling and often personal and engaged narrative, *The Age of Sustainable Development* reads like a "looking glass" through which to glimpse the future—a future still characterized by more questions than answers as to how 9 billion will—together—govern the sustainable development of our planet.

▸ *Under-Secretary-General Achim Steiner of the United Nations*, executive director, United Nations Environment Programme

In this era marked by geopolitical and ecological crises, unspeakable structural violence, and unprecedented inequality, on the one hand, and breathtaking advances in science and technology and untold opportunities, on the other, a new narrative for social, economic, and environmental justice is urgently needed. Prof. Sachs, distinguished thinker, commentator, and activist, charts a clear and compelling path, both moral and scientific, which cuts neatly through the increasing complexity and interdependence to the realization of a holistic vision of sustainable development. A very timely handbook to inform and empower all those who have a stake in the post-2015 debate on the future of our planet.

> ▸ *Mr. Michel Sidibé*, executive director, UNAIDS

It is truly gratifying that an economist as distinguished as Jeffrey Sachs has authored *The Age of Sustainable Development*. His vast knowledge in the field and deep insights on various sectors of economic activity make him uniquely qualified to provide a view of the subject. Given the fact that almost a quarter-century has gone by since the Rio Summit of 1992, it is time for an authoritative volume such as the one Professor Sachs has authored for the benefit of global society.

> ▸ *Dr. Rajendra K. Pachauri*, director-general, TERI; chairman, IPCC

Comprehensive and crystal clear—what we need to make sustainable development a reality, not a dream.

> ▸ *Dr. James E. Hansen*, director, Earth Institute Program in Climate Science, Awareness, and Solutions

THE AGE OF
SUSTAINABLE
DEVELOPMENT

COLUMBIA UNIVERSITY PRESS   NEW YORK

THE AGE OF SUSTAINABLE DEVELOPMENT

JEFFREY D. SACHS

Columbia University Press

Publishers Since 1893

New York, Chichester, West Sussex

cup.columbia.edu

Copyright © 2015 Jeffrey D. Sachs

All rights reserved

Library of Congress Cataloging-in-Publication Data

Sachs, Jeffrey.

The age of sustainable development / Jeffrey D. Sachs.

pages cm

Includes bibliographical references and index.

ISBN 978-0-231-17314-8 (cloth : alk. paper) —

ISBN 978-0-231-17315-5 (pbk. : alk. paper) —

ISBN 978-0-231-53900-5 (e-book)

1. Sustainable development.

2. Economic development—Environmental aspects.

I. Title.

HC79.E5S2155 2015

338.9' 27—dc23

2014034070

Columbia University Press books are printed
on permanent and durable acid-free paper.
This book is printed on paper with recycled content.
Printed in the United States of America

c 10 9 8 7 6 5 4 3 2 1

p 10 9 8 7 6 5 4 3 2

Cover image: Félix Pharand-Deschênes/Globaïa
Cover and book design: Lisa Hamm

References to websites (URLs) were accurate at the time of writing.
Neither the author nor Columbia University Press is responsible
for URLs that may have expired or changed since
the manuscript was prepared.

For Sienna Sachs Beck, Willa Tatum Sachs Beck, and their generation:

Let us build a world of sustainable development

CONTENTS

FOREWORD

Secretary-General Ban Ki-moon of the United Nations

Sustainable development is the central challenge of our times.

Our world is under strain. Poverty continues to plague communities and families. Climate change threatens livelihoods. Conflicts are raging. Inequalities are deepening. These crises will only worsen unless we change course.

That is why world leaders are hard at work on a new development agenda—including a set of concrete sustainable development goals—to help guide humanity to safety and prosperity.

It is critical that we understand how sustainable development can be achieved in practice, on the ground, in all parts of the world. This book by my special advisor Professor Jeffrey D. Sachs provides a deeply informed panorama, outlining opportunities, challenges, and risks. I hope it will serve as a useful resource as countries, cities, businesses, families, and individuals take on the shared responsibilities of this new agenda.

Real solutions exist in the global fight against poverty, hunger, and disease. As we now expand on the lessons of the Millennium Development Goals and transition toward the sustainable development agenda, there is good reason for optimism.

We have the technologies and the know-how to succeed. With the dedicated efforts of each and every one of us, we can be the first generation to end extreme poverty and the last generation to face climate change as an existential threat.

Your energy and ideas can help lead the way to the age of sustainable development. Together, we can build a future of shared prosperity and a life of dignity for all.

PREFACE

We have entered a new era. Global society is interconnected as never before. Business, ideas, technologies, people, and even epidemic diseases cross borders with unprecedented speed and intensity. We share the exhilaration of the new information age and also the fears of global-scale environmental disruption. Business practices, technologies, and the size and age structure of populations are changing rapidly. There are new opportunities and also new risks. For these reasons, I suggest that we have arrived in the Age of Sustainable Development.

As I will describe, sustainable development is both a way of looking at the world, with a focus on the interlinkages of economic, social, and environmental change, and a way of describing our shared aspirations for a decent life, combining economic development, social inclusion, and environmental sustainability. It is, in short, both an analytical theory and a "normative" or ethical framework. Our new era will soon be described by new global goals, the Sustainable Development Goals (SDGs).

Other powerful ideas also express the novelty of our era. Leading geoscientists have coined the term *Anthropocene*, with the Greek roots meaning "human-made" (*anthropo*) and "new" (*cene*). The Anthropocene is our current, unprecedented epoch of the Earth in which the Earth's physical change—climate, biodiversity, chemistry—is mainly driven by human activity. Leading ecologists have adopted the idea of "planetary boundaries" to explain the limits beyond which human activities will tip the Earth into uncharted and dangerous patterns of climate disruption, loss of biodiversity, and change in the chemistry of the air, land, and oceans. Both of these key concepts describe the realities of the new Age of Sustainable Development.

Our geopolitics is changing rapidly as well. Our once "bipolar" world, divided between the rival superpowers of the United States and the Soviet Union, is now a complex multipolar world with many regional powers and nearly 200 countries, many new and with fragile institutions. This new multipolar world must find the means to preserve peace, pursue economic development, and face the unprecedented environmental challenges of our age. New forms of global governance will therefore play a key role in meeting the new SDGs.

I am excited to share these ideas with you as we enter a new era together. Our hopes are high—to end extreme poverty and protect the planet from the side effects of our own actions. Yet the challenges are immense and unprecedented. I hope that this volume will be of help in identifying pathways forward in our new age. And I hope that many young readers and students will soon become the creative and dedicated leaders of sustainable development in our new age.

ACKNOWLEDGMENTS

This book was developed as part of a global massive open online course (MOOC) of the same name, *The Age of Sustainable Development*. I'm very gratified that tens of thousands of students around the world have already taken the course. They have sparked a rich global discussion around the themes of sustainable development, especially the new Sustainable Development Goals.

I want to thank the wonderful team that worked with me to design and implement the MOOC and this book: Claire Bulger and Aditi Shah, who helped lead all stages of the project; Karena Albers and Tad Fettig of Kontentreal, whose creative vision helped bring the videos to life; Amir Jina, an outstanding sustainable development student and young leader who helped design the course material; Erin Trowbridge, Kyu Lee, and Arif Noori, who helped bring the course and book to students around the world; Chandrika Bahadur, who is pioneering the SDSN's online education effort globally; Imer Jasiel del Cid, who helped with key support on copyrights; and with my deep gratitude, Bridget Flannery-McCoy, Patrick Fitzgerald, and Lisa Hamm of the Columbia University Press, whose vision, enthusiasm, hard work, and creativity helped make this book a reality and an effective pillar of the online course.

A book like this, drawing on so many distinct fields of inquiry, is based on the deep research and wisdom of the global sustainable development community. I'd especially like to thank my colleagues at the Earth Institute and the United Nations system for their collegiality and their care in for educating me in these matters. Of course all remaining errors are my own.

Finally, as with all my books, they depend on the support, guidance, love, and patience of my family, including Sonia, Lisa, Matt, Adam, Tatyana, Hannah, Joan, Andrea, Sienna, and Willa.

THE AGE OF
SUSTAINABLE
DEVELOPMENT

1

INTRODUCTION TO SUSTAINABLE DEVELOPMENT

I. What Is Sustainable Development?

Sustainable Development as an Analytical and Normative Concept

Sustainable development is a central concept for our age. It is both a way of understanding the world and a method for solving global problems. Sustainable Development Goals (SDGs) will guide the world's economic diplomacy in the coming generation. This book offers you an introduction to this fascinating and vital field of thought and action.

Our starting point is our crowded planet. There are now 7.2 billion people on the planet, roughly 9 times the 800 million people estimated to have lived in 1750, at the start of the Industrial Revolution. The world population continues to rise rapidly, by around 75 million people per year. Soon enough there will be 8 billion by the 2020s, and perhaps 9 billion by the early 2040s (Sustainable Development Solutions Network [SDSN] 2013a, 2, 5).

These billions of people are looking for their foothold in the world economy. The poor are struggling to find the food, safe water, health care, and shelter they need for mere survival. Those just above the poverty line are looking for improved prosperity and a brighter future for their children.

Those in the high-income world are hoping that technological advances will offer them and their families even higher levels of wellbeing. It seems that the super-rich also jostle for their place in the world's rankings of richest people.

In short, 7.2 billion people are looking for economic improvement. They are doing so in a world economy that is increasingly interconnected through trade, finance, technologies, production flows, migration, and social networks. The scale of the world economy, now estimated to produce $90 trillion of output per year (a sum called the gross world product, or GWP), is unprecedented (SDSN 2013a, 2). By crude statistics, the GWP measures at least 200 times larger than back in 1750. In truth, such a comparison is difficult to make, since much of the world economy today consists of goods and services that did not even exist 250 years ago.

What we know is that the world economy is vast, growing rapidly (by 3–4 percent per year in scale), and highly unequal in the distribution of income within countries and between countries. Ours is a world of fabulous wealth and extreme poverty: billions of people enjoy longevity and good health unimaginable in previous generations, yet at least 1 billion people live in such abject poverty that they struggle for mere survival every day. The poorest of the poor face the daily life-and-death challenges of insufficient nutrition, lack of health care, unsafe shelter, and the lack of safe drinking water and sanitation.

The world economy is not only remarkably unequal but also remarkably threatening to Earth itself. Like all living species, humanity depends on nature for food and water, materials for survival, and safety from dire environmental threats, such as epidemics and natural catastrophes. Yet for a species that depends on the beneficence of nature, or on what the scientists call "environmental services," we are doing a poor job of protecting the physical basis of our very survival! The gigantic world economy is creating a gigantic environmental crisis, one that threatens the lives and wellbeing of billions of people and the survival of millions of other species on the planet, if not our own.

The environmental threats, we shall learn, are arising on several fronts. Humanity is changing Earth's climate, the availability of fresh water, the oceans' chemistry, and the habitats of other species. These impacts are now so large that Earth itself is undergoing unmistakable changes in the functioning of key processes—such as the cycles of water, nitrogen, and carbon—upon which life depends. We don't know the precise scaling, timing, and implications of these changes, but we do know enough to understand that they are extremely dangerous and unprecedented in the span of humanity's 10,000 years of civilization.

Thus we arrive at sustainable development. As an intellectual pursuit, sustainable development tries to make sense of the interactions of three complex systems: the world economy, the global society, and the Earth's physical environment. How does an economy of 7.2 billion people and $90 trillion gross world output change over time? What causes economic growth? Why does poverty persist? What happens when billions of people are suddenly interconnected through markets, technology, finance, and social networks? How does a global society of such inequality of income, wealth, and power function? Can the poor escape their fate? Can human trust and sympathy surmount the divisions of class and power? And what happens when the world economy is on a collision course with the physical environment? Is there a way to change course, a way to combine economic development with environmental sustainability?

Sustainable development is also a normative outlook on the world, meaning that it recommends a set of *goals* to which the world should aspire. The world's nations will adopt SDGs precisely to help guide the future course of economic and social development on the planet. In this normative (or ethical) sense, sustainable development calls for a world in which economic progress is widespread; extreme poverty is eliminated; social trust is encouraged through policies that strengthen the community; and the environment is protected from human-induced degradation. Notice that sustainable development recommends a holistic framework, in which society aims for economic, social, and environmental goals. Sometimes the following shorthand is used: SDGs call for *socially inclusive and environmentally sustainable economic growth.*

To achieve the economic, social, and environmental objectives of the SDGs, a fourth objective must also be achieved: good governance. Governments must carry out many core functions to enable societies to prosper. Among these core functions of government are the provision of social services such as health care and education; the provision of infrastructure such as roads, ports, and power; the protection of individuals from crime and violence; the promotion of basic science and new technologies; and the implementation of regulations to protect the environment. Of course, this list is just a brief subset of what people around the world hope for from their governments. In fact, all too often they get the reverse: corruption, war, and an absence of public services.

In our world today, good governance cannot refer only to governments. The world's multinational companies are often the most powerful actors. Our wellbeing depends on these powerful companies obeying the law, respecting the natural environment, and helping the communities in which they operate, especially to help eradicate extreme poverty. Yet as with governments, reality is often the reverse. Multinational companies are often the agents of public corruption, bribing officials to bend regulations or tax policies in their favor and engaging in tax evasion, money laundering, and reckless environmental damage.

Thus the normative side of sustainable development envisions four basic objectives of a good society: economic prosperity; social inclusion and cohesion; environmental sustainability; and good governance by major social actors, including governments and business. It's a lot to ask for, and there is no shortage of challenges to achieving sustainable development in practice. Yet the stakes are high. Achieving sustainable development on our crowded, unequal, and degraded planet is the most important challenge facing our generation. The SDGs must be the compass, the lodestar, for the future development of the planet during the period 2015 to mid-century.

Before proceeding further, let me give a very brief history of the concept of sustainable development. The term "sustainable" as applied to ecosystems goes back a long way. Fisheries managers, for example, have long used the concept of the "maximum sustainable yield" to denote the maximum fish catch per year consistent with a stable fish population. In 1972, at the UN Conference on the Human Environment in Stockholm, the challenge of maintaining sustainability in the context of economic growth and development was first brought to the global forefront. That same year, the blockbuster book *Limits to Growth*, published by the Club of Rome, argued forcefully that continued economic growth on the prevailing economic patterns would collide with the Earth's finite resources, leading to a future overshoot and collapse.

While 1972 put the challenge of sustainable development onto the global stage, the phrase itself was introduced eight years later, in an influential publication entitled *World Conservation Strategy: Living Resource Conservation for Sustainable Development* (1980). This pathbreaking publication noted in its foreword that

> human beings, in their quest for economic development and enjoyment of the riches of nature, must come to terms with the reality of resource limitation and the carrying capacity of ecosystems, and must take account of the needs of future generations.

The purpose of the document was to "help advance the achievement of sustainable development through the conservation of living resources" (iv).

The phrase was then adopted and popularized in the report of the United Nations Commission on Environment and Development, known widely by the name of its chairwoman, Gro Harlem Brundtland. The Brundtland Commission gave a classic definition of the concept of sustainable development, one that was used for the next twenty-five years:

> Sustainable Development is development that meets the needs of the present without compromising the ability of future generations to meet their own needs. (Brundtland 1987, 41).

This "intergenerational" concept of sustainable development was widely adopted, including at the Rio Earth Summit in 1992. One of the key principles of the Rio Declaration was that "development today must not threaten the needs of present and future generations."

Over time, however, the definition of sustainable development evolved into a more practical approach, focusing less on intergenerational needs and more on the holistic approach linking economic development, social inclusion, and environmental sustainability. In 2002, at the UN World Summit on Sustainable Development (WSSD) in Johannesburg, the WSSD Plan of Implementation spoke of "the integration of the three components of sustainable development— economic development, social development and environmental protection—as interdependent and mutually reinforcing pillars" (World Summit on Sustainable Development 2002, 2). The concept of intergenerational justice remains but is now secondary to the emphasis on holistic development that embraces economic, social, and environmental objectives.

This three-part vision of sustainable development was again emphasized on the twentieth anniversary of the Rio Summit. In the final outcome document for the Rio+20 Summit ("The Future We Want"), the aim of sustainable development was put this way:

> We also reaffirm the need to achieve sustainable development by: promoting sustained, inclusive and equitable economic growth, creating greater opportunities for all, reducing inequalities, raising basic standards of living; fostering equitable social development and inclusion; and promoting integrated and sustainable management of natural resources and ecosystems that supports inter alia economic, social and human development while facilitating ecosystem conservation, regeneration and restoration and resilience in the face of new and emerging challenges. (UN General Assembly 2012, para. 4)

The SDGs, called for in the same outcome document, are to be based on the three-part framework. Here is how the SDGs were announced in "The Future We Want":

> [The SDGs] should address and incorporate in a balanced way all three dimensions of sustainable development and their inter-linkages . . . We also underscore that SDGs should be action-oriented, concise and easy to communicate, limited in number, aspirational, global in nature and universally applicable to all countries while taking into account different national realities, capacities and levels of development and respecting national policies and priorities . . . Governments should drive implementation [of the SDGs] with the active involvement of all relevant stakeholders, as appropriate. (UN General Assembly 2012, paras. 246–7)

I will discuss the SDGs in greater detail in the final chapter. Until then, I will use the concept of sustainable development in the current sense of a three-way normative framework, embracing economic development, social inclusion, and environmental sustainability. In addition, I will refer to sustainable development as an *analytical field of study*, one that aims to explain and predict the complex and nonlinear interactions of human and natural systems. I turn next to this analytical sense of the term.

Embracing Complexity

In addition to being a normative (ethical) concept, sustainable development is also a science of complex systems. A system is a group of interacting components that together with the rules for their interaction constitute an interconnected whole. The brain is a system of interacting neurons; the human body is a system of some 10 trillion individual cells, with those cells interacting in systematic ways in various organ systems (circulatory system, nervous system, digestive system, etc.); the cell itself is a system of interacting organelles; and the economy is a system of millions of individuals and businesses, bound together in markets, contracts, laws, public services, and regulations.

We talk about these systems as complex because their interactions give rise to behaviors and patterns that are not easily discernible from the underlying components themselves. The conscious brain cannot be reduced to a list of its neurons and neurotransmitters; functions such as consciousness depend on highly complex interactions of the component neurons. A living cell is more than the sum of the nucleus, ribosomes, and other components; the systems of metabolism, gene expression, and the like depend on highly complex interactions of the components. A growing economy is more than the sum of its individual businesses and workers. Complexity scientists speak of the *emergent properties* of a complex system, meaning those characteristics that emerge from the interactions of the components to produce something that is "more than the sum of its parts."

Complex systems have many unexpected characteristics. They often respond in a nonlinear way to shocks or changes, meaning that even a modest change in the components of the system can cause a large, perhaps catastrophic change in the performance of the system as a whole. A small change in the cell's chemistry can lead to its death; a small change in the physical environment may cause large and cascading changes to the relative abundance of the species in that environment. The failure of a single business can lead to a financial panic and a global downturn, such as occurred when the Lehman Brothers investment bank failed in September 2008. A single bank failure, or a single infection, or a slight change in Earth's temperature, can lead to a chain reaction, positive-feedback process, which has explosive consequences.

Sustainable development involves not just one but four complex interacting systems. It deals with a *global economy* that now spans every part of the world; it focuses on *social interactions* of trust, ethics, inequality, and social support networks in communities (including new global online communities made possible by revolutions in the information and communications technologies, or ICTs); it analyzes the changes to complex *Earth systems* such as climate and ecosystems; and it studies the problems of *governance*, including the performance of governments and businesses. In each of these complex systems—economic, social, environmental, and governance—the special properties of complex systems, such as emergent behavior and strong, nonlinear dynamics (including booms and busts), are all too apparent.

Complex systems require a certain complexity of thinking as well. It is a mistake to believe that the world's sustainable development problems can be boiled down to one idea or one solution. A complex phenomenon such as poverty in the midst of plenty has many causes that defy a single diagnosis or prescription, just as in the cases of environmental ills or communities torn asunder by mistrust and violence. Medical doctors are trained to understand and respond to the complex system known as the human body. Medical doctors know that a fever or a pain can have countless causes. Part of the job of a skilled medical doctor is to make a differential diagnosis of the specific cause of a fever in a particular patient. A skilled sustainable development practitioner needs to be a complex-systems expert in the same way, acknowledging the complexity of the issues and looking to make a specific diagnosis of each specific case.

The Role of Technological Change

The Maglev in Shanghai (figure 1.1) is a remarkable piece of technology that carries people at speeds of more than 200 miles per hour to and from Shanghai City and its international airport. It is a product of joint work between engineering companies from Europe and China and has been operating for the past decade. It is an example of how new technology can help to achieve sustainable development by improving transport services and energy efficiency, and eventually enabling a shift to a clean, low-carbon energy system. The maglev, unlike earlier generations of rail, is powered by electricity rather than by coal or petroleum. If and when the electricity that powers the maglev is eventually produced with a low-carbon primary energy source, rather than the coal that today dominates electricity generation in China, the electric-powered intercity rail will also promote the shift from fossil fuels to safer low-carbon energy sources such as wind and solar power (which are much less polluting and do not result in human-induced climate change, as discussed later).

Throughout our study of sustainable development we will note three aspects of technology. First, technological advances are the main driver of long-term global economic growth. The rapid growth of the world economy since 1750 is the result of 250 years of technological advances, starting with the steam engine and steam-powered transportation, the internal combustion engine, electrification, industrial chemistry, scientific agronomy, aviation, nuclear power, and today's ICTs.

1.1 The maglev train in Shanghai

"The Shanghai Transrapid maglev train," Lars Plougmann, Flickr, CC BY-SA 2.0.

Without these advances, the world economy and world population would have stopped growing long ago.

Second, technological advances often have negative side effects, even when their direct effects are enormously positive. The burning of coal is both the emblem of the Industrial Revolution and the root of our current environmental crisis. One can say that coal enabled modern civilization through the invention of the steam engine and the harnessing of fossil fuels for motive force. Yet coal is now used on such a scale, and with such dire side effects, that it endangers civilization itself. In 2010, humanity emitted around 14 billion tons of carbon dioxide (CO_2) through coal burning, close to half of the world's total emissions of CO_2 due to fossil fuels. Unless coal is phased out rapidly or used with new technologies (such as carbon capture and sequestration, discussed later), the damage to the planet and the global economy will be overwhelming.

Third, technological advance is, at least to some extent, under human guidance. Sometimes technological advance is portrayed as a great lottery, determined by the luck of the draw or the skill of individual inventors and scientists. Alternatively, technological advance is sometimes described as merely following the demands of the market. Companies invest in research and development (R&D) in order to pursue profits. We end up with research on challenges sought by the marketplace, not necessarily those of vital importance for the poor or for the environment. Yet there is another side to technological change, the idea that it can be *directed* toward human goals through a deliberate, goal-based interaction of public and private R&D efforts.

We are used to the idea that governments steer technology for "reasons of state"—that is, for military purposes. Governments have long hired engineers and inventors to design and build new weapons and defenses, many involving pioneering breakthroughs in technology. World War I heralded major advances in aviation; and World War II brought advances in computers, radar, semiconductors, rocketry, antibiotics, communications, semiconductors, and countless other advances led by state-supported research, including America's Manhattan Project, which brought together world-renowned physicists to design and construct the first atomic bombs.

Of course, we should greatly prefer to achieve technological advances through peaceful means. And indeed, there is a distinguished track record of government support for civilian technological advances (though often governments had military purposes also in mind even in these civilian breakthroughs). In recent decades, the Internet, information technology, aviation, space technology (such as global positioning systems), genomics, nanotechnology, and countless other areas of technological advance owe their development in significant measure to government support. In the age of sustainable development, we will need such directed technological change in order to develop new technologies for sustainable energy, transport, construction, food production, health delivery, education, and more. Governments will rely on many policy tools to drive innovations in a targeted direction, including the public financing of R&D, direct research in public laboratories, regulations, prizes for new inventions, and modifications of patent laws (e.g., to encourage R&D on specific diseases).

Sustainable Development as a Normative Approach

Sustainable development is a way to understand the world as a complex interaction of economic, social, environmental, and political systems. Yet it is also a normative or ethical view of the world, a way to define the objectives of a well-functioning society, one that delivers wellbeing for its citizens today and for future generations. The basic point of sustainable development in that normative sense is that it urges us to have a holistic vision of what a good society should be. The easy answer for many people is that a good society is a rich society, one in which higher incomes are the ultimate purpose of economic and political life. Yet something is clearly too limited in such a view. Suppose a society was rich on average because one person was super-rich while the rest were in fact very poor. Most people would not regard that as a very attractive society, one that brings wellbeing to the citizenry. People care not only about the average income but about the income distribution as well.

There are at least five kinds of concerns about the distribution of wellbeing. The first is extreme poverty. Are some people still exceedingly poor in the midst of plenty? The second is inequality. Are the gaps between the rich and poor very wide? The third is social mobility. Can a poor person today hope to achieve economic success in the future, or are the practical barriers to advancement too high?

The fourth is discrimination. Are some individuals such as women, racial minorities, religious minorities, or indigenous populations disadvantaged by their identity within a group? The fifth is social cohesion. Is the society riven by distrust, animosity, cynicism, and the absence of a shared moral code? Sustainable development takes a view on these issues, calling on society to aim for the end of extreme poverty; the reduction of glaring gaps of wealth and poverty; a high degree of social mobility, including good life chances for children born into poverty; the absence of discrimination including by gender, race, religion, or ethnicity; and the fostering of social trust, mutual support, moral values, and cohesion. We can summarize these objectives with the term *social inclusion*.

Another aspect of a good society is being a good steward of the natural environment. If we break the physical systems of water and biodiversity, if we destroy the oceans and the great rain forests, we will lose immeasurably. If we continue on a path that fundamentally changes the Earth's climate, we are going to face grave dangers. Therefore, from a normative perspective, environmental sustainability certainly seems right and compelling if we care, as we should, about the wellbeing of our children and our children's children and future generations.

We also care about how our government functions. Good governance and the rule of law create a sense of security and wellbeing. On the other hand, corruption, lawlessness, untrustworthy politicians, unfair government services, significant discrimination, insider dealing, and so forth create a lot of unhappiness. Careful studies have confirmed that across the world people feel happier and more satisfied with life when they trust their government. Unfortunately, in many places in the world, people do not trust their governments to be honest and fair and to keep them basically secure, and they have all too many valid reasons for that distrust.

From a normative perspective then, we could say that a good society is not only an economically prosperous society (with high per capita income) but also one that is also socially inclusive, environmentally sustainable, and well governed. That is my working definition of the normative objectives of sustainable development. It is the point of view endorsed by the SDGs adopted by the UN member states. The fundamental question is how to take our knowledge of the interconnections of the economy, society, the environment, and governance and apply it to determine how to produce prosperous, inclusive, sustainable, and well-governed societies; that is, how do we achieve the SDGs? We shall see that there are indeed some powerful ways to achieve sustainable development as a shared set of goals for the planet.

Trade-offs Versus Synergies in Economic, Social, and Environmental Goals

The conventional view is that there are important trade-offs in pursuing economic, social, and environmental goals. For example, it is conventionally believed that society can aim to be rich, or it can aim to be equal; but if it aims for more equality, it will end up less rich. In such a view, income and equality are *substitutes*. In colloquial terms, the debate is often over whether to "grow the economic pie" or to "divide the pie more equally." A similar trade-off is often perceived to exist regarding the environment. A poor society, it is said, must choose between growth and the environment.

Economists often use the terms "efficiency" and "equity" to describe such choices. Efficiency means the absence of waste in the economy. There is no way to raise one person's income or wellbeing without lowering someone else's. The pie, in essence, is as large as possible. Equity means fairness in the distribution of the pie, remembering that standards of fairness may differ across individuals. To restate the common view described above, there is a trade-off between efficiency and equity. Societies that aim to be fairer, in that conventional view, inevitably introduce inefficiencies into the economy, leading to a waste of resources. For example, a tax on rich people to distribute income to the poor may lower the work incentive of both the rich (who must pay part of their income in taxes) and of the poor (who have less incentive to work). The result may be fairer but at the cost of efficiency and lower output.

That view is much too pessimistic. We will see throughout this book why investing in fairness may also be investing in efficiency, and why attention to sustainability can be more fair and more efficient at the same time. Here are two easy examples. Suppose the tax on the rich is used not for consumption by the poor but for the education and health of the poor. The investments in health and education may well have a very high return for the poor, enabling them to be much more productive. If the work effort of the rich is little affected by the tax, while the productivity of the poor is strongly boosted, the result may easily be more efficiency and more equity. Similarly, an investment in pollution control may raise productivity of the workforce by cutting disease and absenteeism, especially of the poor who are living in the most polluted conditions. Pollution control thereby achieves three aims: high output, more fairness, and more sustainability. In these cases, sustainable development offers *synergies* rather than trade-offs in the pursuit of efficiency, equity, and sustainability.

II. An Introduction to Economic Growth

Measuring the Size of the Economy

Economists typically summarize a country's overall economic development by gross domestic product (GDP) per person. The GDP measures the market value of total production within the country in a given time period, usually a year. Gross domestic product per capita (GDP per person) is simply GDP divided by the population. Since GDP is the size of the overall economic pie, the GDP per capita is the size of the average slice per person. Of course, the actual income distribution in any country will be uneven. Some households will have a very large slice of the pie, while others may receive no more than mere crumbs. Nonetheless, the average slice, the GDP per capita, is fairly closely though imperfectly correlated with other measures of national wellbeing, such as life expectancy, levels of education, quality of infrastructure, and levels of personal consumption spending.

There are a few quick points to mention about the measurement of GDP. First, GDP measures the production inside the boundaries of a country. This is different from the income earned by residents of the country. Suppose the country is an oil exporter, and the government owns two-thirds of the oil, while foreign companies own one-third. The GDP would count all of the oil produced within the country, but national income would include only the two-thirds of the oil owned by the government. We give the name gross national product (GNP) to the income-based measure. In this example, GNP would be less than GDP.

Second, the GDP measures output at market prices. For each output in the economy, such as bushels of grain, production of automobiles, sales of haircuts, and rentals of apartments, the quantity produced is multiplied by the price per unit, to calculate the value of production. These are summed to calculate the GDP. At this level, the GDP of each country is expressed in the national currency, such as dollars, pesos, euros, yen, yuan, won, and others. To compare across countries, the national currency units are converted to U.S. dollars, using the market exchange rate. We then have a common standard for comparing the GDP across countries. Dividing by each nation's population, we find the GDP per capita, which gives an indication of the relative living standards across countries (remembering that living standards will vary within each country based on the distribution of income across households).

Yet there is a problem with this comparison. The prices of individual products differ across countries, even when expressed in U.S. dollars. Suppose that in the first country, barbers sell $50 million worth of haircuts, while in the second, they sell $25 million. If the price of haircuts is the same across the two countries, we would be right to conclude that the first country enjoys twice the number of haircuts sold as the second. Yet if the market price of haircuts is twice as high in the first country, then the number of haircuts is actually the same, even though the sales volume in the first country is twice as high.

When we compare GDPs we want to compare the real volume of goods and services, not the difference merely due to market prices. In order to make a good comparison of GDP across countries, therefore, statisticians have decided to use a common set of "international prices" to sum up the production and consumption in each country. This adjusted measure is called the GDP at purchasing power parity (PPP). The use of a common set of international prices assures us that $1 of GDP in every country, when measured at PPP (or at international prices), has an equal purchasing power in terms of actual goods and services.

Third, we must also note GDP measures only the goods and services transacted in the market economy, not those that take place outside of the marketplace, such as production that occurs within the home. When a mother looks after her own children, the home day care is not counted as GDP. If the mother looks after the neighbor's child for a fee, however, that day care is counted as part of the GDP.

Also, GDP does not measure the "bads" or harms that often accompany production, such as the costs of industrial pollution or destruction by war. Therefore, GDP per person is only a rough indicator of true economic wellbeing per person. Plenty of terrible things—pollution, natural disasters, war—may afflict people in high-income countries without GDP reflecting those costs to society.

Defining Economic Growth

Ask an economic policy maker almost anywhere in the world about the country's main economic goal, and the answer will typically be "economic growth." Every day, the newspapers recount the recent growth rates of the major economies, as well as commentaries about the prospects for future growth. Yet what exactly is being measured by economic growth?

Economic growth, in simplest terms, measures the change in the GDP over a given period, for example, the current year relative to the past year or the current quarter of the year (January–March) compared with the preceding quarter (October–December). Economic growth signifies an increase in GDP.

Once again, we must immediately highlight some details. If the GDP rises by 100 percent (i.e., doubles in size), but the population also doubles, then the size of the average slice of the economic pie remains unchanged. Our interest in growth is therefore typically in the rise of GDP per capita rather than GDP by itself.

Moreover, we are interested in the rise of output of actual goods and services, not just the prices of goods and services. Here is an example. If the country produces one ton of steel, at $500 per ton, the contribution to GDP is $500. If the price of steel goes up to $1,000 per ton, while production remains at one ton, the contribution of steel to GDP rises to $1,000, even though there is no change in the actual production of the economy. Therefore, we generally are interested not in GDP at current prices (whether domestic or international), but GDP at constant prices. For example, we might decide that for the next several years, every ton of steel will be measured at a constant price of $500, even if the actual market price fluctuates. We call this GDP at constant prices. For the reasons described, we are in fact typically interested in "GDP at constant international prices" or "GDP at PPP in constant prices."

Why are we so interested in GDP per capita at constant international prices? As mentioned earlier, that measure tends to be related to several other indicators of prosperity. When GDP per capita increases, economic wellbeing tends to rise. Richer countries—those with a higher GDP per person—tend to have higher material wellbeing on average than poorer countries. People in richer countries tend to have higher consumption levels, greater food security, longer lives, and greater protection from diseases and environmental catastrophes. Chances of violence and war are lower. And people living in richer societies also tend to express greater satisfaction when asked for subjective assessments of their lives, as discussed in the next chapter.

Yet for many reasons, some already mentioned and others that will be mentioned later, the rise in GDP per capita is *far from a perfect measure of wellbeing*. It is quite conceivable that GDP per capita rises but that many people in the country end up being worse off. That could be true, for example, if only a small part of the society is recipient of the higher production. It could also be true if the rise in market-based output is offset by "bads" occurring outside the market, for example, environmental destruction such as air and water pollution.

Still, let us focus on the long-term trajectory of GDP per capita measured at constant international prices. The good news is that the world economy in total has tended to grow over the course of many decades. This means that if we add up the GDP (at constant international prices) of every country, and call the result the GWP, and then divide by world population to find GWP per person, we find that GWP per capita has been rising fairly consistently by around 2 to 3 percent per year. In turn, this global growth, reflecting the growth of national economies as well (using GDP per capita as the measurement), has been associated with many other gains in material wellbeing, such as improved health, better education, and more food security (though also more obesity, alas).

A handy rule of thumb for economic growth, and indeed for any kind of growth, is called the "rule of 70." The idea is the following. Consider the growth rate of the world economy, say a 2 percent per year increase of the GWP per person. If we take 70 divided by the annual growth rate, in this case 70 divided by 2, or 35, we determine the number of years it takes for the economy to double in size. So an economy growing at 2 percent per year will double in 35 years (= 70/2);

if the global growth heats up to 4 percent per year, the doubling time therefore drops by half, to 17.5 years (= 70/4).

Now the key point is that the world economy has been growing consistently since the start of the Industrial Revolution in the middle of the eighteenth century. Angus Maddison, the late economic historian, did a great service for the economics profession by estimating the GDP per person over the long time period from the start of the Common Era (1 C.E.), with the most detailed data after 1820. He measured GDP in each period and country using the same standard: international prices of 1990. By that measure, the GWP rose from $695 billion in 1820 to around $41 trillion by 2010. During that same period, the world population rose from around 1.1 billion to 6.9 billion. Therefore, the GWP per capita (in constant 1990 international dollars) increased from $651 to $5,942 in Maddison's estimates (Maddison 2006).

How fast is that growth on an annual basis? Note that there are 190 years between 1820 and 2010. Therefore, we can find the average growth rate between 1820 and 2010 by solving the following equation:

$$(\$GWP \text{ per person in } 2010)/(\$GWP \text{ per person in } 1820) =$$
$$\$5942/\$651 = (1 + g)^{190}$$

Solving for the growth rate g, we find that $g = 1.1$ percent per annum is the average annual growth rate since 1820. If we make exactly the same calculation for the years 1970 to 2010, we find that the more recent growth rate is even higher, at 1.5 percent per annum.

Figure 1.2 shows an admittedly rough estimate of GWP per person, measured at constant international dollars, over a *very long* time period, specifically from 1 C.E. to 2010. Of course, the actual GWP per person in earlier centuries is based on rough estimates rather than precise data. Still we see something absolutely extraordinary about this graph. During most of the history of the past two millennia, there was little or no economic growth. GWP per person only started to rise around 1750 and then only very gradually at the beginning. (Note that Maddison presents estimated world output for 1700 and 1820, but not for the year 1750.) The whole story of economic growth in human history is a recent one, stretching over little more than two centuries! Economic historians call the period since 1750 the "age of modern economic growth." This period is the central period of our study. One can see from figure 1.2 that even though the Industrial Revolution began in Britain sometime in the middle of the eighteenth century, it became noticeable at the global level only in the nineteenth century (hence Maddison's more detailed data starting only in 1820).

We can say the following now, to be elaborated later. For most of human history, output per person was at a very low level, just around the level needed to survive. Most of humanity lived on farms and grew food for their own subsistence. In most years, the food was enough to keep them alive. In bad years, with droughts or floods or heat waves or pests, the harvest might fail, and people would die, sometimes in large numbers. Poor harvests might also make the population more susceptible to infectious diseases, since malnourishment weakens the body's immune system. Starting around 1750, something fundamentally new began to occur: positive economic growth. We will see that economic growth started only in a few places, including Great Britain and the United States. Eventually it spread around the world, though quite unevenly.

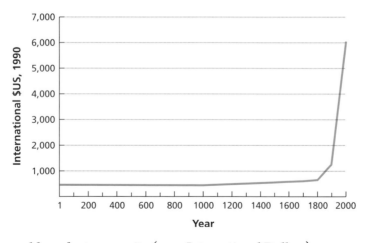

1.2 Gross world product per capita (1990 International Dollars)

Source: Bolt, J., and J. L. van Zanden. 2013. "The First Update of the Maddison Project: Re-Estimating Growth Before 1820." Maddison Project Working Paper 4.

The rise in GWP was first associated with the rise of industry, such as coal mining, steelmaking, and textile production. Indeed, we usually call the first takeoff of economic growth from around 1750 to 1850 the Industrial Revolution, with a capital "I" and "R." More recently, after 1950 or so in the high-income countries, the rise in GWP has been associated with the rise of services, such as the banking system. The overall result is that the world's output per person, or GWP per capita, lifted above the subsistence level and over a period of roughly 250 years, grew by a factor of around 30. In some countries, the increase has been a factor of around 100.

Figure 1.3 shows something else that is also astounding and that seems to follow a similar course. This graph looks quite like figure 1.2, but instead of measuring GWP per person, it measures the world population over a very long stretch of time, in this case all the way back to the presumed beginning of civilization, around 12,000 years before the present day (sometimes called 12,000 B.P.). This is the time when human beings shifted from hunting and gathering their food to growing it in one place; the change from nomads shifting locations to find food to farmers living in fixed villages. The period before agriculture is known as the

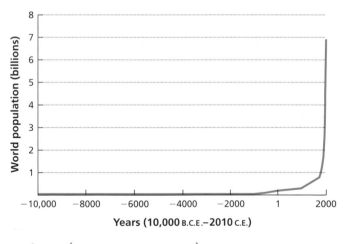

1.3 Global population (10,000 B.C.E.–present)

Source: Bolt, J., and J. L. van Zanden. 2013. "The First Update of the Maddison Project: Re-Estimating Growth Before 1820." Maddison Project Working Paper 4.

Paleolithic Era (*Paleo* = old + *lithic* = stone). The period after the start of agriculture is known as the Neolithic Era (*Neo* = new + *lithic* = stone).

What we see is that just like GWP per capita, the global population changed fairly little over very long stretches of time, always remaining well under 1 billion people. From 10,000 B.C.E. to around 2000 B.C.E., the human population was well under 100 million. Around 1 C.E., at the time of the Roman Empire, the world population according to Maddison's estimate was around 225 million. As of 1000, it was 267 million, on Maddison's best estimate; in 1500, around 438 million. It reached 1 billion around 1820. The world population therefore rose perhaps 4 times in the 18 centuries between 1 C.E. and 1820, implying an annual growth rate of just 0.08 percent per year. For most people in most of history, population seemed relatively unchanged over the course of a lifetime, indeed many lifetimes. The only changes were due to mass deaths from wars, famines, and plagues, followed by subsequent recoveries of population to more "normal" levels.

Then, in the same era as the Industrial Revolution, population broke free of its ancient restraints. At that point in history, the population curve turns up remarkably steeply. Around 1820 or so, humanity reached the great milestone of a billion people on the planet, then from 1820 to around 1930, in roughly one century, the second billion was added. Then the numbers really started to soar. In just 30 years, from 1930 to 1960, the third billion was added. The world went from 3 billion in 1960 to 4 billion in 1974, 5 billion in 1987, 6 billion in 1999, and 7 billion in the year 2011. Notice that the recent increments of 1 billion have occurred in roughly dozen-year intervals!

One clear reason for the rise in the world's population is the increased ability to grow more food and feed a rising population. Just as humanity learned to harness technology for industrialization, it learned to harness technology to raise food production. Since 1750 or so, farmers have been able to grow more food thanks to better seed varieties; better farming techniques (such as rotating crops through the years to maintain soil fertility); chemical fertilizers to boost soil nutrients; and machinery to sow seeds, harvest crops, process foodstuffs, and store and transport food to cities.

We are on track to reach 8 billion around 2024 or 2025, and 9 billion sometime in the early 2040s. After that, the numbers become far more uncertain but will probably continue to rise, at least on present patterns of fertility (childbearing) and mortality (deaths). The rise in population since the early 1800s is absolutely astounding, unprecedented in human history, just as with GWP per person. The fundamental fact is that modern economic growth and global population increases have tended to come side by side, although the relationship between the two is complicated, as we shall learn.

The age of modern economic growth is one of rising output per person combined with rapid overall population growth. Together, those two dynamics, more income per person and more people on the planet, have meant a massive expansion of total economic activity. Indeed, it is an obvious relationship that total output in the world, the GWP, is equal to the output per capita multiplied by the world's population:

$$GWP = GWP \text{ per capita} \times \text{world population}$$

Figure 1.4 shows Maddison's estimates of GWP production expressed in constant 1990 international dollars. Since GWP per capita and population both have

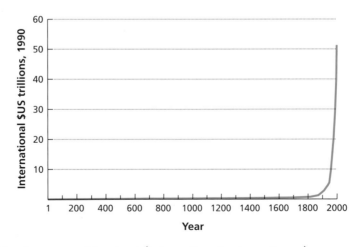

1.4 Growth of gross world output (international prices of 1990)

Source: Bolt, J., and J. L. van Zanden. 2013. "The First Update of the Maddison Project: Re-Estimating Growth Before 1820." Maddison Project Working Paper 4.

the same astounding pattern of nearly no change during 1 C.E.–1800, and then a sharp upturn, the graph of GWP has the same characteristic shape. World output has soared around 240 times since 1800. This has been a huge boon to average wellbeing (e.g., longer life expectancy), industrialization, urbanization, and yes, environmental threats as well.

The Recent Growth of China

Let's look at what growth really means in one very important case. There has been no exemplar of rapid economic growth more remarkable than China. China's growth is superlative in every aspect. As the world's most populous country, with 1.3 billion people, anything major that happens in China is earthshaking, but since 1978, China has also been among the fastest-growing economies in world history. When Deng Xiaoping came to power at that time, China undertook some basic market reforms that put the country on a trajectory of extraordinary economic growth, averaging roughly 10 percent per year in GDP growth.

Remember the rule of 70. A growth of 10 percent means that China has been doubling its GDP roughly every 7 years ($= 70/10$). This is absolutely astounding. Since China has grown at this torrid rate for almost 35 years, that is roughly 5 doublings ($= 35$ years/7 years per doubling). That in turn implies that the economy has grown roughly by a factor of 2^5 (or 32 times) since Deng Xiaoping opened the Chinese economy to market forces and international trade! In per capita terms, the growth is only slightly less impressive, at roughly 9 percent per annum, or 11.8 times overall between 1978 and 2013.

What does such extraordinary growth signify? To appreciate China's accomplishment, take the example of Shenzhen, which is a city very close to Hong Kong in southern China. In 1980, Shenzhen was a small, mainly rural village of some 30,000 people, as seen in figure 1.5.

Compare that with Shenzhen today, shown in figure 1.6.

Now, with nearly 12 million people, this modern metropolis is incomparable with its roots just three decades earlier. This kind of torrid growth is characteristic of China's eastern seaboard, where coastal cities became powerhouses of international trade. More than 200 million people have flocked from the countryside to

1.5 Shenzhen, China, in 1980

"Looking northwest . . .," Leroy W. Demery, Jr., Flickr. Used with permission.

1.6 Shenzhen, China, in 2002

Reuters.

the cities in search of new jobs in industry and services. China has become the world's largest trading country and the industrial workshop of the world.

China's experience has the hallmarks of modern economic growth, albeit in a turbocharged version. The economy has proceeded from rural to urban, from agricultural to industrial and service-oriented. It has gone from high fertility rates (many children per woman) to low fertility rates, and from high child mortality to low child mortality. Life expectancy has soared, public health has improved, and educational attainment has gone up steeply. With its vast population and strong educational orientation, China now turns out more PhDs per year than any other place in the world. And all of this has happened in the span of just a bit over three decades. This is the kind of remarkable experience that inspires many countries to aim to end poverty within their borders.

We must not leave the impression that all is well with China's economic growth. There have been at least three serious downsides. First, the rapid transition from rural to urban, and from farming to industry and services, has disrupted the lives of hundreds of millions of people, causing mass migration within China and disrupting families, as fathers and mothers often went off to find work in the cities and left their children with grandparents in the countryside. Second, the inequality of income has soared, as urban workers have advanced in living standards, while the incomes of those left behind in the countryside have often stagnated. Third, the physical environment has been devastated, with massive pollution accompanying China's massive industrialization. Indeed, as we shall see, the pollution has become so bad that it is causing widespread disease and premature deaths, especially from heart and lung diseases, stroke, and cancers, slowing China's gains in life expectancy. China, in short, has achieved rapid economic growth but has not yet achieved sustainable development, meaning growth that is also socially inclusive and environmentally sustainable.

Improvements in Global Health

Global growth in GWP per capita has been accompanied by another positive development: the improvement in public health. Higher incomes have meant improved food security for many (as well as unhealthy diets leading to obesity). Advances in technology in agriculture and industry have also been accompanied by rapid advances in health technologies, including medical advances such as antibiotics, vaccinations, diagnostics, and vast improvements in surgery, as well as advances in other fields with major health benefits, such as improved provision of water supplies, sewerage, and household sanitation.

Around 1950, for every 1,000 children who were born an estimated 134 would die before their first birthday. That number, 134 deaths under 1 year per 1,000 births, is called the infant mortality rate, or IMR. It signifies the proportion of newborns that won't survive to their first birthday, in this case, 13.4 percent (134/1,000). It is heartening that the IMR has been coming down sharply to an estimated 37 per 1,000 today. But we must remember this means 37 of every 1,000 newborns (3.7 percent) still do not survive to their first birthday, dying of malaria, pneumonia, diarrheal disease, or other preventable diseases of infants. These are tragedies that continue to occur around the world, with around 5 million children, almost all in developing countries, succumbing by age 1, and around 6 million dying each year under the age of 5. Still, the drop from 134 to 37 in infant mortality is a tremendous accomplishment of economic development and public health systems (including improved medical care, greater food security, greater access to safe water and sanitation, and other contributors to improved health). The decline in mortality rates at all ages has improved the quality of life and certainly eliminated a lot of the tragedy and anguish that was part of humanity's existence up until the improvements in public health and medical care in the past century.

With more children surviving and with health improving at older ages as well, the good news is that our life expectancy is also rising considerably. A good measure of this is life expectancy at birth. Life expectancy measures the average life span, taking into account the risks of death at each age. In the middle of the last century, in the five-year period from 1950 to 1955, the average life expectancy for the entire world population was around forty-seven years. Today, the estimated life expectancy at birth is roughly seventy-one years, and it is as high as eighty years in high-income countries. This tremendous increase in longevity is another benefit of economic growth and material progress and exemplifies the broad trend of improvement being achieved in most parts of the world.

The first major economic lesson of recent history is that the first pillar of sustainable development—prosperity achieved through economic growth—is achievable on a large scale, and indeed is being achieved across large parts of the planet. Most parts of the world have been benefiting from a rise in GDP per person. That increase in GDP per person has been accompanied by several structural changes in society: from rural life as peasant farmers to urban life with employment in industry or services. There are fewer tragic deaths of young children and greater health and longevity for most of us, with life expectancy now several decades higher than what it was in the middle of the twentieth century.

China's experience, repeated somewhat less dramatically in many other countries, shows that high per capita incomes need not be the preserve of a small, narrow part of the world (covering the United States, Canada, Europe, Japan, Australia, and New Zealand, but very few other places) as it was until recently, but can in fact be achieved almost everywhere. However, as we noted briefly in the case of China, even rapid economic growth is not sufficient to ensure wellbeing. We must ensure that the economic growth is inclusive and does not leave millions of people behind. We must ensure that economic growth is environmentally sustainable, so that progress does not undermine the Earth's life-support systems of high biodiversity, soil productivity, a safe climate, and productive oceans. Unless we combine economic growth with social inclusion and environmental sustainability, the economic gains are likely to be short-lived, as they will be followed by social instability and a rising frequency of environmental catastrophes.

III. Continuing Poverty in the Midst of Plenty

In many ways we already live in a world of plenty. Economic growth has produced incredible wealth, and most parts of the world have escaped from extreme economic hardship. Countries like China that were once very poor are now middle-income countries. Yet despite these advances, parts of the world remain stuck in extreme poverty. Perhaps the most urgent economic challenge on the planet is to help populations still living at the edge of survival to achieve economic growth and escape from poverty.

What is extreme poverty today? Figure 1.7 shows a smallholder farmer living in northern Ethiopia, in the Koraro village of Tigray Province, the site of a

1.7 Smallholder farming life in Northern Ethiopia

Photo courtesy of John Hubers.

Millennium Village. The farmer is hidden behind a great bale of grain carried by his donkey. There is no modern transport, no electricity grid. The land is parched. This is a dry region of poor farm households eking out a living and trying to ensure enough annual food production to feed themselves and their families. If they are lucky, they may produce a small surplus of grain to bring to market for a bit of cash income.

Figure 1.8 shows a street in Nairobi's Kibera slum, the urban face of extreme poverty. Hundreds of millions of people live in urban slums around the world. Often urban poverty abuts right up against great urban wealth. Looking closely, we see an unpaved muddy road that is not really passable by vehicles. As the photo shows, people living in this slum may see power lines overhead, but they may be

1.8 Kibera slum, Nairobi, Kenya

"Scenes from the Kibera slum in Nairobi," Karl Mueller, Flickr, CC BY 2.0.

too poor to be connected to the grid. These people also probably get by without modern sewerage or household sanitation, often having to defecate in open places. They are perhaps buying their water from a water truck, because there is no piped water to individual households and perhaps not even a shared public water stand for the community.

In short, even though these slum dwellers are living in an urban area of several million people, they are, like their counterparts in northern Ethiopia, mostly unable to secure basic needs, to access emergency health care, electricity, adequate nutrition, clean cookstoves, safe water, and sanitation. They may barely eke out a living in informal employment. They may earn just enough to buy a minimum of food, water, clothing, and shelter.

Extreme poverty is a multidimensional concept. Poverty is typically described as the lack of adequate income, but extreme poverty should be understood in more general terms as the inability to meet basic human needs for food, water, sanitation, safe energy, education, and a livelihood. Extreme poverty means lacking modern energy for safe cooking, such as natural gas, with the household instead relying on wood-burning stoves that cause chronically smoke-filled homes and subsequent respiratory diseases in the children. Extreme poverty often means that households cannot secure decent schooling for their children. There may be no school nearby, or no qualified teacher, or a school that charges tuition beyond the household's income.

People living in extreme poverty are, simply put, people who cannot meet their basic needs. Life is a daily struggle for dignity, and even for survival. While the numbers living in extreme poverty around the world have been declining, and the proportion of the world population living in extreme poverty has been shrinking even faster in recent decades, the number of people still struggling in extreme poverty is staggering. Depending on one's estimate and the exact definitions used, more than 1 billion people, and perhaps as many as 2.5 billion people, can be categorized as living in extreme poverty. It is probably accurate to say, and shocking to think about, that around one billion people struggle each day for their mere survival. They worry about whether they will they have enough to eat; they worry that unsafe water will cause a life-threatening disease; they worry that a mosquito bite transmitting malaria will take the life of their child, because they cannot afford the 80-cent dose of medicine needed to treat the infection.

This struggle for survival occurs in both rural and urban areas. It is still predominantly rural (perhaps in a ratio of 60:40), but it is increasingly taking on an urban face in the world's slums. Where is this extreme poverty? One shortcut is to look at the GDP per person around the world. As a general principle, economies with low GDP per capita also tend to be places where households live in extreme poverty. Figure 1.9 is a color-coded map of the world in which the colors denote the GDP per capita measured in purchasing power–adjusted terms (in 2011 prices). The map shows the huge variation in per capita GDP around the world. The countries in blue have GDP per capita above $30,000. There are not too many of them: the United States and Canada, most of western Europe, Japan,

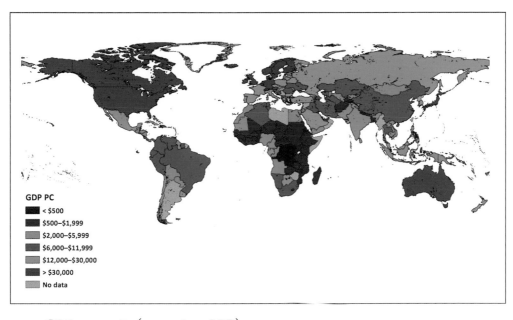

1.9 GDP per capita (2011 prices, PPP)

Source: World Bank. 2014. "World Development Indicators."

Australia, and a few small, oil-rich states in the Middle East. By and large, extreme poverty has been completely eliminated from those countries.

Next come the countries colored lighter blue, with GDP per capita between $12,000 and $30,000, still high by world standards. These include Israel, Korea, New Zealand, Russia, and several countries in central Europe. Contrast that with the red parts of the world. These are the places of very low GDP per capita, less than $2,000 in PPP terms, and also the highest concentration of populations living in extreme poverty. It is clear from this map that the poorest countries in the world are concentrated in tropical sub-Saharan Africa, those countries lying south of North Africa and north of the southern tip of Africa. Many of these tropical African countries are very poor, with around half of the population living in extreme poverty. The next poorest region, also home to vast numbers of people living in extreme poverty, is South Asia—including India, Pakistan, Nepal, and Bangladesh. Even though the GDP per capita is typically higher in South Asia than in tropical Africa, the South Asian economies have vast populations and many people living in extreme poverty. In both Africa and South Asia, the proportions of households living in extreme poverty have been falling, but there is still a huge challenge in ending extreme poverty, a subject we will consider in detail in a later chapter.

Notice as well a few other places with pockets of poverty, such as landlocked Bolivia in South America and landlocked countries in central Asia such as Mongolia. These are countries where poverty is high and geography is difficult. We shall see that being landlocked makes economic growth more difficult. Economic growth often depends on international trade, but international trade is very hard for countries that are hundreds or even over a thousand kilometers from a port, with the port possibly in another country to boot. (Coastal countries with ports are often relatively hostile to their landlocked next-door neighbors, since they have sometimes fought wars over access to the sea.)

Figure 1.10 maps another aspect of extreme poverty: the mortality rate of infants (IMR) (deaths of children under 1 year per 1,000 births), shown for data from 2013. Infants living in extreme poverty face a burden of disease and much higher risks of mortality than non-poor children. Once again, where is the concentration of child mortality? Tropical Africa and parts of South Asia are again the epicenters of the global challenge.

Even in countries where the vast majority of the population has escaped from extreme poverty, there can still be very significant pockets of poverty. Brazil is a case in point. Most of the poor in Brazil are able to meet their basic needs (and hence should not be described as living in "extreme" poverty), but they are still vastly poorer and vastly disadvantaged compared with their richer urban neighbors. Sometimes the starkness of the divisions of income and social status are right in front of our eyes, and the eyes of the poor. Take, for example, the view of Rio de Janeiro in figure 1.11, with its contrast of favelas (slums) and modern high-rises.

As always with sustainable development, there is hope for the extreme poor and for those living in relative poverty as in Rio. There are practical approaches, things that can be done, to help even the poorest of the poor to meet their basic needs and to help them succeed in the daily struggle for survival. We will be examining such approaches in detail later in the book. One that I find most exciting

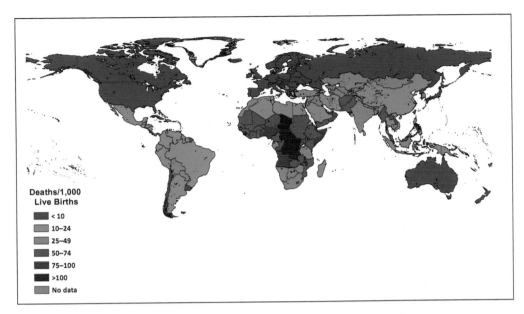

1.10 Global infant mortality rates (deaths under 1 per 1,000 births)

Deaths/1,000
Live Births
- < 10
- 10–24
- 25–49
- 50–74
- 75–100
- >100
- No data

1.11 Wealth and poverty in Rio de Janeiro

is the idea of Community Health Workers (CHWs) in poor villages and slums who bring health care to people who otherwise would be disconnected from the health system. We will see how modern technologies have made the CHWs especially effective in recent years.

We have noted that the extent of poverty has a strong geographical pattern. The highest proportions of extreme poverty are in tropical Africa and South Asia. We will study some of the reasons for this geographical pattern. It is not a coincidence. Geography shapes many things about an economy, including the productivity of farms, the burdens of infectious diseases, the costs of trade, and the access to energy resources. We will examine such geographical factors later in the book. Fortunately, geography is not destiny. Even if a particular region is vulnerable to specific diseases (such as malaria), modern technologies offer modern solutions. Geographical reasoning helps us to identify the high-return investments that can help the poorest of the poor to escape from poverty.

IV. Global Environmental Threats Caused by Economic Development

One of the most important messages of the field of sustainable development is that humanity has become a serious threat to its own future wellbeing, and perhaps even its survival, as the result of unprecedented human-caused harm to the natural environment. Gross world product per person, now at $12,000 per person, combined with a global population of 7.2 billion people, means that the annual world output is at least 100 times larger than at the start of the Industrial Revolution. That 240-fold increase in world output (or even a thousandfold increase on particular dimensions of economic activity) results in multiple kinds of damage to the planet. Large-scale economic activity is changing the Earth's climate, water cycle, nitrogen cycle, and even its ocean chemistry. Humanity is using so much land that it is literally crowding other species off the planet, driving them to extinction.

This crisis is felt by rich and poor alike. In late October 2012, police cars floated down the street in Manhattan during Superstorm Sandy, one of the strongest storms to hit the Eastern Seaboard in modern times (see figure 1.12). Even if

1.12 Flooding in Manhattan during Superstorm Sandy, October 2012

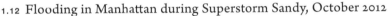

"Hurricane Sandy Flooding Avenue C 2012," David Shankbone, Wikimedia Commons, CC BY 3.0.

scientists can't determine whether the storm's remarkable ferocity was due in part to human-induced climate change, they can determine that human-induced climate change greatly amplified the *impact* of the storm. As of 2012, the ocean level off the Eastern Seaboard of the United States was roughly one-third of a meter higher than a century earlier, the result of global warming causing a rise in ocean levels around the world. This higher sea level greatly exacerbated the flooding associated with the superstorm.

Superstorm Sandy wasn't the only climate-related shock to the United States that year. Earlier in the year, U.S. crops suffered major losses as the result of a mega-drought and heat wave in the Midwest and western grain-growing regions (see figure 1.13). Drought conditions have continued to burden some parts of the U.S. West since then, with California in an extreme drought as of 2014.

Halfway around the world from New York City, also during 2012, Beijing experienced massive flooding that followed especially heavy rains. Bangkok

1.13 Corn fields in Iowa drought (2012)

"Iowa County Drought," CindyH Photography, Flickr, CC BY-SA 2.0.

experienced astounding floods in October 2011 (see figure 1.14). Indonesia experienced heavy flooding in early 2014, while Australia suffered another devastating heat wave. All of these events were huge setbacks for both the local and global economy, with loss of life, massive loss of property, billions or even tens of billions of dollars of damage, and disruptions to the global economy. The floods in Bangkok, for example, flooded automobile parts suppliers, shutting down assembly lines in other parts of the world when the parts failed to arrive.

The particular disasters are varied, but it is clear that one broad category—climate-related catastrophes—is rising in number and severity. One major class of climate shocks is known as "hydrometeorological disasters." These are water- and weather-related disasters, including heavy precipitation, extreme storms, high-intensity hurricanes and typhoons, and storm-related flood surges such as those that swept over Manhattan, Beijing, and Bangkok. Massive droughts

1.14 Bangkok floods (2011)

"USS Mustin provides post-flood relief in Thailand," Jennifer Villalovos, U.S. Navy.

cause deadly famines in Africa, crop failures in the United States, and a dramatic increase in forest fires in the United States, Europe, Russia, Indonesia, Australia, and other parts of the world. Other climate-related catastrophes include the spread of diseases and pests that threaten food supplies and the survival of other species.

The frequency and severity of these threats have risen dramatically and are likely to increase still further. Indeed, the reshaping of the Earth's physical systems—including climate, chemistry, and biology—is so dramatic that scientists have given our age a new scientific name: the *Anthropocene*. This is a new word that comes from its Greek roots: *anthropos*, meaning humankind, and *cene*, meaning epoch or period of Earth's history. The Anthropocene is the era—our era—in which humanity, through the massive impacts of the world economy, is creating major disruptions of Earth's physical and biological systems.

In the language of the scientists, human-induced changes are "driving" the Earth's physical and biological changes. To a layperson, the word "driving" might suggest that somebody is in control. That's not what the scientists mean. They mean that humanity is *causing* changes that are large, serious, and highly disruptive, with most of humanity, including most political leaders, having little scientific understanding of the dangers ahead.

The study of sustainable development requires a deep understanding of these human-induced changes, most importantly so we can change course and protect ourselves and future generations. One of the main drivers of change is humanity's massive use of coal, oil, and natural gas, the primary energy sources we call fossil fuels. When we burn coal, oil, and gas to move vehicles, heat buildings, transform minerals into steel and cement, and produce electricity, the combustion process produces CO_2 that is emitted into the atmosphere. The rising concentration of CO_2 in the atmosphere is the main, though not the only, source of human-induced climate change.

Figure 1.15 tells a remarkable story. It depicts the fluctuating levels of CO_2 in the atmosphere over the past 800,000 years. The distant past is on the left-hand side of the figure; the present is all the way to the right. The vertical axis measures CO_2 in the atmosphere. The measurement unit is the number of molecules of CO_2 for every 1 million molecules in the atmosphere. As of today, there are around 400 CO_2 molecules per million, or 400 parts per million (ppm). That doesn't seem like very much: just 0.04 percent. Yet even small changes in this concentration have a big effect on the climate.

Start on the left-hand side of the graph. 800,000 years ago, the CO_2 concentration was around 190 ppm. We see that it rose to peak around 260 ppm before falling to a low of around 170 ppm around 740,000 years ago. In general, CO_2 rises and falls like the teeth of a saw. These fluctuations are natural. They are "driven" (i.e., caused) mainly by slight changes in the Earth's orbital patterns around the sun; changes involving the shape of the orbit; slight variations in the Earth's distance to the sun; and the fluctuations in the tilt of the Earth relative to the plane of the Earth's orbit, causing slight changes in the pattern of the seasons. When the orbit changes slightly in ways that tend to heat the Earth, a feedback process tends to cause the release of CO_2 dissolved in the oceans, which then escapes into

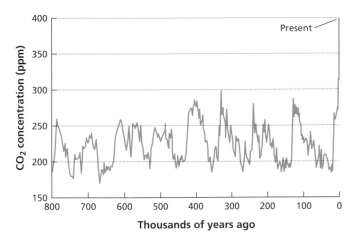

1.16 CO_2 in the atmosphere over the past 800,000 years

Reprinted by permission from Macmillan Publishers Ltd: Nature, Lüthi, Dieter, Martine Le Floch, Bernhard Bereiter, Thomas Blunier, Jean-Marc Barnola et al. "High-resolution Carbon Dioxide Concentration Record 650,000–800,000 years Before Present," copyright 2008.

Note: Ice core data before 1958; Mauna Loa data after 1958.

the atmosphere (just as CO_2 gas bubbles escape if one heats a pot filled with soda water). In turn, as the CO_2 rises in the atmosphere, the result is to warm the planet even more. We say that the increase in CO_2 is a "positive feedback." The change in the orbit slightly warms the planet; that releases CO_2 into the atmosphere, which in turn causes a further rise in temperature.

Scientists have shown that whenever the atmospheric concentration of CO_2 was high, the Earth tended to be warm (mostly because of the CO_2). Whenever CO_2 was low (because the atmospheric CO_2 was reabsorbed into the ocean), the Earth tended to be cold. Indeed, in the low phases of the natural CO_2 cycle, the Earth was actually cold enough to produce an ice age, with much of the Northern Hemisphere covered by a thick sheet of ice. By relating the concentration of CO_2 to the Earth's temperature (determined by other means), scientists have found a systematic relationship of high CO_2 and high Earth temperatures.

The far right-hand side of the graph shows that in the last blink of an eye in geological time, really in the past 150 years, the concentration of CO_2 has shot up like a vertical rocket. This is not because of natural changes in the Earth's orbit. This time, the rise in CO_2 has a human cause: the burning of fossil fuels. Notice the key and alarming point: humanity has pushed the level of CO_2 in the atmosphere to 400 ppm, higher than at any time in the past 800,000 years. Indeed, the last time the CO_2 concentration was so high was 3 million years ago, literally off the chart! And when the CO_2 level was that high 3 million years ago, the Earth was vastly warmer than today.

Why worry, you might wonder. The reason is that all of our civilization—the location of our cities, the crops we grow, and the technologies that run our industry—is based on a climate pattern that will soon disappear from the planet. The Earth will become much warmer than it has been during the entire period of civilization; the sea level will be come much higher, threatening coastal cities and low-lying countries; the crops that feed humanity will suffer many devastating harvest failures as a result of high temperatures, new kinds of pests, droughts, floods, losses of biodiversity (such as pollinating species), and other calamities. We will study these threats in detail.

A few years ago, a group of scientists noted that what humanity is doing, including producing carbon emissions but also much more, is disrupting not just the climate but several of Earth's natural systems. These include the depletion of freshwater sources (such as underground aquifers); the pollution from heavy use of chemical fertilizers (applied in order to improve crop productivity); the change in ocean chemistry, mainly the increasing acidity of the ocean resulting from atmospheric CO_2 dissolving into ocean water; the clearing of forests to create new pastureland and farmland; and particulate pollution caused by many industrial processes, especially those involving the combustion of coal. All pose deep threats to the Earth and the wellbeing of humanity. These scientists argued that the extent of the damage is so large that humanity is leaving the "safe operating conditions" for the planet (Rockström et al. 2009). It is as if we are driving the car right off the road and into the ditch, or worse, right over the cliff.

The scientists argued that it is urgent to identify the safe operating limits for the planet or, put another way, to define the "planetary boundaries" beyond which humanity should not venture. For example, pushing CO_2 to 400 ppm might be dangerous, but pushing CO_2 to 450 ppm (through continued heavy use of fossil fuels) could be reckless. Depleting some groundwater could be inconvenient. Depleting major aquifers could be devastating. Raising the ocean's acidity slightly could be bad for shellfish. Raising the ocean's acidity dramatically could kill off a massive amount of marine life, including the species of fish and shellfish that humanity consumes as a vital part of our food supply.

Figure 1.16 offers the scientists' visualization of these planetary boundaries (Rockström et al. 2009, 472). Starting at 12 o'clock and moving clockwise around the circle we see the ten major planetary boundaries that humanity is in danger of exceeding, starting with climate change, ocean acidification, and so forth. The red shaded area shows the scientists' assessment of how close the world is to exceeding each of these boundaries. In the case of nitrogen flux (from fertilizer use) and biodiversity loss, the entire wedge of the circle is red. We have already exceeded these planetary boundaries. For other threats, we are still some way from the boundaries, although the red-shaded portions of each slice of the pie

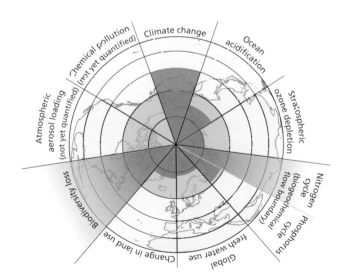

1.16 Planetary boundaries

Reprinted by permission from Macmillan Publishers Ltd: Nature, Rockström, Johan, Will Steffen, Kevin Noone, Åsa Persson, F. Stuart Chapin, Eric F. Lambin, Timothy M. Lenton et al. "A Safe Operating Space for Humanity," copyright 2009.

are increasing rapidly. During the twenty-first century, the entire circle will likely turn red unless there is a fundamental change of strategy. Put another way, humanity will exceed the safe operating limits unless the world adopts a strategy to achieve sustainable development.

V. Pathways to Sustainable Development

The first part of sustainable development—the analytical part—is to understand the interlinkages of the economy, society, environment, and politics. The second part of sustainable development—the normative part—is to do something about the dangers we face, to set SDGs, and to achieve them! Our overarching goal should be to find a global path, made up of local and national paths, in which the world promotes inclusive and sustainable economic development, thereby combining the economic, social, and environmental objectives. This can only be accomplished if a fourth objective—good governance of both governments and businesses—is also achieved. Good governance, I shall repeatedly emphasize, means many things. It applies not only to government but also to business. It means that both the public sector (government) and the private sector (business) operate according to the rule of law, with accountability, transparency, responsiveness to the needs of stakeholders, and with the active engagement of the public on critical issues such as land use, pollution, and the fairness and honesty of political and business practices.

In the coming chapters, I will constantly refer to a comparison. On the one hand, we will consider the implications of humanity continuing on the current course. For example, suppose that the world economy continues as today to be run mainly on fossil fuels, so that the CO_2 concentration in the atmosphere continues to rise rapidly. Or suppose that farmers continue to overuse groundwater so much that the aquifers are depleted. These scenarios will be called *business as usual*, or BAU for short. Such scenarios will be compared with a dramatic change of course for humanity, one in which the world quickly adopts new technologies (e.g., solar power to replace coal-fired electricity generation or more efficient water use to avoid depleting the aquifers). The alternative path, one that aims not only for economic growth but also for social inclusion and environmental sustainability, will be called the *sustainable development* path, or SD for short.

We will examine and contrast the BAU and SD trajectories. If we continue with BAU, what would happen? Certainly there would continue to be many kinds of progress. Science and technology won't stand still. The poor will benefit from advances in ICTs, such as access to higher education through free, online learning. Poverty would continue to fall in many places. The rich might continue to become richer for another decade or two. Yet eventually, the negative consequences of rising inequality and rising environmental destruction will come to dominate the positive tendencies. Progress will peak. Calamities, both social and environmental, will start to dominate. More than 200 years of progress could be choked off, and even sacrificed to war.

What about SDGs? Can we find alternatives to fossil fuels, groundwater, pasturelands, and the like, to meet human needs without destroying the physical environment? Some of the key solutions are likely to be more expensive in the short term, such as buildings specially designed to use less energy for heating through better design, insulation, materials, and overall systems strategy; or electric vehicles with battery power that are still expensive compared with normal gas-guzzling internal combustion engines. Some fear that we can't afford the SD path; that the SD path might "save" humanity at the cost of ending economic progress; and that SDGs are therefore unrealistic, even impossible to achieve. A major task of this book is to examine this claim. Without giving away the entire plot, I'll say at the outset that if we are clever and apply ourselves to the study and design of new sustainable business practices and technologies, sustainable development is both feasible and affordable. Indeed, it is business as usual that eventually would impose the truly devastating costs.

The essence of sustainable development in practice is *scientifically and morally based problem solving*. We indeed have a lot of problems. We have continued life-threatening poverty in the midst of plenty. We have built up inequalities of wealth and poverty, and we have deployed technology systems that are now crossing planetary boundaries. We are going to need a coordinated global effort in a focused and relatively short period of time, a matter of decades rather than centuries, to move from the BAU to the SD trajectory. In order to accomplish the SDGs, every part of the world will have to be involved in problem solving, in brainstorming, and in determining new and creative ways to ensure inclusive and sustainable growth. This book aims to contribute to that problem solving. We will describe the challenges, identify the best candidates for SDGs, and determine how those SDGs can in fact be achieved.

2

AN UNEQUAL WORLD

I. Incomes Around the World

Sustainable development has three major aspects: economic development, broad-based social inclusion, and environmental sustainability; all supported by good governance. But what do we mean by economic development? How do we measure it, and what is the state of play in today's very complicated and diverse world?

There are many different dimensions of economic development, and therefore many measurements are needed to assess a country's development process. Nonetheless, we tend to rely heavily on a single measurement called the gross domestic product (GDP) of a country. The GDP means the total production occurring within the geographic boundaries of a country, typically expressed for a one-year period. *Gross* in the GDP means measuring every market transaction within the country. *Domestic* signifies that the measurement is of economic activity within specific geographic boundaries; usually a country, but sometimes a city, region, or even the world. *Production* signifies that what we're measuring is not the trade in preexisting capital (such as the resale of a house) but the flow of new output in a given time period.

There is one more critical point about measuring GDP. In general we are interested in getting a sense of a country's standard of living. To do that, we take the total production in the country over a given time period and divide it by the population to derive the GDP per person, or GDP *per capita*. Larger countries have more people and more workers and therefore produce more. If we simply compared countries in terms of the total production, we would find that highly populated countries have higher production, but we would not know whether the living standards of the larger countries are really higher than those of the smaller countries that produce less in total but more per person.

GDP per person is really not a comprehensive measure of economic development, because there are many other important indicators of wellbeing that it does not precisely capture, including the health and education of the population. Still, as a shorthand, GDP per person is a reasonable first indicator of where things stand on many dimensions, all of which tend to be related to output per capita, even if they are not precisely measured by it. Indeed, GDP per capita is what the World Bank and other international organizations use to summarize a country's current level of development. The United Nations and the World Bank keep systematic tabulations on GDP per person and classify countries on this basis.

The World Bank places countries into three main categories: high-income, middle-income, and low-income. The classification is based on the country's GDP per capita. On the current criteria, a country is low-income if its GDP per capita is below $1,035 per person per year, or about $3 per day. A middle-income country is in a band between $1,035 and $12,615 per person per year. The high-income countries are above the $12,616 per person threshold. There are then refinements. The middle-income group, which is quite big, is split between the upper-middle-income and the lower-middle-income, with the dividing line at $4,085 per person per year. Note that the precise boundaries are readjusted periodically with changes in global market conditions.

Figure 2.1 shows the high-income countries, shaded blue: the United States, Canada, western Europe, Japan, South Korea, Australia, New Zealand, and a few other parts of the world. This is about 1 billion of the roughly 7 billion people on the planet, around 15 percent of the world's population. The large middle group of green colors covers quite a wide expanse of the world. Indeed, five out of seven in the world's population are in the middle-income category, with approximately 2.5 billion people in both the upper-middle-income and the lower-middle-income

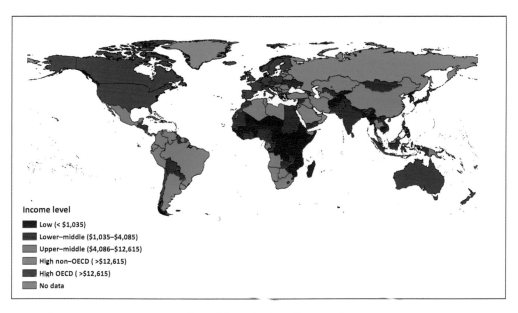

2.1 Country income groups (World Bank classifications)

Source: World DataBank.

categories. The countries shaded red are the low-income countries, with approximately 1 billion people. There is a strong geographical pattern in these classifications. The low-income countries are heavily concentrated in two regions: tropical Africa and South Asia, with a few other low-income countries scattered in other parts of the world. (Note that tropical Africa signifies the part of Africa between the Tropic of Cancer at 23.45 degrees north latitude and the Tropic of Capricorn at 23.45 degrees south latitude. It excludes the countries of North Africa and those of southern Africa.)

There is one more very important UN category. There is a subgroup within the low-income countries that is in rather desperate shape. Not only are they poor; in most cases the levels of disease, education, and social instability are also unusually bad. Many are relatively isolated, poor island economies. Moreover, many of the countries in this group are very vulnerable to droughts, floods, conflict, and violence. The United Nations has classified this group as the *least-developed countries* (LDCs). Figure 2.2 shows the fifty countries currently on the list. Among the

LDCs, the poorest of those are again heavily concentrated in tropical Africa and Asia (in this case spanning central, South, and Southeast Asia).

The map also suggests something very interesting and telling. Asia, Afghanistan, Nepal, Bhutan, and Laos are all landlocked countries. In Africa, the LDCs include several more landlocked countries: Burkina Faso, Mali, Niger, Chad, the Central African Republic, Uganda, Rwanda, Burundi, and Malawi. This very high presence of landlocked countries on the LDCs list is not a coincidence. Economic development depends heavily on international trade, which is significantly more difficult for landlocked countries. Being landlocked does not sentence a country to extreme poverty (just think of Switzerland and Austria!), yet it is an added hurdle to development, especially when the landlocked country is surrounded by poor coastal economies. Note that the LDCs also include a number of small island economies, which can be quite vulnerable. They are subject to extreme climate catastrophes (such as tropical cyclones, storm surges, and droughts) and are often relatively isolated with small populations and high shipping costs to major ports.

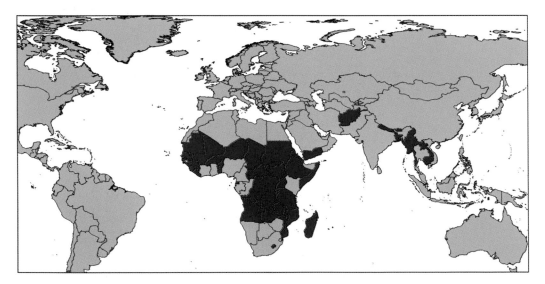

2.2 The least-developed countries (LDCs)

Source: World DataBank.

There are two more important details in terms of GDP measurement. Since countries typically have their own national currencies (or are part of a regional currency such as the euro), each nation's transactions are measured in its own currency. To make a common standard, the national currencies must be converted to a common currency using an exchange rate. Almost always, the common standard is the U.S. dollar, with each country's exchange rate to the dollar used to convert the GDP to a common U.S. dollar base. Mexico's 2012 GDP per capita, for example, is 135,000 Mexican pesos per dollar. The peso-dollar exchange rate in 2012 was 13.2 pesos per dollar. Therefore, the Mexican GDP equaled $10,200 per person in U.S. dollars.

There is one more translation that can also be enormously useful to make one more adjustment for the difference of costs or prices in different countries. If one buys an automobile or a television set almost anywhere in the world, the price will be fairly similar, because these are goods that are traded internationally and therefore have a similar price. Yet for many other goods and services, such as apartment rents, foods that are produced and consumed locally, or personal services (such as haircuts or movie tickets), prices can vary enormously across countries even when converted to dollars by the market exchange rate. The price of a haircut in Mogadishu might be one hundredth of what one would pay in a salon in Paris. We must take such price differences into account in order to compare living standards across countries.

Suppose that in one country the GDP per capita (converted to U.S. dollars at the market exchange rate) is $6,000 per person, while in a second country the GDP per capita is $3,000 per person. It might appear that the first country is twice as rich as the second. Yet if the average price level in the second country is also half of that in the first, that is, if the goods and services on average cost only half as much, the actual living standards of the two countries would be comparable.

To take price-level differences into account, the GDP per capita is sometimes measured using a common standard of international prices, such as prices for foodstuffs, rent, haircuts, movie tickets, legal fees, and the rest. This unit of account is called GDP per capita at international prices (expressed in dollars), or GDP per capita at purchasing power parity. Purchasing power parity is sometimes abbreviated as PPP, so one will often see this measurement called GDP per capita (PPP).

Consider once again the Mexican example. In 2012, the Mexican GDP per capita was around $10,200, while the U.S. GDP per capita was around $51,000.

It would seem that average living standards (based on GDP per capita) were roughly five times higher in the United States. Yet Mexico's price level was lower than in the United States, so Mexico's living standards did not really lag behind those of the United States by a factor of five. On one set of measurements, used by the International Monetary Fund (IMF), Mexico's overall price level was roughly two-thirds of the U.S. price level (taking an average over a large number of goods and services). Thus, while the GDP per capita was $10,200, Mexico's GDP per capita at PPP was roughly $15,400. Thus, the United States exceeded Mexico's average GDP per person not by a factor of five, but by a factor of 3.3, not quite so large.

The PPP adjustment is largest for the poorest countries (since the price level of goods and services in the poorest countries tends to be lowest—think haircuts!). In a typical poor African country, for example, the GDP per capita in PPP units tends to be three to four times larger than the GDP expressed at market prices. Consider poor, landlocked Malawi, for example. For 2012, the IMF records its GDP per capita at market prices at $250, less than $1 per day. Adjusting for differences in price level, Malawi's GDP per capita (PPP) was $848, still a very low living standard but not quite as low as suggested by the GDP at market prices.

The big picture is that we classify economic development in shorthand by the GDP per person, adjusting for population, currency, and price level. We can then study key questions. Why are countries at different levels of economic development? How do these levels of development relate to things like health, wellbeing, and happiness, concepts not directly measured by GDP per capita? What can the low-income countries do to raise living standards and achieve economic growth fast enough to narrow the gap significantly with the wealthier countries? This final question, of course, is one of the key policy challenges of sustainable development.

II. Urban-Rural Inequality

We have looked at GDP per person as a summary measure of how to classify countries in their respective levels of economic development, but we should note that countries have great variations of living standards within them as well as between them. It is very important, especially in view of sustainable development's commitment to social inclusion and broad-based prosperity, to understand the variations and inequalities of living standards within countries. Perhaps the starkest kind of variation within countries is the difference between rural and urban life.

Before the Industrial Revolution, virtually the entire world population, roughly 90 percent, were living in rural areas, eking out an existence as smallholder farmers trying to grow enough food for their families and perhaps a little for the marketplace, at least in good years. When we think about the preindustrial era, we might think about bucolic England, with shepherds guiding their flocks on the hillsides while farmers toil in the fields below. That kind of rural scene is still very familiar in many parts of the world, especially in Africa and Asia. However, the world is very quickly becoming more urbanized. The process of urbanization is changing lives in fundamental ways and leading to vast differences within countries as well.

In almost every country in the world there are both rural and urban populations, often with very different qualities and types of lives. (The key exceptions are a number of small city-states, such as Hong Kong and Singapore, where there is no rural life.) Since many countries are in a transition from predominantly rural to predominantly urban life, it is quite important to understand the differences between rural and urban living and what these differences mean in terms of wellbeing, income levels, and kinds of economic activities.

It is important to start by clarifying the definition of "urban." Interestingly, there is no official international definition of what it means to be an urban area. The United Nations relies on national definitions, which differ across countries. Generally speaking, an urban area is a place where at least several thousand people live in a relative densely settled area. Of course there can be millions of people living in a single densely settled area (called an *urban agglomeration*), but it is the minimum threshold that defines an urban area versus a rural village, and that threshold varies by country. Some countries use a threshold of at least 2,000 people living in a dense settlement as the dividing line between a rural village and a small urban area. Other countries use 5,000 as the threshold.

Either way, the difference between rural and urban has some pretty basic characteristics that are very important to the process of economic development and to the nature of inequality within countries. The first consequential difference is what people do to earn a living. Agriculture is the mainstay of rural areas, whereas industry and services are the mainstays of urban areas. As countries experience rising proportions of their populations in urban areas, this also signifies a rising proportion of the labor force in jobs in industry and services and a declining share in agriculture. This is a nearly universal trend as part of the process of GDP per capita rising.

Generally speaking, income per person tends to be higher in urban areas than in rural areas, which encourages the flow of the rural population to the urban areas. (Such a pattern is common but not universal.) The locations of rural villages and urban areas generally also differ. Rural populations are settled in good food-growing areas. Urban areas tend to be located at coasts or along rivers, where it is easier to engage in sea-based trade. The balance of rural and urban is thus also typically a balance of interior and coastal orientation of the country. As countries urbanize, they tend also to become more coastal in orientation, that is, with a higher proportion of the population living near the sea.

Population densities—the numbers of people residing per square mile or square kilometer—also tend to be very different in rural and urban areas. Population densities are generally low in rural areas, since each farm family needs a high ratio of land to people. The population density is typically below 100 people living within a square kilometer (km^2), although higher ratios are recorded in crowded rural Asia. By contrast, urban areas are often packed with thousands of people per square kilometer.

As a result, the quality of public services also tends to differ. It is much harder to provide electricity, piped water, and sewerage systems in rural areas where populations are quite dispersed, relative to urban areas where populations are tightly settled. This is one of the reasons why income levels, health standards, and overall living standards tend to be higher in urban areas.

One other quite notable difference between rural and urban areas is that fertility rates (the average number of children per woman) tend to be higher in rural areas, or to put it differently, rural families tend to be larger. There are many reasons for this difference, but one key reason is that young children are often seen as productive farm workers, whereas children in urban areas are often seen as "expensive," because they go to school rather than do chores on the farm. The result is that when households move from rural to urban areas they also often choose to have fewer children. Other causes include: the higher income levels of urban households; the higher level of mothers' education in urban areas; the greater access of urban mothers to family planning and modern contraceptives; and the higher child survival in urban areas, so urban families have less fear about the survival of their young children. All of these factors contribute to the lower fertility rates in urban areas.

Global Trends in Urbanization

Figure 2.3, which shows the proportion of each country's population in urban areas, looks a bit like the map of income per person. The richer parts of the world tend to be more urban; the poorer parts of the world tend to be more rural. The Americas are very highly urbanized societies, with generally 80 percent or more

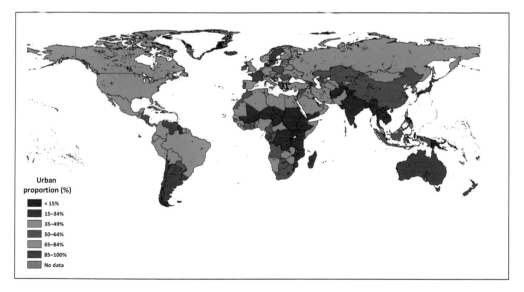

Urban proportion (%)
- < 15%
- 15–34%
- 35–49%
- 50–64%
- 65–84%
- 85–100%
- No data

2.3 Proportion of the population in urban areas

Source: United Nations Department of Economic and Social Affairs Population Division (DESA Population Division). 2012. "World Urbanization Prospects: The 2011 Revision."

of the population living in urban areas. Meanwhile, tropical Africa is still quite rural, with around 25–35 percent of the population living in urban areas. One sure fact, however, is that almost everywhere in the world, urbanization is proceeding rapidly and is a critical part of the economic development process.

Figure 2.4 shows the growth rates of the urban areas, demonstrating the annual proportionate increase of populations living in urban areas. It is actually in Africa, today's most rural region, where urbanization rates are extraordinarily high. Africa is catching up in urbanization. Africa's urban areas are often growing at around 5 percent per year, in which case it takes around 14 years for the urban areas to double in size (using the rule of 70, we find that 14 years = 70/5). In that case, a large urban agglomeration of 5 million today would become a mega-city of 10 million in just 14 years.

The worldwide trend is toward urbanization. We expect the world's population to reach 8 billion by 2025 and 9 billion by 2040. All of that increased population

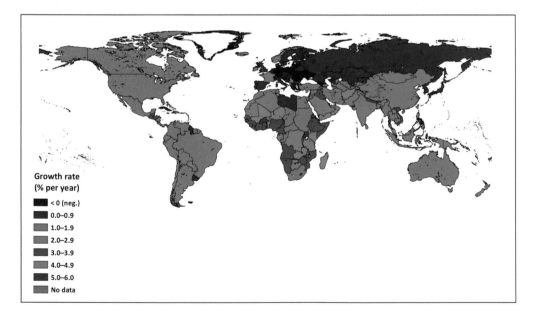

2.4 Urban growth rates

Source: United Nations Department of Economic and Social Affairs Population Division (DESA Population Division). 2012. "World Urbanization Prospects: The 2011 Revision."

is expected to live in urban areas, since the total rural population is expected to remain roughly constant at around 3.3 billion by 2035 and then to decline slightly to 3.2 billion by 2050. This means that all future population growth on the planet is urban population growth. The proportion of the world living in urban areas is going to rise from around 53 percent in 2013 to around 60 percent by 2030 and 67 percent by 2050. Prosperous, healthy, and resilient cities are going to be a core challenge of sustainable development.

As the process of urbanization takes place, the countryside is also transformed. For example, in the United States the farm population is now under 1 percent of the entire labor force and the rural population is only 19 percent, even though 95 percent of U.S. land is classified as rural. Farms in the United States are very large and efficient, whereas in today's poorer rural countries there are still many smallholder farmers working very small plots. As the population moves to urban areas, a number of those farms will eventually consolidate and become larger per farm household.

In sum, in the process of economic development the arrow is pretty strong from rural to urban. Urbanization is associated with higher incomes, better public services, better education, and declining fertility rates. We see that in many parts of the low- and middle-income regions of the world. Along the way, societies and cultures are deeply divided between rural and urban interests, politics, and ways of life.

III. Income Inequality Within Countries

The variation in incomes across households and individuals within a country can also be very large. We must take into account not just a country's average level of income, but its distribution of income. To use an old quip, a man whose feet are in the fire and whose head is on ice is asked, "How's the temperature?," and he replies, "On average just fine." The inequality of fire and ice is devastating, and societies can also be like that in their divisions of income, wealth, and opportunity. On average, income can be just fine, but if the average income is "just fine" because a few people are fantastically rich and the rest of the country is excruciatingly poor, things are not so fine after all.

Therefore, in addition to measuring the GDP per person, we want to measure the inequality of income within the country. Several indicators are used. We can look at the ratio of incomes of those at the top of the income distribution to those at the bottom, sometimes comparing the average incomes of the richest (top) 20 percent and poorest (bottom) 20 percent of households. Another useful, widely used measure is the Gini coefficient. The Gini coefficient varies between 0.0 and 1.0, with 0.0 meaning complete equality of income (every person or household has the same income), and 1.0 signifying complete inequality (all income is owned by one person or household, with all the rest having no income). Real societies are of course somewhere in between. Societies regarded as rather equal, with a broad middle class, like Sweden, Norway, or Denmark, have a Gini coefficient around 0.25. Countries that are much less equal by comparison, with both a lot of wealth at the top and a lot of poverty at the bottom, have a Gini coefficient of 0.4 or higher. Figure 2.5 maps the Gini coefficient around the world with the best recent comparative measurements. Note that measuring the Gini coefficient is difficult, with countries using survey data in different years and often with different definitions.

The lowest inequality (the blue countries) tends to be in western Europe and especially in Scandinavia, with a Gini of around 0.25. In comparison with Scandinavia, the United States is shaded green, as the United States is quite unequal in income distribution, with a recent Gini of 0.45. The United States has an estimated 442 billionaires and a remarkable estimated 13 million households with a net worth of more than 1 million dollars. Yet the United States also has tens of millions of very poor people, with very low incomes and almost no net worth at all (or, in fact, net debt). America's poor are not as excruciatingly poor as one would find in the LDCs, but they are poor indeed, having difficulty keeping food on the table. African countries, for those with Gini data, are also rather unequal. China was rather equal in pervasive poverty fifty years ago, but with its recent economic development and a stark divide between the wealthier urban areas and poorer rural areas, the inequalities in China have risen to levels similar to those of the United States.

Looking at the Gini coefficient among the high-income countries shows there are very different pathways to economic development. Getting richer does not mean necessarily becoming more unequal, but nor does it guarantee becoming

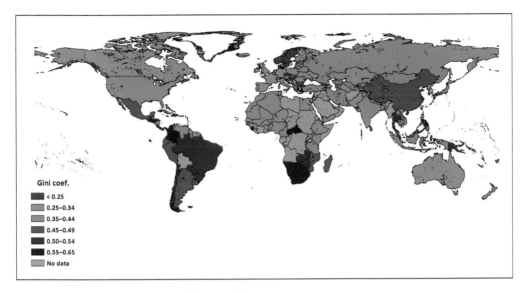

Gini coef.
- < 0.25
- 0.25–0.34
- 0.35–0.44
- 0.45–0.49
- 0.50–0.54
- 0.55–0.65
- No data

2.5 Gini coefficients around the world

Source: The World Factbook. CIA.

more equal. There are different pathways for development. Northern Europe has chosen a pathway of becoming wealthier with considerable social equality, whereas the United States has been on a path of rising incomes alongside rising inequality, as is shown in figure 2.6.

How do we explain these gaps? This is a very complicated, contentious, and much-debated topic. There are many reasons for inequality. History, geography, and government all play a big role in a society's relative equality or inequality. Traditionally, when most wealth was farm wealth, the size of landholdings made a big difference. Some countries, especially in the Americas, had huge farms and haciendas that were often taken by the Europeans who came to the Americas, displaced indigenous populations, and used slaves to work the large farms. These inequalities in the Americas continue today, though in more subtle ways.

In today's world where industry and services are much more important, variations in education levels are also a very significant source of inequality. Young

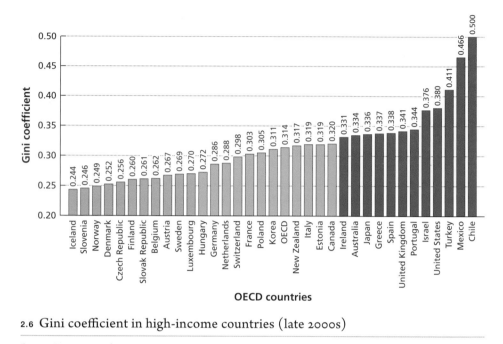

2.6 Gini coefficient in high-income countries (late 2000s)

Source: Organization for Economic Co-operation and Development. 2014. "Income Inequality." In OECD Factbook 2014: Economic, Environmental and Social Statistics. *OECD Publishing. doi: 10.1787/factbook-2014-24-en.*

people who are able to obtain a higher education are also usually able to translate their higher education into higher income levels. Children who are not able to obtain a higher level of education, often because of poverty, may end up with very low-paying jobs. Education can become an equalizer if everybody has the same educational opportunity, but it can also become a source of inequality as well if only the children of the rich are able to obtain a quality education.

As I noted, the rural-urban divide is another key to inequality. Families that move to urban areas often find better employment prospects and higher incomes, whereas those left in villages as smallholder farmers often barely survive.

Discrimination still matters tremendously. Women all over the world are not given the same chance as men in the labor market and do not earn the same incomes, even when they are doing the same job or better. Racial, ethnic, and religious minorities often face terrible conditions that hinder their access to education and to quality jobs in the labor market. Consequently, they are unable to get the kind of employment they deserve.

Government policies can make a huge difference to promoting equality or favoring inequality. Corrupt governments that use their revenues for a very small class of insiders tied to the government can create a huge amount of inequality. Often countries that live off a few mining resources such as oil, gold, or diamonds are very unequal because of the ways the government revenue is allocated among the favored insiders. The result can be what is known as the "natural resource curse," the paradox that a country with abundant natural resources ends up poor and undeveloped with massive corruption and inequality. (There can be other causes of a natural resource curse, such as when a resource boom causes an overvalued exchange rate that hurts the more technologically dynamic sectors in agriculture and industry.)

At the same time, governments can be great equalizers. If governments use revenues to ensure widespread access to education and health care, they can narrow the income inequality while also raising overall economic efficiency (by ensuring that the poor, like the rich, can invest in their own lifetime productivity). In Scandinavia, poor families are given financial support to ensure that their children too will still have a good chance to succeed in life. The result is a very low level of poverty, a high overall prosperity, and a very low degree of income inequality across households.

There is a lot of choice about what can be done to promote greater equality, and we are living in a period when we face that choice more starkly than ever. Practicing sustainable development means both understanding the nature and sources of inequality and setting the goal of a high degree of social inclusion in economic development. This is our battle and our challenge, to understand the inequalities of income within societies and to ensure that all children, rich and poor, have the chance of prosperity. If we find ways to ensure a big middle class and a decent chance for a child born into a poor family, as the impressive Scandinavian societies have done, that can become the path to social inclusion, a key pillar of sustainable development.

IV. Measuring Wellbeing

We are all interested in the overall quality of our lives, sometimes described as "life satisfaction," "wellbeing," or "happiness." Part of this wellbeing is determined by the ability to meet our material needs and aspirations, and therefore depends on income. Part depends on the social services provided by the government. We are also affected by the extent of personal security or insecurity (e.g., living in a war zone). Of course, extreme poverty is an affront to wellbeing and life satisfaction. Impoverished families go hungry, lack safe water and sanitation, and cannot access health care when they need it. Children die young and tragically. Extreme poverty is a burden that deserves our highest priority.

Yet we also know that "man does not live by bread alone." Material possessions are not everything. In fact their importance diminishes in proportion to income. For a poor family, income may loom largest as the determinant of wellbeing. An extra dollar of income can be the difference between a meal or no meal. For a rich person, an extra dollar of income will have essentially no effect at all on wellbeing.

We therefore know that simple measures like GDP per person give only a rough reflection of the overall level of wellbeing of an individual or a nation. But for sustainable development we are interested in raising human wellbeing, not just in raising income, still less in a mad race for more riches for people who are already rich. Therefore, it is important to ask how we can best measure wellbeing (or life satisfaction) beyond GDP per capita.

What then are some of the options for measuring human wellbeing that go beyond the simple calculation of the GDP per person? One important innovation, championed by the UN Development Programme (UNDP) during the last quarter century, is the Human Development Index (HDI) (UNDP 2013b). It tries to give a more holistic account of human development by taking into account the important things that empower people and help them meet their capacities. The Human Development Index takes income per person as one of the three basic dimensions of wellbeing. Instead of measuring income per capita directly, it uses the *logarithm* of income per capita. Using the logarithm, each higher level of income boosts the HDI by a smaller increment.

The HDI also uses indicators of educational attainment, such as mean and expected years of schooling, and indicators of health, notably the life expectancy at birth. By taking the weighted average of income, education, and health, UNDP creates the HDI. The map of the HDI in figure 2.7 is similar to the map of GDP per person, but it is by no means the same. Note that tropical Africa is once again the epicenter of the development challenge. Just as the World Bank categorizes countries as high-, middle-, and low-income, UNDP categorizes countries as exhibiting high, middle, and low levels of human development.

The GDP per capita and HDI are related but not the same. There are countries that are relatively low on income per capita but do quite well on the HDI, because they have favorable outcomes on life expectancy and educational attainment; and there are countries that are very rich on paper according to GDP per capita, yet their populations suffer poor standards of health and education and hence a level of human development far lower than would be suggested by income alone.

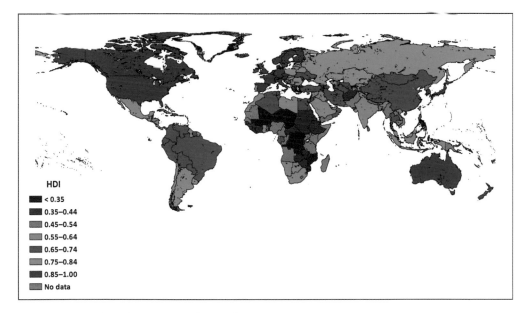

HDI
- < 0.35
- 0.35–0.44
- 0.45–0.54
- 0.55–0.64
- 0.65–0.74
- 0.75–0.84
- 0.85–1.00
- No data

2.7 Human Development Index rankings

Source: United Nations Development Programme. 2013. Human Development Report 2013. *New York: United Nations Development Programme.*

It is helpful to take a look at some examples of countries that are rather rich in terms of their GDP per capita but rather poor on their human development indicators. Equatorial Guinea is such a case. Equatorial Guinea was an utterly impoverished country in West Africa until major resources of oil and gas were discovered and developed by international oil companies. Those hydrocarbon resources can be hugely beneficial to the people of Equatorial Guinea, but only if those oil and gas earnings are used wisely and invested in raising over-all wellbeing. Yet so far, Equatorial Guinea's GDP per capita has run far ahead of the outcomes on literacy, life expectancy, and other dimensions of life. The great boost of oil income has yet to reach the large masses of the population in terms of discernible increases in quality of life. As it stands today, Equatorial Guinea ranks rather high on income per person, at 41st in the world; yet it ranks 144th in the HDI.

Other countries, by contrast, rank much higher on human development than on their average income per capita. A solid case in point is South Korea, one of the world's greatest development success stories in modern history. South Korea has had enormous economic growth over the last half century and has become one of the richest countries in the world. Part of that has resulted from its remarkable focus on raising educational standards and improving the health of the population. South Korea now ranks roughly thirtieth in GDP per capita on world standings, while it ranks even higher in HDI, at fifteenth in the world.

Table 2.1 includes some other interesting examples. The reader can suggest reasons for these discrepancies.

Subjective Wellbeing

There is another fascinating way that we can attempt to assess wellbeing. Why not ask people directly about the quality of their lives? Most frequently, surveys use what is called the Cantril ladder. People are asked to place their lives on a ladder with ten rungs, where the top (tenth) rung is the best possible life and the bottom rung is the worst. Specifically, the international survey firm Gallup International asks the following:

Table 2.1 Rankings of GDP per Capita and HDI for Selected Countries, 2013			
Country	GDP per capita rank	HDI rank	Difference
Some countries with HDI higher than GDP per capita			
United Kingdom	23	14	9
New Zealand	21	7	14
Slovenia	34	25	9
South Korea	30	15	15
Some countries with GDP per capita higher than HDI			
Qatar	3	31	−28
Kuwait	17	46	−29
Equatorial Guinea	41	144	−103
Gabon	60	112	−52

Source: HDI from UNDP 2014; GDP per capita from IMF (constant USD).

Please imagine a ladder with steps numbered from zero at the bottom to 10 at the top.

The top of the ladder represents the best possible life for you and the bottom of the ladder represents the worst possible life for you.

On which step of the ladder would you say you personally feel you stand at this time?

In recent years there has been an enormous, eye-opening global effort to assess in that straightforward way. The Organization for Economic Co-operation and Development and Gallup International have been in the forefront of this movement. The recent global results have been published in the *World Happiness Report*, which can be downloaded for free (Helliwell, Layard, and Sachs 2013).

Psychologists and other survey experts distinguish between two quite distinct dimensions of happiness and two very different kinds of questions to elicit these dimensions of happiness. One is to ask about a person's recent emotions. "Did you have a good day yesterday?," "Were you happy?," "Did you smile?" This is sometimes called emotional or *affective* happiness. The other way is to ask about overall life satisfaction, as is done with the Cantril ladder. "Are you satisfied with your life as a whole?," "Where would you put yourself on the ladder of life?" This is called *evaluative* happiness, since it seeks something more permanent than yesterday's emotions. It seeks an actual evaluation of life as a whole. Both measures are important, but evaluative happiness is the one used to get an overall sense of life satisfaction in a country.

Figure 2.8 shows the distribution of reported evaluative happiness around the world, based on the Cantril ladder. It is quite fascinating. Richer countries do tend to be happier, but some middle-income countries are close to the top of the

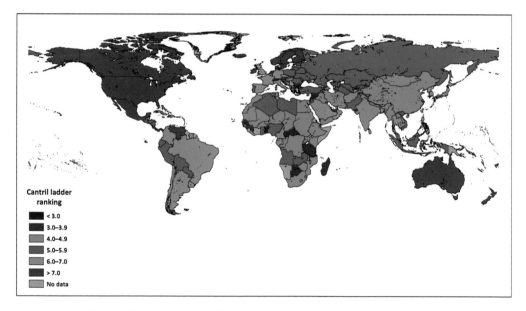

2.8 Subjective wellbeing around the world

Source: Helliwell, John, Richard Layard, and Jeffrey D. Sachs. 2013. World Happiness Report. New York: Sustainable Development Solutions Network.

charts in happiness, and some richer countries are not so happy. We can learn a lot from this. The Americas, western Europe, Australia, and New Zealand rank highly on happiness, while some of the poorer regions, perhaps not surprisingly, rank much lower.

What do we learn when we study the differences in life satisfaction around the world in this evaluative sense? (I will use the word "happiness" as shorthand for "life satisfaction" in this discussion.) We find out that income per person (GDP per capita) indeed matters, but as just one aspect of happiness. A second major reason for happiness versus unhappiness is "social capital," or the quality of the social environment and community. Do people have good support networks of friends and colleagues? Do they trust others in their community? Do they trust their government to be honest? The quality of social life is an enormously important determinant of whether people report high or low life satisfaction. As Aristotle said more than 2,000 years ago, "Man is by nature a social animal." Our happiness depends fundamentally on our relations with others.

Not surprisingly, physical and mental health play a very important role. We should underscore that mental health—for example, the presence of depression or anxiety disorders—can cause havoc in a person's life, lowering life satisfaction dramatically. The availability of mental health services is therefore a key intervention in raising life satisfaction for people suffering from mental health disorders.

Finally (though not exhaustively), each individual's values and the values of society at large are related to happiness. People who report a very strong orientation to materialistic values (such as an intense drive to earn more money or to accumulate more consumer goods) do not report as much life satisfaction as those with a less materialistic orientation. Buddha, Aristotle, and other sages made this point in ancient times, and it has been verified in careful modern psychological studies and opinion surveys. Those individuals who report that generosity is very important to them, through volunteering, philanthropy, or other forms of altruism, also report greater happiness. The role of altruism not only in helping the recipient but in helping the giver as well is obviously important in thinking about the path to sustainable development.

If we as individuals simply pursue income per capita as our main goal in life, we will lose out on many counts. Our societies will tend to become highly unequal.

The rich will use their political influence to gain further income and wealth. The environment will come under great threat, since short-term desire for income will tend to trump long-term concern for the environment and the wellbeing of future generations. In the end, if our societies are driven overwhelmingly by the goal of increased incomes and consumerism, we are unlikely to achieve the kind of happiness and life satisfaction that we desire. With a more balanced and holistic approach, as sustainable development bids us to undertake, greater happiness can arise. Yes, societies should pay attention to higher income per person (especially when they are poor!), but they should also focus on health, social inclusion, honest government, and networks of social support and altruism. Societies will benefit if they find ways to promote the values of generosity, compassion, and volunteerism rather than the values of individualistic materialism.

In the end, the overriding goal is not income but life satisfaction and wellbeing. We have seen again and again the importance of a holistic perspective. Fortunately, as we have seen, we have more and more tools to measure, to assess, and ultimately to help promote improvements in wellbeing.

V. Convergence or Divergence?

One of the most important questions in the study of economic development is whether today's poor countries have the chance, and indeed are on a path, to close the large gap in GDP per capita and other indicators of wellbeing with the high-income countries. That successful transition would improve not only income per person but also other important aspects of life such as health, life expectancy, educational attainment, and life satisfaction. Throughout this book, and throughout the pursuit of sustainable development, we are interested in determining how poor countries can narrow and eventually close the development gap with richer countries.

Economists use a couple of important terms for this concept. The term *convergence* is used to convey a narrowing of the proportionate gap of a poor country and a richer comparison country. (The proportionate gap signifies the ratio of the poor country to the rich country in the variable of interest.) The opposite of convergence is *divergence*, which means that the poorer country, in relative terms, is becoming even poorer when compared with the rich country. The study of whether countries are converging or diverging tells us a lot about whether overall differences of material life, life expectancy, health, educational levels, and the degree of urbanization are tending to narrow or instead are tending to widen. This is a very complicated question, and no one story line fits all cases. It is fair to say, however, that in broad terms the first phase of modern economic growth, roughly from 1750 to 1950, was characterized mainly by divergence and since the middle of the twentieth century, the forces of convergence have tended to gain the upper hand.

Modern economic growth, we shall see shortly, took off in the Industrial Revolution, roughly dated from around 1750 to 1850. Until the Industrial Revolution, most of the world was poor and rural, and so the gaps of rich and poor countries were quite narrow, not like the huge gaps today. The Industrial Revolution sparked a takeoff in GDP per capita in a relatively small part of the world, starting in England and then spreading to all of Great Britain, much of western Europe, across the Atlantic to the United States and Canada, and across the Pacific to Oceania (Australia and New Zealand). Initially, very few other places experienced early industrialization. The overall economic process viewed globally was one of divergence. The rich in Great Britain, the United States, and a few other places, were becoming richer because they were industrializing. The poor remained as they always were, eking out their living in peasant farming in most of the world, relatively untouched by the new inventions of the steam engine, railroad, telegraph, or other technological advances.

Then came a major new phase in the political phenomenon of imperialism. As western Europe became industrially and militarily dominant, it also seized political control over more and more of the world, notably in Africa, Asia, and the Middle East. This was a big political setback to the potential of convergence. Colonized peoples were not able to undertake the key steps crucial for economic development, such as improving infrastructure and raising educational levels. The imperial masters were generally not interested in the overall economic and social development of their colonial possessions but were focused on extracting as many of the resources of those countries as they could for the benefit of the home country and its industries.

In the two decades after World War II, a quite different political development, one very important for global economic development, occurred: the end of imperial rule. Countries around the world rapidly gained political independence, often through political and military struggle, and this newly won independence gave them a much greater ability to undertake economic development on their own. Newly independent countries were finally able to start building infrastructure to provide a basis for industrial development, and they began trying to attract both foreign and domestic investors. With that massive global political change, faster economic growth in the poorer countries became much more of a typical condition than it had been in the decades prior to the 1950s. Further technological developments such as better transport, better communications, and the new information age enabled the poor countries of the world to pick up the pace of their economic development. The past half century has thus seen a tendency toward convergence.

One of the crucial goals of sustainable development is that *all of today's low-income countries*, especially the LDCs, should make that transition successfully through convergence to at least middle-income status. It is therefore critical for us to understand how that process can work and to address and overcome some of the remaining barriers to convergence. The world today provides many examples of countries that were once poor achieving very rapid convergence. China is the exemplary case, with an astounding economic performance over the roughly thirty-five years since 1978, the watershed year in which China undertook many important economic reforms that put it on a path of convergent growth.

Yet we also see many parts of the world still stuck in poverty. That is a poverty trap so tight that countries are not yet achieving economic convergence. Niger is an example of such a country. It is a landlocked country in the Sahel region of Africa, just south of the Sahara Desert. It is one of the world's poorest countries. It is also at the bottom of the world's HDI, meaning that not only is it income poor, but its health and education conditions are dire as well. Unlike China, Niger has been stuck in a poverty trap for a long period and has not been achieving economic convergence. The numbers are illuminating. Looking at GDP per person in PPP-adjusted terms, which take into account differences in price levels across countries, the U.S. per capita income in 1980 in the prices of those days was about

$12,000 per person. In China, it was about $250 per person. In Niger it was about $450 per person (IMF 2014).

Now fast-forward to the year 2010. What has happened? We have already noted in the previous chapter the growth of Shenzhen, China, from a village to a modern metropolis. China experienced more than three decades of double-digit economic growth, with the economy doubling roughly every seven years on average. That meant that by the time 2010 came around, China was no longer at $250 per person; it was now at nearly $10,000 per person.

Niger, unfortunately, is still stuck at below $1,000 per person and is still among the least developed countries. China therefore went from being around 2 percent of the U.S. GDP per person to being around 20 percent. It is still far below the U.S. income level, but it is narrowing the gap very quickly. Niger, sadly, started out at around 4 percent of the U.S. per capita level, but by 2010 was less than 2 percent. In other words, it has been falling behind and experiencing divergence rather than convergence. One of the most important objectives in the coming chapters will be to examine the processes of convergence versus divergence to try to understand the underlying factors so that today's poorest countries still stuck in the poverty trap, like Niger, can get themselves (with the help of the rest of the world) on to a trajectory of convergence. When that happens, these countries too can enjoy the improvements in material life and related benefits (in health, education, and more) that we know are key components of improved wellbeing.

3

A BRIEF HISTORY OF
ECONOMIC DEVELOPMENT

I. The Age of Modern Economic Growth

The world we've seen is divided—startlingly so—with 55 high-income economies (1.3 billion people), 103 middle-income countries (4.9 billion people), and 36 low-income countries (0.8 billion people). How did these vast differences across the world come about? How is it that there are countries like the United States at more than $50,000 per person per year of income, and countries like Niger at under $500 per person per year, less than 100th the income levels of the high-income countries when measured at market exchange rates? This huge gap certainly did not exist two centuries ago.

Just before the great takeoff of modern economic growth—before the start of the Industrial Revolution around 1750—the world was fairly equal in income levels. To be more precise, the world was nearly equal in its poverty. Just about every part of the world was rural, with smallholder peasant farmers trying to eke out a survival for their families. A bad harvest could mean famine and death in any part of the world. Today's rich countries in Europe were certainly not exempt, as the hunger that fueled the French Revolution demonstrates.

The story of today's inequality therefore is also the story of the era of modern economic growth, the period since the start of the Industrial Revolution. It is only in this period that some parts of the world experienced sustained increases of gross domestic product (GDP) per person over long stretches and thereby transformed themselves from rural to urban, from peasant agriculture to high-yield agriculture, and from cottage industry (e.g., spinning and weaving) to modern industry and by now to a modern high-tech, knowledge-intensive industrial and service economy. It is only in this modern period of roughly 250 years that the vast gaps in income between the rich and the poor opened up.

How did this happen? And why did modern economic growth take off early only in some places in the world? If all countries started poor, why did some get rich while others were left behind? We need to understand the nature of modern economic growth and why it has varied so much across the world. And as sustainable development is also a set of goals including the end of poverty, we must consider what can be done to unlock rapid economic growth in today's low-income countries and especially in the least-developed countries (LDCs).

This takeoff of modern economic growth is a new event from the point of view of the long haul of human history. Our modern species, *Homo sapiens*, has been around for roughly 150,000 years. Our civilization, based on settled agriculture, is roughly 10,000 years old. During almost all of that period, economic change was so gradual that life seemed unchanged from one generation to the next, aside from wars, famines, and other temporary shocks. The idea of sustained economic progress simply did not exist. There was no evidence for it.

We can see that very starkly in the depiction in figure 1.4 (as best it can be estimated) of the growth of the world economy over the very long haul of human history from 1 C.E. The estimate of the total world output is essentially flat for almost 1,800 years (and would be flat for even longer if we went back further in time). The economic takeoff begins around 1750, and then world output shoots up sharply and dramatically. That steep upward sloping line, looking almost vertical in recent years, continues upward today because the world economy is growing very strongly today, although at different rates in different parts of the world.

The total output of the world, which is the sum of the GDP in each country, has two parts. One is the GDP per person in the world, and the other is the world population. The product of the two gives us total world output. Was the world economy's takeoff around 1750 due to rising output per person, or was it due to a rise in the number of people? In fact, both factors have played a huge role. The world population was fairly stable for thousands of years, below half a billion people at the time of the Roman Empire, though of course with significant fluctuations such as during bad periods like the Black Death. Starting in the middle of the eighteenth century, the world population turned steeply upward, as figure 1.3 shows (Maddison 2006, 242). This population increase was largely enabled by changes in economic and technological know-how, most importantly in the ability to grow food and thereby sustain a larger global population.

Output per person also soared, starting at roughly the same time (the onset of the Industrial Revolution) (Maddison 2006, 262). The graph in figure 1.2 of world output per capita is also nearly flat for centuries. If there was a long-term rise in output per person, the progress that occurred over decades, even over centuries, was nearly imperceptible before the middle of the eighteenth century. Only then did output per person begin its steep climb.

These numbers are of course rough estimates, the best possible reconstructions using various kinds of evidence. The world before 1750 was a world of poverty; one that could nonetheless produce beautiful treasures for human history, like the Egyptian pyramids, the Acropolis, the Great Wall of China, the Hagia Sophia in Constantinople, and Notre Dame Cathedral. Yet for all of those grand monuments, most people in most ages lived difficult rural lives, always on the edge of famine, disease, and early death.

One of the greatest economists of modern history, the British economist John Maynard Keynes, wrote a quite remarkable description of this long period of near stasis from the time of the Roman Empire until the onset of the Industrial Revolution. Keynes wrote:

> From the earliest times of which we have the record, back say to 2000 years before Christ, down to the beginning of the 18th century there was no very great change in the standard of life of the average man living in the civilized centers of the earth. . . .
>
> This slow rate of progress, or lack of progress, was due to two reasons—to the remarkable absence of important technical improvements and to the failure of capital to accumulate. The absence of important technical inventions between the prehistoric age and the comparatively modern times is truly remarkable. Almost

everything which really matters and which the world possessed at the commencement of the modern age was already known to man at the dawn of history. Language, fire, the same domestic animals which we have today. Wheat, barley, the
vine and the olive, the plow, the wheel, the oar, the sail, leather, linens and cloth,
bricks and pots, gold and silver, copper, tin and lead—and iron was added to the
list before 1000 B.C.—banking, statecraft, mathematics, astronomy, and religion.
(Keynes 1930, 2)

Keynes's point is that technology is crucial for the long haul of economic development. For a very long period of history, technology was relatively unchanging,
so much so that farmers in Roman times and in early seventeenth-century England would have experienced similar conditions: the same techniques, similar living standards, a world little changed over a span of seventeen centuries!

Then, dramatically, everything changes. The curves of population, output per
capita, and technological advancement start soaring out of sight. That is the next
subject we will tackle: how the Industrial Revolution began and how it changed
human history and human destiny.

II. The Industrial Revolution Begins in England

Modern economic growth began in England. This unique phenomenon started in
one particular place on the planet. We can watch it, and therefore we can understand how this came about. It is a bit like a biologist being able to watch the start
of life itself. Life as we know it appeared just once, and from there it has evolved.
Modern economic growth also has a kind of DNA. It also came together from a
number of different materials, and something took off. This was an extraordinarily
unusual occurrence that happened in England in the middle of the eighteenth
century. If it were so easy to create economic life, it would have happened in many
places. But as Keynes rightly pointed out, this did not happen. What happened in
the middle of the eighteenth century in England was a unique coming together of
various forces that allowed economic life to take off and eventually spread to the
entire world economy.

What is it about the Industrial Revolution? We can take a hint from the word *industry* itself. For the first time, a society moved beyond agriculture as the economic base to one in which industry was the economic base. This required a fundamental change in know-how and technology. Just as life depends on the complex interactions of many components of a living cell, so too, the life of a modern economy requires the interactions of many parts. New technologies—the steam engine, mechanized spinning and weaving, large-scale steel production—were certainly vital, but many complex economic interconnections were needed as well. Rural areas needed higher food productivity to produce a surplus for the industrial workforce (which obviously was no longer growing its own food). Transport was needed to carry food from farms to industrial towns, and industrial goods such as linens and apparel from the factories to the countryside. New ports and global shipping carried manufactured goods abroad as exports, to be traded for the primary commodities needed for industrial production. A worldwide supply system began to take hold. And these increasingly complex transactions required markets, insurance, finance, property rights, and other "software" and "hardware" of a modern market-based economy.

Thus, the emergence of modern economic growth reflected the unique confluence of several factors, and England was the place where these factors first aligned. One sees several notable things happening in the 1600s and 1700s in England, including many social and technical innovations imported from the Netherlands across the North Sea. First, agricultural productivity started to rise. There was more urbanization. More trade. A more sophisticated market economy began to take hold. Property rights deepened in complexity and flexibility (e.g., in the formation of new companies or the protection of patents for new discoveries). The rule of law deepened. And of course there was the wonder of the Scientific Revolution of the 1500s and 1600s. Galileo had paved the way for a new physics and opened the path for the incredible discoveries of perhaps the greatest physicist of all time, Isaac Newton. Newton changed the way humanity looked at the world, in terms of timeless and discoverable laws of nature. Sir Francis Bacon, writing before Newton, predicted that science and technology could dramatically transform the world for human benefit. In this, he proved among the most prescient thinkers of history.

One of the great technological breakthroughs came in 1712, with the invention of a steam engine by Thomas Newcomen. Newcomen's new invention burned coal to create motive force that could be used to pump water out of the shafts of coal mines. And then came a wonderfully creative and fiercely targeted genius, who realized that Newcomen had, however, made a couple of design mistakes. James Watt, working in a Glaswegian university lab, improved upon Newcomen's steam engine, and the Watt steam engine came to life in 1776. From a technological point of view, this was the most important breakthrough of the industrial era and the technological trigger of most that followed (McCord and Sachs 2013, 3). It was now possible to harness massive amounts of coal-powered energy efficiently and economically. James Watt was after profits and the patent; his aims included intellectual property, glory, and riches. He was working in an environment in which he could succeed, because the beginnings of commercial law existed in England, as opposed to many other places on the planet where such property rights had not yet been recognized.

These are some of the components that uniquely came together in England. And for all of the genius of Newcomen and Watt, if there had been no coal and iron ore in England, there never would have been a steam engine or Industrial Revolution! The coal and iron ore deposits were also transportable, thanks to the favorable transport conditions in England via rivers, canals, and roads. Topography, river ways, canals, ports, and mineral deposits therefore all helped, in combination with market incentives, rule of law, and a scientific outlook fostered by great universities. These are the special conditions—nature and nurture one can say—that converged in mid-eighteenth-century England to make possible the Industrial Revolution.

The first individual to describe this phenomenon was Adam Smith, the author of *An Inquiry into the Nature and Causes of The Wealth of Nations* and rightly known as the father of modern economics. He published *The Wealth of Nations* in 1776, the same year James Watt produced the modern steam engine, the American colonies declared their independence, and Edward Gibbon published *The History of the Decline and Fall of the Roman Empire*. Quite a year for a takeoff! Adam Smith was the first economist to explain the workings of a modern economy in terms of specialization and the division of labor. He gave us the idea of the Invisible

Hand, by which individuals acting out of their own narrow self-interest, trading in the marketplace, bring about a rise in productivity and therefore the "wealth of nations." As one of Smith's many wonderful observations in *The Wealth of Nations* explains:

> It is not from the benevolence of the butcher, the brewer or the baker that we expect our dinner, but from their regard to their own interest. We address ourselves not to their humanity, but their self-love and never talk to them of our necessities but of their advantages. (Smith 1776, 19)

In other words, it is the motivation of meeting our own wants and needs, via market transactions, that gives rise to the division of labor and the workings of the modern economy.

We also know the images of the early modern industrial era created by James Watt's steam engine, shown in figure 3.1: new factory towns with the massive coal burning and smoke pouring out of the new high chimneys. Factories that until

3.1 Depiction of Industrial Revolution–era factory town

Karl Eduard Biermann: Borsig-Maschinenbauanstalt zu Berlin. 1847.

recently had been powered by human or animal traction, or by wind and water, could now operate with vastly more powerful steam engines. The scale of industrial activity began to soar.

The steam engine made possible new forms of transport as well, including steam-powered railroads and steam-powered ocean freighters. The far greater energy also allowed a far greater scale of industrial transformation of materials than ever before. The production of steel soared, and this in turn made possible the massive expansion of cities, industries, and infrastructure of all kinds.

The transformation of life was dramatic, and often traumatic. One of the fiercest critics of the harshness of early industrialization was of course none other than Karl Marx. Marx and his coauthor Friedrich Engels wrote *The Communist Manifesto* in 1848 as a kind of ironic tribute to the power of the new industrial economy and the breakthroughs in industrial technology. Marx and Engels described this new world in vivid language:

> Modern industry has established the world market, for which the discovery of America paved the way. This market has given an immense development to commerce, to navigation, to communication by land. This development has in its turn reacted on the extension of industry and in proportion as industry, commerce, navigation, railways extended in the same proportion, the bourgeoisie, the new capitalist class developed, increased its capital and pushed into the background every class handed down from the Middle Ages. (Marx and Engels 1848, 15)

A new world indeed had arrived, one that would overrun the old and create a new global age, in part based on European colonial domination powered by the new industrialization. As Marx and Engels famously put it:

> The bourgeoisie, by the rapid improvement of all instruments of production, by the immensely facilitated means of communication, draws all, even the most barbarian, nations into civilisation. The cheap prices of commodities are the heavy artillery with which it batters down all Chinese walls, with which it forces the barbarians' intensely obstinate hatred of foreigners to capitulate. It compels all nations, on pain of extinction, to adopt the bourgeois mode of production; it

compels them to introduce what it calls civilisation into their midst, i.e., to become bourgeois themselves. In one word, it creates a world after its own image.

III. The Great Waves of Technological Change

By the early nineteenth century, the new era of modern economic growth was now underway. Markets and technological advance drove this process, first in a highly uneven way and eventually to nearly the entire world (as Marx foresaw). This is the period that Simon Kuznets, the great Nobel laureate, economic historian, and conceptualizer of economic development, called the era of modern economic growth.

We have defined economic growth as the sustained increase of GDP per person. For a global average, we sum the national GDPs to find the gross world product (GWP), which we divide by the world population. For more than 200 years now, the era of modern economic growth, the GWP per capita has increased on a sustained basis, though in a very uneven way across different regions of the world. A few of the world's poorest countries still have not yet achieved the takeoff of modern economic growth that other countries experienced two centuries ago.

We need to understand that global growth process. In order to do so, we need to make a quite basic distinction between two kinds of economic growth. Each one is characterized by a sustained increase of output per person, but each has a very distinct underlying dynamic. One kind of growth is the growth of the world's technological leaders. In the early nineteenth century that was certainly England; in the middle to end of the nineteenth century, it was Germany and the United States; in the twentieth century the United States was by far the most technologically dynamic country in the world. The "technological leaders" had a very particular kind of economic growth driven by relentless technological advance, in which advances in one technology tend to spur advances in other technologies as well, through new innovations and new combinations of processes. For example, after James Watt invented his improved steam engine in 1776, it was taken up in textiles, mines, steam-powered rail, steam-powered ships, steel production, and countless other areas. Each of these sectors became its own site of technological advances, which then spurred further technological breakthroughs.

Economists have given a name to this kind of growth: *endogenous growth*. "Endogenous" means something that arises from within a system, rather than from the outside. Endogenous growth means economic advancement that emerges from the internal workings of the economy. In its simplest description, a technological breakthrough raises GDP, which in turn raises the incentives for innovation more generally, since a higher GDP offers the prospect of higher profits for new products and processes. These new innovations raise GDP still further, spurring yet more innovations. And the innovations combine in novel ways, giving rise to new kinds of equipment, machinery, industry, and manufacturing techniques.

There is a second kind of economic growth, the growth of a "laggard" country that for whatever reasons of history, politics, and geography lagged behind as the technological leaders charged ahead. China, for example, did not industrialize in the nineteenth century. At some point countries like China and today's emerging economies begin to catch up by drawing on the technologies and organizational systems of the leaders. This kind of growth is very different from endogenous growth. It is sometimes called *"catch-up" growth*. The technologies that fuel it come from outside the economy engaged in rapid catching up. The essence of the strategy is to import technologies from abroad rather than develop them at home. It is still true that even imported technologies need to be adapted to local conditions, but they do not need to be invented and tested anew.

Catch-up growth can be considerably faster than endogenous growth. Technological leaders have tended to grow at around 1–2 percent per capita, while the fastest catching-up countries (like South Korea and China) have enjoyed per capita GDP growth of 5–10 percent per annum. No technological leader has ever sustained such rapid growth rates, and no laggard country has sustained them after the point of catching up with the leading countries. Super-rapid growth is about closing gaps, not about inventing wholly new economic systems or technologies.

These two different mechanisms of growth, the first one based on continuing innovation and the second one based on closing a gap by adopting (and adapting) the technologies of those countries already ahead, are the two major ways that economic growth proceeds in the world. *The failure to recognize the fundamental differences between endogenous growth and catch-up growth has led to all sorts of confusion in the discussion of economic development.* For example, the kinds of institutions that countries need in order to innovate and spur endogenous growth are typically quite different from the institutions needed to promote rapid catch-up growth. The first is based on innovation; the second on rapid adoption and diffusion of existing (though mostly foreign) technologies.

For catch-up growth, a strong role of government (as in China, South Korea, and Singapore), for example, can often be a major spur to the rapid adoption of advanced technologies from abroad. Innovation per se is not as important as the rapid development of infrastructure and the ability to link the domestic economy with high-tech foreign companies from abroad. This can be done, for example, by enticing foreign companies to invest in high-tech production in the catch-up economy, both to serve the fast-growing home market and also to produce goods at low cost for exports to the world market.

Let us focus on endogenous growth, the growth of technological leaders, in this section of the text, and then turn to catch-up growth in the following section. Economists sometimes call endogenous growth a process of "dynamic increasing returns to scale" or a chain-reaction economy. Innovations spur further innovations, keeping the growth process alive, just like in a nuclear chain reaction. The basic mechanism is the following. A new innovation causes the growth of GDP. This in turn increases the purchasing power of the market for further innovations. Other potential inventors therefore scale up their own research and development (R&D) in search of profitable innovations. Some of these R&D efforts prove to be successful, increasing the GDP still further, and thereby spurring even more R&D. The process continues in a chain reaction of innovation, economic growth, and then further innovation. And the innovation process is helped by the fact that various innovations can be *combined* to produce new innovations. The Industrial Revolution, for example, began with steam power, and then with advances in steel production, and those two sectors permitted an explosion of innovation in other kinds of heavy machinery, including the railroads, ocean steamers, and eventually the breakthrough of automobiles based on the internal combustion engine.

Ever since the onset of the Industrial Revolution there have been waves of tech-nological change, often bunched together because of the incentives of a growing marketplace and the R&D potential of combining new technologies. We speak loosely of the steam age, the age of electricity, the age of the automobile, the age of aviation, and so on. There have also been many theories of these technology waves. Perhaps the most influential in economic history has been that of the Russian economist Nikolai Kondratiev, who worked at the time of the Russian Revolution and whose greatest masterwork, *The Major Economic Cycles*, was published in 1925.

Kondratiev's main idea was that economic development was propelled by waves of major technological change dating back to the Industrial Revolution. He regarded these long waves of technological change as the main drivers of economic advancement and also as the sources of economic crisis when the growth dynam-ics of one cycle reach their conclusion while the next technological wave has not yet gathered force. Followers of Kondratiev today generally identify roughly four to six such long waves of technological change. One such illustrative classifica-tion, with five such waves, is shown in figure 3.2. I would like to emphasize here that different researchers in the Kondratiev tradition come up with somewhat dif-ferent timing and labels for the technology waves.

In the classification in figure 3.2, the first of the "Kondratiev waves" puts the steam engine at the core, from 1780 to 1830, roughly from the time of James Watt's invention to its widespread application. This classification seems to be unim-peachable; the steam engine truly defines the first breakthrough of modern eco-nomic growth.

The second of the technological waves is the great burst of railway building and steel production, dating to about 1830. These are crucial applications built upon the steam engine, the growing metals industry, and the development of precision engineering. These technologies transformed national economies and the world economy by dramatically reducing transport costs and thereby linking distant markets. Primary commodities (such as coal and ore deposits, or grain and timber production overseas) could now be profitably shipped and traded in international markets.

The third of the technology waves is the age of electricity, which itself had a few major subphases. Major discoveries of the physics of electricity date back to

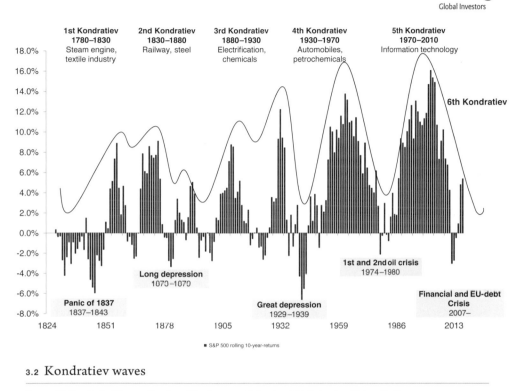

3.2 Kondratiev waves

Source: Shiller, Robert J. 2005. Irrational Exuberance. Princeton, NJ: Princeton University Press.

the end of the eighteenth century and the first half of the nineteenth century, to Benjamin Franklin, Michael Faraday, and the initial understanding of electromagnetism and electromagnetic induction. Then toward the end of the nineteenth century, Thomas Edison, George Westinghouse, and others applied the growing scientific knowledge of electricity to give us electric lighting and incandescent bulbs on city streets and then electricity in homes and factories. Electricity generation through coal-fired steam turbines and hydropower created the new power-generation industry.

The fourth technological wave is said to be from 1880 to 1930, the age of the automobile, which dramatically expanded mass transport and allowed the growth of major cities, and the chemical industry, which brought new materials, including explosives, chemical fertilizers, dyes, and eventually polymers, including plastics. One could add to this wave the age of modern aviation of the first half of the twentieth century. While the underlying technologies for the automobile, including the internal combustion engine, began in the second half of the nineteenth century, the dramatic scale-up began in the early years of the twentieth century with the Model T in 1908, built at low cost with Henry Ford's process innovation of the modern assembly line. The mass production of automobiles and trucks deeply transformed the way we live, where we live, how we produce goods, and, of course, how we ship and trade goods in the economy.

The fifth wave in this classification dates to around 1970, but again, with roots that go back much earlier. This is the wave of information and communications technology (ICT) made possible by the digital revolution. Fundamentally, the digital revolution is built on the realization that complex information can be stored as 0s and 1s (bits) and that these bits of information can be processed and transmitted with unimaginable speed and precision through new inventions such as transistors (to process and store the information) and fiber optics (to transmit massive amounts of information).

The age of ICT has given rise to the new "knowledge economy," in which massive amounts of data can be stored, processed, and transmitted globally for use in just about every sector of the economy (education, health, finance, entertainment, production, logistics, agriculture, and much more). The invention and spread of mobile phones, and now smartphones and other handheld devices, has made the ICT revolution also a mobile revolution, wherein information can readily reach every nook and cranny of the planet. Combined with advances in space science, notably satellite systems, ICT is also enabling breakthroughs in geopositioning, mapping, spatial planning, and countless other applications of geographic information.

The ICT revolution, the wave we are living through now, builds on waves of scientific and technological innovations. Intellectual giants such as Alan Turing, John von Neumann, and Claude Shannon pioneered the basic concepts of digital information and computation in the 1930s and 1940s. World War II gave a major spur to countless technologies, including semiconductors, radar, digital communications, computers, coding, and others. The invention of the transistor at the end of the 1940s was the next crucial step in the ICT revolution, and it in turn led to the concept of the integrated circuit in the late 1950s, which in turn spurred the modern computer revolution.

Beginning in the late 1950s, the newly invented integrated circuit gave rise to an imaginably dynamic process of technological advance based on the ability to pack more and more transistors into an integrated circuit and thereby giving rise to an upward spiral of the capacity to store, process, and transmit bits of information. In 1965, Gordon Moore, then the CEO of Intel, noted the phenomenon that the transistor count on an integrated circuit was doubling roughly every 18–24 months and had been doing so since the late 1950s. He predicted that this doubling process would continue for years into the future. Indeed, it has continued until now. There have been roughly 58 years of Moore's law, which means around 30 doublings of transistors on an integrated circuit. Thirty doublings, or 2^{30}, equals 1,073,741,824. The capacity to manage bits and bytes of information has increased by roughly 1 billion times since the mid-1950s!

Add to that the ability to transmit that information through satellites, fiber optics, and microwaves, and we have arrived at the mobile information revolution. In the 1980s, almost all telephony was carried on fixed landlines, and most of the world was still without a telephone. As of 1990, there were around 50 million cell phone subscribers, all in the high-income world. As of 2014, there are roughly 7 billion mobile subscribers and around 1 billion smartphone users. Mobile telephony now reaches the world's most remote villages. By 2020, almost all of the world will be within range of wireless broadband. The Internet, with its own technological marvels (such as asynchronous packet switching) and its globally shared protocols and standards means that the world of online information is now commonly accessible (or at least potentially accessible) by nearly all parts of global society.

Will there soon be a Kondratiev sixth wave of technological change? The one we really need now is a wave of sustainable technologies—ways to produce and mobilize energy and to transport ourselves and transport goods, to relieve the massive human pressures and human-caused destruction of the Earth's ecosystems. Indeed, one could say that spurring this sixth wave, the wave of sustainable technology, is a core part of achieving sustainable development. We now need to promote the next great wave: sustainable technologies. Fortunately, many of the advances and insights of the fifth wave will be helpful for the sixth. Energy efficiency, sustainable materials, nanotechnology, and breakthroughs in sustainable chemistry and food production will all benefit enormously from the recent advances of computation science and information technology.

IV. The Diffusion of Economic Growth

We have seen how modern economic growth burst forward in England in the middle of the eighteenth century and how subsequent waves of technological change have kept the process of endogenous economic growth continuing now for well over two centuries. This process describes well how the technological leaders have continued to forge new advances in economic life and transformation and how GDP per capita has continued to rise for two centuries.

During this period, the United States has been the main technological leader for well over a century and has been near the technological forefront dating back to 1820 or so. Since that time, per capita growth of the U.S. economy has averaged around 1.7 percent per year (Maddison 2006, 186). This does not seem very dramatic, especially since many developing countries are achieving growth rates of up to 10 percent per year. Yet 1.7 percent per year achieved for two centuries is a great achievement indeed. The United States in 1820 had a GDP per capita on the order of $2,000 measured in 2014 U.S. dollars. With 1.7 percent growth over around 194 years, from 1820 to 2014, the U.S. economy has expanded by a factor of roughly 26. Thus, the economy of $2,000 per capita in 1820 is now an economy of around $52,000 per capita. (Note that 1.017, which is 1 plus the growth rate, taken to the 194th power, equals 26.3)

But economic growth has another crucial dimension. For most of the world, economic growth has been about catching up with the technological leaders. This second type of growth might also be called a process of *diffusion*, because diffusion means that something spreads from one place to another. Think of the ripples in the pond when you throw a stone, moving away from the center where the stone hits the water. If the point of contact of the stone and the pond is where the endogenous technological growth is taking place, the ripples signify the diffusion of those technologies—and the modern economic growth that goes along with them—to more and more of the world.

How does that ripple effect work? Why is it that some places in the world are able to follow close on the heels of a technology leader, whereas other parts of the world seemingly have not yet been able to take advantage of advances in technologies that are already more than a century old? Perhaps one billion people or more do not have access to electricity in the twenty-first century, a technology that was developed and adopted by the technological leaders at the end of the nineteenth century.

What has stopped the ripples from reaching those places? The patterns of diffusion lie at the heart of the study of economic development. Yes, part of the study of development is the study of the technological leaders. Yet an equally if not more important part is the study of diffusion of technologies and rising GDP per capita from the leaders to the rest.

Economists have been thinking about this puzzle for a long time. In *The Wealth of Nations*, Adam Smith talked about the fact that diffusion would take considerable time and that economic growth would typically start at the coast of a country and move to the interior only after a considerable lag in time. Why at the coast? Because conditions for trade, specialization, and market dynamics are much easier at the coast. And why would the lag time to reach the interior be long? That, noted Smith, is because transport costs are very high to bring goods and services into the interior of a country or continent, except for those places well served by rivers or easily reached by man-made canals. In book 1 of *The Wealth of Nations*, Smith explains:

> Since such therefore are the advantages of water carriage [water-based transport], it is natural that the first improvements of art and industry should be made where this conveniency opens the whole world for a market to the produce of every sort of labor and that they should always be much later in extending themselves into the inland parts of the country. (Smith 1776, 23)

Today, more than 200 years after Smith wrote, landlocked countries like Bolivia, Chad, Niger, Kyrgyzstan, and Nepal still face the disadvantages of high transport costs. Of course, many technological advances since Adam Smith's time, including rail, trucking, and now Internet-based services, have allowed even the distant parts of the world to integrate more closely with the global economy.

There are several general factors that are conducive to the movement of those ripples from the center of the technological forefront out to the rest of the world. A poor country that is close to a rich country is likely to receive the ripples sooner than a poor country that is very distant from the high-income economies. Proximity matters, just as the ripples move outward from the spot where the stone hit the water. In the nineteenth century, western European countries that were geographically close to Great Britain had an advantage for their own economic development. The proximity meant that they could more easily access Britain's advanced technologies and could sell their own production into Britain's booming market. In the twentieth century, proximity to the United States made a difference. In the early twenty-first century, proximity of poor Asian countries to Japan, South Korea, and now China, has made a difference in speeding catch-up growth.

Favorable agricultural conditions are also a huge benefit for receiving the ripples from abroad. Countries with high agricultural potential (e.g., farmlands that are easy to irrigate or that can grow multiple crops in a year) are more likely to make rapid advances in farm yields that in turn free up labor for work in industry and services.

Places with their own energy resources, be it coal, oil and gas, hydroelectric power, or now solar and wind potential, have a huge advantage in catch-up growth. While it is generally possible to export goods and import primary energy in return (as South Korea and Japan do), it is generally very hard to get that process started in a place without any domestic low-cost sources of primary energy. In the nineteenth century, regions with coal had an advantage. In the twentieth century, regions with oil and natural gas had the advantage. In the twenty-first century, perhaps the desert regions, with massive potential for low-cost solar energy, will now have the advantage!

A physical environment conducive to human health is also important for receiving the technological ripples from abroad. A disease-ridden environment, one that is burdened by malaria, worm infections, and other terrible infectious diseases, can be a serious impediment to the diffusion of economic growth. I will note later that the "excess disease burden" of parts of the tropics, notably tropical Africa, has definitely been one factor in holding back the catch-up growth of these poor, laggard, tropical regions.

And finally, though by no means least, is politics. If the politics are dysfunctional; if a colonial power dominates the society; if a dictator rules; or if chaos and violence grip a country, then catch-up growth is not possible. During the nineteenth century and until the 1960s and 1970s, many countries could not catch up simply because they were under foreign rule. European empires held most of Africa and much of Asia in economic stagnation. The imperial powers were not interested in the economic development of their colonies. They were more interested in the exploitation of the primary commodities—from the mines, oil wells, forests, farmlands, and fisheries—of those countries. In the late twentieth century, the political problems have often been internal rather than international. Despots and dictators have often "run" the economy for their personal or tribal benefit, not for the benefit of economic growth of the entire country.

Historical Patterns of Catch-Up Growth

We can apply these general insights to understand the actual ripples that have spread over the world economy since the start of the Industrial Revolution. My colleague Gordon McCord and I have found it interesting and worthwhile to ask the following question: When did each economy in the world first escape from extreme poverty?

This is like asking when the ripples of global economic growth first arrived in each national economy. To measure extreme poverty for this purpose, we use the threshold of GDP per capita of $2,000, measured in PPP prices.

The first country in history to reach the $2,000 threshold was Great Britain, the home of the Industrial Revolution. That is where the proverbial stone (of endogenous economic growth) first hit the water, around 1820. Then the ripples started to spread from Great Britain, and by now, two centuries later, have reached most of the world. Within Europe, the closer the country is to England, the faster the ripples reached that country. For example, western European countries reached $2,000 earlier than eastern European countries. Belgium, France, and the Netherlands reached the $2,000 threshold ahead of Spain and Scandinavia. Since Europe is relatively compact, diffusion reached almost the whole continent in the nineteenth century.

For the rest of world, the story is obviously much different. The ripples have to travel much longer distances, face far more complex conditions, and encounter barriers such as malaria, desert conditions, landlocked regions, and so forth, which have blocked rapid catching-up growth. Moreover, politics got in the way—big time. Europe's conquest of far-flung colonies in the nineteenth century set back the economic prospects of those places, often for a century or more. It was only upon political independence of those colonies that national governments were able to start investing in the education and infrastructure needed for catch-up growth.

Figure 3.3 shows an approximate timing of the takeoff. The first major economic advance (measured by achieving $2,000 per capita GDP) outside of Europe occurs in places settled by Britain itself, such as the United States and Australia. These British offshoots had several favorable conditions for catch-up growth: vast arable land and energy resources, good coastlines for trade, strong connections with British industry, and technological knowledge. These countries had achieved modern economic growth by 1860.

The next group of countries, which achieved the $2,000 threshold by 1900, includes Argentina, Uruguay, Chile, and Japan. All of these are temperate-zone countries with favorable conditions for agriculture. Japan became the first Asian economy to achieve catch-up growth. Looking at the map, we can see that Great Britain and Japan have lots of geographical similarities. Both are islands off the main Eurasian landmass. Both have been relatively protected from invasions from the mainland. Both have been able to trade heavily with the mainland. Both are temperate-zone economies with relatively high-yield agriculture. Both have relatively healthful environments, free of the burdens of massive tropical diseases. Both had achieved substantially urban, literate, and politically stable societies by the nineteenth century.

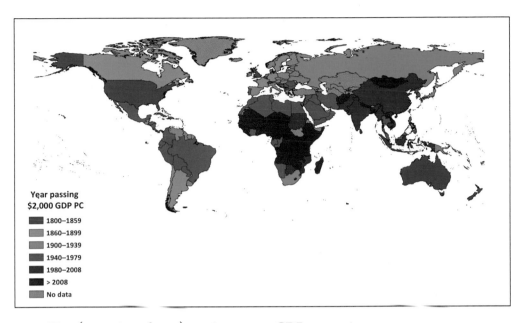

Year passing
$2,000 GDP PC

- 1800–1859
- 1860–1899
- 1900–1939
- 1940–1979
- 1980–2008
- > 2008
- No data

3.3 Year (or projected year) passing $2,000 GDP per capita

Source: McCord, Gordon, and Jeffrey Sachs. 2013. "Development, Structure, and Transformation: Some Evidence on Comparative Economic Growth." NBER Working Paper No. 19512. Washington, DC: National Bureau of Economic Research.

Much of the rest of the world did not have the economic good fortune of Europe, the United States, Canada, Japan, Australia, and the Southern Cone of South America (Argentina, Chile, and Uruguay). Most of the rest of the world had to wait until after 1950 to have the ripples of modern economic growth reach their economies! As I have emphasized, large swaths of the world were held back by imperial conquests. The European empires did not pursue broad-based modern economic growth in most of their colonies. (There were a few exceptions, such as Hong Kong and Singapore, where the colonies served as trading posts rather than as sources of raw materials.) By the end of the nineteenth century, India, much of Asia, and virtually all of Africa (figure 3.4) was under European colonial rule. Most of the colonized regions did not experience modern economic growth until decolonization in the 1940s–1960s.

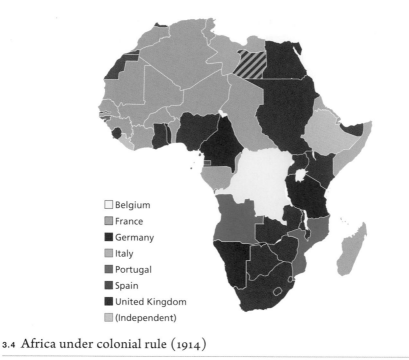

- ☐ Belgium
- ☐ France
- ☐ Germany
- ☐ Italy
- ☐ Portugal
- ☐ Spain
- ☐ United Kingdom
- ☐ (Independent)

3.4 Africa under colonial rule (1914)

"Colonial Africa 1914 map," Declangraham et al., Wikimedia Commons, CC BY-SA 3.0,2.5,2.0,1.0.

V. Economic Development Since World War II:
The Making of Globalization

As of the beginning of the twentieth century, the world economy could be described as follows. On the whole, it was a miraculous economic age, unprecedented in the long sweep of human history. Waves of technological change had led to unprecedented breakthroughs in the ability of humankind to produce goods and services, meet material needs, extend life spans, solve long-standing problems of public health, and make breakthroughs in quality of life in countless ways through electrification, modern transport, and mass industrial production. Yet by 1900 the world was also one of unprecedented gaps between the rich and poor. Modern economic growth had come to Europe and to a few other temperate-zone countries (the United States and Canada, the Southern Cone of South America, Japan, Australia, and New Zealand), but not yet to the rest of the world.

At the end of World War I, John Maynard Keynes looked back to the period just before World War I and described the unique global circumstances this way (in his famous work, *The Economic Consequences of The Peace*):

> What an extraordinary episode in the economic progress of man that age was which came to an end in August, 1914! . . . The inhabitant of London could order by telephone, sipping his morning tea in bed, the various products of the whole earth, in such quantity as he might see fit, and reasonably expect their early delivery upon his doorstep; he could at the same moment and by the same means adventure his wealth in the natural resources and new enterprises of any quarter of the world, and share, without exertion or even trouble, in their prospective fruits and advantages; or he could decide to couple the security of his fortunes with the good faith of the townspeople of any substantial municipality in any continent that fancy or information might recommend. . . . But, most important of all, he regarded this state of affairs as normal, certain, and permanent, except in the direction of further improvement, and any deviation from it as aberrant, scandalous, and avoidable. (Keynes 1920, chap. 2, pp. II.4)

Of course Keynes was speaking as a brilliant and highly privileged Englishman. He was the one sitting in bed, sipping his tea and ordering commodities from all parts of the world. Those under the fist of colonial rule obviously could not do the same.

Still, Keynes was also expressing the uniqueness of an era in which modern economic growth had taken hold in many parts of the world and had already created a global market economy (as Marx, we recall, had predicted in 1848). That global economy succumbed tragically and unexpectedly to war and chaos, for no good reason, with the outbreak of World War I in 1914. That war, in turn, unleashed worldwide chaos, millions of deaths from violence, millions more from infectious diseases such as the 1918 flu pandemic, and also the upheavals of revolution, most importantly the 1917 Bolshevik Revolution that gave birth to Soviet-era

Communism. World War I unleashed tremendous political and financial crises that led to enormous monetary and financial instability in the 1920s and that in turn played a key (though complex) role in the onset of the worldwide Great Depression in 1929. And of course the Great Depression gave rise to another wave of political horrors, including the rise to power of Adolf Hitler in Germany in early 1933 and the rise of Fascism in Japan in the 1930s as well. One can say, in the shortest of shorthand, that World War I unleashed mass deaths after 1914; economic chaos in the 1920s; the Great Depression in the 1930s; and the onset in 1939 of World War II, which engulfed the world until 1945.

By the end of World War II, many cutting-edge technologies (radar, semiconductors, computers, space science, aviation, nuclear energy, and many more) had continued to advance rapidly, though many of the prewar technological leaders, including Germany and Japan, were in ruins. But the world's main technological leader, the United States, was certainly not in ruins in 1945. Other than the 1941 Japanese attack on Pearl Harbor, the United States had passed the war largely unscathed. By the end of World War II the United States was far and away the world's leading economy and would remain so till the end of the century.

By 1945 the world economy was roughly divided into three parts. The first (called "the First World") included the United States, western Europe, and Japan, the market-oriented industrial world that operated within a security system led by the United States. The second (called "the Second World") included the Communist countries led by the Soviet Union and, after 1949, including China. The third included most of the newly independent countries just escaping from colonial rule. Some of these postcolonial countries signed up to the U.S. security umbrella. A few joined the Soviet bloc. Many, however, declared themselves to be unaligned. These constituted the new "Third World." By the 1960s, a fourth informal term crept into the global parlance, "the Fourth World," signifying the poorest of poor countries. With the end of the Cold War in 1991, this jargon of First, Second, Third, and Fourth World countries has largely been abandoned.

The world economy evolved under these geopolitical divides for several decades. The First World recovered from the damage of World War II remarkably quickly during the 1950s. Endogenous technology-driven economic growth took hold and living standards rose rapidly in the high-income countries. The post–World War II period for these countries was briefly a period of rebuilding, and then a period of dynamic, endogenous growth. In the Second (Communist) World, industrialization seemed to be rather dynamic for a time, but by the 1960s the Second World was already facing a crisis of economic stagnation. By the 1970s economic development under Communist systems was basically screeching to a halt, which prompted some of the Second World countries to begin reform. China was the first great reformer of the Communist group, when Deng Xiaoping came to power in 1978 and opened China to a market system and to international trade and investment. Those reforms unleashed China's own catching-up growth with remarkable success, to the point where China became the fastest-growing major economy in history.

Other parts of the Communist world took longer to break free because the Soviet Union refused to make similar reforms for a long time, until Mikhail Gorbachev came to power in 1985 and began his own market reforms. After that came the democratic and economic revolutions of Eastern Europe in 1989. With the end of the Soviet Union itself at the end of 1991, the Second World finally became part of the world economy.

The so-called Third and Fourth Worlds included dozens of countries, each with their own economic history, politics, and strategies. A few of the countries were soon interested in integrating with the First World economies. These countries realized that the arrival of economic ripples could lift them into a very special kind of catch-up industrialization. This new form of "late industrialization" took the form of local factories producing for multinational companies as part of global production systems. For example, a company in South Korea or Taiwan would begin to produce electronics goods or clothing for retailers in the United States and Europe, according to the technology designs and intellectual property of the U.S. and European companies. The early adopters of this strategy for catching up were called the "Asian Tigers," and included South Korea, Taiwan, Hong Kong, and Singapore. By the 1960s these four economies were growing extremely rapidly through integration of their new industrial base with the high-tech industries of the First World. As their success became evident, other developing countries also took notice and began to open their economic doors to trade and foreign investment in order to attract new multinational companies and to catch the ripples of global technology-based growth.

This is how our own era of globalization came into being, step by step, after World War II. The new catch-up growth took off in countries that opened their borders to trade and foreign investment. New global production systems, centered around large multinational companies, used the poorer countries as places for low-wage, labor-intensive parts of their production systems. The global value chain of production (for a car, a shirt, a home computer for global sales) was increasingly divided up among many countries to take advantage of different wage levels, local skills, and transport conditions. Poor countries were able to become part of global production systems when they offered good infrastructure, transport, and low-cost and reasonably skilled labor.

This new globalization of production was facilitated by many breakthroughs in technology and transport, including standardized containerization of trade in 20-foot containers that allowed easy transport from ships to trucks. Other key technologies included computer-assisted design and manufacturing (CAD/CAM), the Internet, and mobile telephony. These ICT technologies revolutionized the ability of companies to engage in sophisticated, dispersed global production systems and to thereby create globally integrated companies, often with hundreds of thousands of employees operating in more than 100 countries. The world's large multinational companies thereby became the main agents for the continuing transmission of economic ripples around the world and the diffusion of modern economic growth.

Japan was one leader in this process, and it developed a wonderful metaphor: the flying geese model. When geese fly in formation (figure 3.5), one goose flies in front, and then in back are others following the lead. This is how economic development in Asia proceeded as well, with the industrialization first of Japan (with endogenous technological change), and then flying in formation just behind came South Korea, Taiwan, Hong Kong, and Singapore. Behind them came Indonesia, Malaysia, and Thailand; and then behind them, China and Vietnam; and now, Cambodia, Laos, and Myanmar.

Figure 3.6 shows where the world's multinational textile and clothing production was located as of 1999. Every red dot is a production site. Note that virtually

3.5 Flying geese formation

"Canada Goose," Joshua Mayer, Flickr, CC BY-SA 2.0.

**Number of
foreign affiliates**

- ◦ 1–4
- • 5–11
- ◦ 12–18
- • 19–30

3.6 Global distribution of textile and clothing production sites (1999)

Source: United Nations Conference on Trade and Development. 2001. World Investment Report 2001: Promoting Linkages. *New York: United Nations.*

every dot in Asia is on the coast, just as Adam Smith predicted and explained in 1776, well before the advent of such global production chains. Again, we see how geography interacts with technology to account for the ripples of global growth.

Figure 3.7 shows a map of China's foreign direct investment (FDI) during the growth spurt of 1978–2000. After Deng Xiaoping opened China to the world in 1978, foreign investment made China an export base for world manufacturing production. China became the industrial workshop of the world, typically using industrial technologies and processes brought from the outside, often through FDI of multinational companies. Once again, we see the wave going from coastal provinces, where FDI is the highest, into the interior, just as Adam Smith had told us it would.

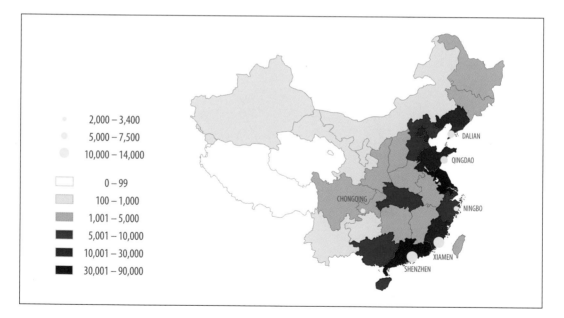

3.7 Distribution of foreign direct investment in China (millions of dollars) (1999)

Source: United Nations Conference on Trade and Development. 2001. World Investment Report 2001: Promoting Linkages. New York: United Nations.

Note: Yellow dots represent FDI stock by major city; provinces are shaded according to FDI.

The economic development that started as a local phenomenon in England and then spread across into western Europe and the other temperate-zone nations had finally spread to Japan by the late nineteenth century, and to the postcolonial world after World War II. By now, the ripples of modern economic growth have reached almost the entire planet.

There are still a few places where modern economic growth has not yet reached. These are generally places facing great geographical difficulties. They include places far in the interior of continents, high in the mountains, and relatively isolated as distant islands in oceans. These places have many burdens and few benefits. In the coming chapters, we will discuss how poverty reduction and economic growth can reach those places that have still not benefited much from the era of modern economic growth.

4

WHY SOME COUNTRIES DEVELOPED WHILE OTHERS STAYED POOR

I. The Idea of Clinical Economics

We have seen how modern economic growth diffused throughout the world during the past 250 years. The Industrial Revolution began in England in the mid-eighteenth century. By the middle of the nineteenth century, only a handful of countries had reached even $2,000 per capita (measured at purchasing power parity in 1990 prices). As late as 1940, the $2,000 threshold had been reached only by the United States, Canada, Europe, the Soviet Union, Australia, New Zealand, Japan, and the Southern Cone of South America (Argentina, Chile, and Uruguay), but still not by most of the world.

We have begun to explore why economic growth occurred earlier in some places than others, and almost not at all in a few places. We have noted that within Europe, industrialization spread roughly from the northwest (Britain) to southeast (Balkans) in the course of the nineteenth and early twentieth centuries. The year that each European country reached $2,000 per capita is well explained by its proximity to Britain: the closer to Britain, the earlier the date of reaching $2,000 per capita (McCord and Sachs 2013, 14). Hence, among the continental European countries, the Netherlands was the earliest and the Balkan states the latest (not reaching $2,000 per capita until the twentieth century).

We have also noted the reasons for early or late industrialization in other parts of the world (McCord and Sachs 2013). Climate zone mattered: development generally came earliest in temperate-zone regions, such as the Southern Cone of South America. Development came first to coastal countries and generally reached landlocked countries (such as Afghanistan, Bolivia, and Mongolia) much later. Geopolitics certainly mattered. Domination by a European or Asian imperial power set back the process of industrialization in countries in Africa (such as Ghana and Kenya), South Asia (such as India), Southeast Asia (such as Malaysia), and Northeast Asia (such as Korea). Disease burden mattered too. Since development depends on a healthy, well-educated population, it's not a surprise that regions beset by a heavy disease burden are held back. In the United States the South, with hookworm, malaria, and yellow fever, was at a disadvantage compared to the North, which suffered much less from these diseases.

There are three main points here. First, modern economic growth was a diffusion process, starting in one small part of the world (Britain) and gradually diffusing and evolving all over the planet. Second, the patterns of diffusion are discernible. They are not just a mystery. Third, many different kinds of factors have been at play during the past 250 years, and the relative importance of those factors continues to change, especially as technologies evolve. In the past malaria was a profound barrier to development in Africa. With recent advances in malaria and treatment technologies, this particular barrier is close to being eliminated for the first time in history.

There seems to be a misguided desire for overly simple explanations of complex economic dynamics. In many places one will read that economic growth depends on "economic freedom," or on "inclusive institutions," or on "controlling corruption." Factors like economic freedom, political institutions, and corruption may play a role, but they certainly do not play the only role, or even the main role, in many places and times of history. These individual factors taken alone neither explain the patterns of development across the globe and over time, nor do they help us predict future development.

The problem is that in a complex process such as economic transformation, many things can go wrong. Think of the complexity of the global economic system like the complexity of the human body. While it is true that in the past doctors and spiritual leaders often blamed a person's disease on one factor (god's punishment of a sinner) or a few factors (the imbalances of the bodily humors), these explanations failed to understand the complexity of human pathophysiology. In a complex system like the human organism, literally thousands, indeed tens of thousands of things can go wrong. A horrific disease like sickle cell anemia, which was long a death sentence in premodern times, is now scientifically explained by a change of one single nucleotide on one gene on one chromosome: a single change among roughly 1 billion base pairs in the human genome. There are many diseases caused by such a single-site polymorphism, as it is called. There are other diseases caused by environmental factors, pathogens such as viral or bacterial infections, trauma, unhealthy behaviors, and many other possible factors.

The art of modern clinical medicine has advanced beyond pronouncing a single cause of disease ("you have sinned") or a single prescription ("take two aspirin and call me in the morning") or a single referral ("go to the emergency room"). The modern doctor is expected to diagnose the specific causes of a specific patient's illness and to offer a specific prescription that is accurately honed to that patient's conditions and needs. The modern economist should do the same in diagnosing the persistence of poverty. Instead of offering one simplistic diagnosis ("stop your corruption"), one prescription ("cut government spending"), or one referral ("go the International Monetary Fund for treatment"), the effective development practitioner should make a diagnosis that is accurate and effective for the conditions, history, geography, culture, and economic structure of the country in question.

I have been lucky in my own work as an economist in thinking about the need for careful diagnosis and prescription, because I have been able to watch a wonderful medical diagnostician do her work. That clinician is my wife Sonia, a clinical pediatrician. When she sees a young child with a fever, she doesn't immediately think she knows what the problem is, that all fevers have the same cause. Her training, knowledge, and experience inform her that there could be thousands of reasons for the fever. To treat the patient effectively, she needs to make a diagnosis of the actual cause of the disease in this particular patient. She begins to ask questions. Her first question, in general, is whether the baby's neck is stiff.

If it is, there is the possibility of cerebral meningitis, a relatively rare but potentially fatal condition. If the mother answers that the baby's neck is stiff, then indeed my wife's next sentence is, "I'll meet you in the emergency room." It is more likely, of course, that the condition is not cerebral meningitis. The list of questions continues, all aiming to ferret out whether the condition is viral, bacterial, environmental (e.g., poisoning), or something else altogether.

Doctors call the process of homing in on the actual cause of a disease a *differential diagnosis*. I have come to the view that in economic development, and indeed in sustainable development more generally, we also need to have a clinical approach based on differential diagnosis. In my book *The End of Poverty*, I called this approach "clinical economics," and I said that the role of a good practicing clinical economist is to make a differential diagnosis for the economic case at hand, just as a good medical doctor diagnoses each individual patient (Sachs 2005, chap. 4). Medical doctors like my wife go through a systematic checklist, asking the relevant questions about potential causes of illness and asking them in a particular order (rule out meningitis first, or move fast to the emergency room!). They look at the evidence and lab results; conduct interviews; try to understand from the parents and from the child what's actually happening; and then on the basis of a wide range of information and evidence, they make a diagnosis and a plan for treatment. The treatment may then proceed according to plan or it may prove to be ineffective, in which case a further round of diagnosis will be necessary. (Fans of the medical television series, *House*, a personal favorite of mine, will know the process.)

Practitioners of sustainable development also need to make such a differential diagnosis. In *The End of Poverty*, I created such a checklist for a particular "disease" that persists into the twenty-first century—extreme poverty. I reasoned that although most of the world had by now escaped extreme poverty, roughly 1 billion people are still trapped in extreme poverty. They are heavily concentrated in tropical Africa and South Asia, but there are pockets of extreme poverty in other parts of the world as well (in countries as varied as Haiti, Afghanistan, and Laos). I presented a diagnostic checklist with seven main categories and many more subcategories (Sachs 2005, 84).

Let's consider the seven headline items of the poverty checklist.

First, the underlying condition could be what I call a *poverty trap*—when the country is too poor to make the basic investments it needs to escape from extreme material deprivation and get on the ladder of economic growth.

Second, the poverty could result from bad economic policies, such as choosing the wrong kind of investment strategy, closing the borders when international trade would make more sense, choosing central planning when a market system would be better, and so forth.

Third, the poverty could reflect the financial insolvency of the government. If a government has a past history of overspending and over-borrowing, it can reach a state of financial bankruptcy. The government then owes so much to its creditors that it is unable to find the money to build roads, schools, or clinics, or to hire doctors, teachers, and engineers.

Fourth, the poverty might be the result of some aspects of physical geography. The country may be landlocked, far from trade; it may be high in the mountains, unable to farm or engage in low-cost manufacturing; it may face an endemic disease burden of malaria or other burdensome diseases; or it might be highly vulnerable to repeated natural disasters such as earthquakes, tsunamis, tropical cyclones (hurricanes and typhoons), droughts, floods, and other crippling conditions. A few countries, such as Haiti and the Philippines, are unusually buffeted by a large number of such conditions.

Fifth, the country might be suffering from poor governance as opposed to poor policies. On paper, the economic policies look good. In practice, they may be riddled with corruption, inefficiency, incompetence, or all of the above. Poor governance is, of course, a matter of degree. There is corruption nearly everywhere. I sometimes say that, "Yes, corruption is real and damaging, and not only in Washington, DC." Corruption is often attributed to poor countries but not to rich countries. Still, corruption is not the all-purpose explanation it is often taken to be. Many countries with moderate levels of corruption have achieved economic development. Corruption can be a problem, both moral and practical, without being a devastating barrier to development. Yet if carried to extremes, then it certainly can stop economic growth.

A sixth factor in continuing poverty may be cultural barriers. As one important example, some societies continue to discriminate harshly against women and girls. Girls may still have little or no chance to attend school and may be expected to marry very early and to bear many children, even when the household is too impoverished to raise these children with proper health, nutrition, and education. Such cultural patterns can be inimical to long-term economic development.

The seventh factor is geopolitics, a country's political and security relations with its neighbors, foes, and allies. Geopolitics can make a big difference. If a country is physically secure from attack, enjoys national sovereignty, and is able to trade peacefully with other countries, geopolitics is the friend of economic development. If, on the contrary, the country is dominated by a foreign power (as in the colonial era) or is part of a proxy war of the great powers, the country can be undermined or even physically destroyed by the actions of more powerful countries. Think of Afghanistan. It has a hard enough time developing in view of being landlocked and vulnerable to many climate shocks such as droughts and floods. Yet since 1978, it is has been subjected to war, incursions, invasions, terrorist cells, and destructive great-power politics. It is no wonder that Afghanistan remains one of the poorest places in the world. The outside powers have hindered rather than advanced Afghanistan's economic development.

There is one overriding point about these seven factors and the many subfactors in each of the seven categories. The problems do not apply equally to every country. Indeed, some categories are relevant for some places, and other categories are relevant for others. There is no single explanation of the persistence of extreme poverty. Local circumstances, history, and context are all important.

In my own experience of almost thirty years of working with countries all over the world, it has struck me how different parts of the world in different times have had extremely different conditions to confront in order to get out of the poverty rut. Always receiving the same doctor's prescription would be a disaster for a medical patient, and the same is true for an economy. I worked in Bolivia in the middle of the 1980s to help end a hyperinflation. Prices were rising by thousands of percent per year. When one made the differential diagnosis for Bolivia, one could see the government was broke and was printing money to pay its bills, causing the hyperinflation. What was required most of all was to get the budget under control in short order so that this fever of hyperinflation could be broken. To end the large budget deficits involved several kinds of actions, including changes in public-sector prices (such as the oil that the government sold to the public), budget outlays, and payments to foreign creditors on Bolivia's mountain of debt. One part of ending the hyperinflation involved canceling around 90 percent of Bolivia's external debt, thereby easing the pressures of debt servicing on the budget (Sachs 2005, chap. 5).

Four years later, I was asked to help Poland overcome its very different crisis. In 1989, Poland was transitioning from Communism to a market economy and from a dictatorship to a democracy. Once again, Poland's malady—high inflation and collapsing output—required a differential diagnosis. Poland's greatest need in my view was to enable supply and demand to work, because the Communist-era central planning had collapsed. Therefore, at the request of the new leadership, I helped to develop an economic strategy to restore market forces, supply and demand, international trade, and budget balance. Poland ended its high inflation and soon resumed economic growth, indeed quite rapid economic growth based on its new and growing economic trade and investment linkages with western Europe (Sachs 2005, chap. 6).

When I began working in tropical Africa in the mid-1990s (first in Zambia and then in numerous parts of Africa), the underlying conditions and causes of poverty were obviously completely different from those of Poland or Bolivia, or indeed other parts of the world. Africa, uniquely, was in the midst of a massive AIDS pandemic and was also suffering from the resurgence of malaria (because the standard malaria medicine was losing its effectiveness as drug resistance spread). Many parts of Africa were so poor that the most basic infrastructure— roads, power, water, and sanitation—did not even exist. Nonetheless, I found some economic officials from international institutions (such as the International Monetary Fund and World Bank) prescribing exactly the same medicine they had been prescribing earlier in Poland. They were asking the impoverished African nations to cut budgets and even to privatize health services. These were ludicrous and destructive prescriptions for impoverished regions suffering from massive disease pandemics. Africa needed its own diagnosis and prescription, not one repeated by rote from another part of the world, much less from Washington, DC.

For most of tropical Africa, I felt that the first category on the checklist—a poverty trap—was the most accurate diagnosis. African governments knew what

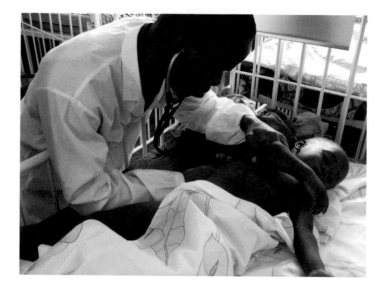

4.1 Child with malaria in Ruhiira, Uganda

Photo courtesy of Kyu-young Lee.

they wanted to do: expand health coverage; improve education; build roads, ports, and power grids; ensure access of the poor to safe water and sanitation; and so forth. These governments recognized their responsibilities. They even had investment plans on the shelf. But they lacked the financial resources to carry out those plans. A health plan might call for outlays of $60 per citizen per year for public health operations, yet that sum, as modest as it is for rich countries, is simply out of financial reach for the poorest countries. This is the poverty trap.

There are two main ways to break this poverty trap. One way is for the government to borrow the funds it needs for public investments and then to count on future economic growth to fill the government's coffers in order to repay the loans. The second way is for foreign governments, businesses, foundations, and international institutions to provide temporary assistance to finance the urgent needs. As economic growth subsequently occurs, the aid can be gradually reduced in scale and eventually eliminated. Such assistance goes by the technical term *development assistance*. It includes all varieties of help on terms better than market-based borrowing of funds. When governments and official agencies provide the help, it is called *official development assistance,* or *ODA.* When nongovernment organizations (NGOs) and private foundations provide the help, it is called *private development assistance.*

Since 2000, with the adoption by the member states of the United Nations of global goals to fight poverty known as the Millennium Development Goals (MDGs), several special institutions have been created to channel ODA to effective purposes. One of the most notable has been the Global Fund to Fight AIDS, Tuberculosis and Malaria (GFATM). Donor governments and private philanthropies and businesses give money to the GFATM, which in turn distributes funds to poor countries suffering from the three diseases addressed by the fund. The program has been highly effective. All three diseases are coming under control. Yet even so, public and professional resistance to ODA remains strong, in part because some economists continue to believe that poverty is caused by single factors (corruption or lack of economic freedom) not addressed by ODA. That may be true in some places, but it seems not to be the case in tropical Africa.

II. A Further Look at Geography—Transport, Energy, Disease, and Crops

Poor countries and individual poor people are often blamed for their plight, even when external factors are the major obstacle to development. It is too easy to say that the poor are still poor because of corruption or bad culture or lack of direction. All such interpretations are rife in public debate. Yet experience on the ground often suggests a very different reality.

Physical geography is the fourth item on the poverty diagnostic checklist. Strangely, practitioners often overlook the most basic realities of physical geography as contributors to poverty. I have learned about geography from the ground up, so to speak, not originally from the classroom (where I was taught very little in graduate school about physical geography). Working in Bolivia, Mongolia, Uganda, Rwanda, Kyrgyzstan, Burkina Faso, Mali, Tajikistan, Zambia, Malawi, and other places has given me a solid appreciation of the extra challenges that an economy faces when it is landlocked, and the difficulties that it faces on top of that when it is both landlocked and mountainous. The view from the Altiplano of Bolivia is strikingly beautiful: blue skies, snowcapped mountains, dry scrub, and rolling hills as far as the eye can see. Yet the difficulty of running a factory up there is just as striking, in view of transport costs to the Pacific ports that are among the highest in the world.

We have already noted that gross domestic product (GDP) per capita (figure 2.1) has very strong geographical correlates. Tropical countries are generally poorer than temperate-zone countries. Disease burden and crop productivity may help to account for such differences. Landlocked countries are generally poorer than coastal countries. Countries facing the hazards of earthquakes and typhoons in the Caribbean and the Asia-Pacific rim (e.g., the Philippines) seem to pay a long-term price for that vulnerability. And so forth.

Another hint of the power of geography is found in figure 4.2, which shows a map of the world's largest urban areas. While there are large cities found in most parts of the world, there is a high proportion of large cities along the coasts of the continents. Those large cities in the interior of the continents are very often along major rivers (such as Chongqing on the Yangtze River in China), so they have waterborne trade. Being on the coast, near ports, or near major rivers has long been a key to a vibrant economy with a sophisticated division of labor that promotes high productivity and allows for a high degree of global trade and economic growth. This proximity to sea-based trade allows exports to world markets at competitive costs and also enables the economy to obtain inputs from the rest of the world at low costs for processing or local production and consumption.

In figure 4.3, countries are color-coded according to an average distance of the economy to the closest seaport. Countries in western Europe, the United Kingdom, and the Arabian Peninsula, among others, are very close to ports and are therefore highly advantaged by very-low-cost transport conditions. Great Britain had a huge advantage in its eighteenth-century economic takeoff as a coastal country with many good seaports; and London, as a city on the Thames River, was able to engage in great international trade. Moreover, England's topography also favored the low-cost construction of canals in places where rivers did not reach. These canals enabled coal to be mined and shipped widely to factory cities in Great Britain.

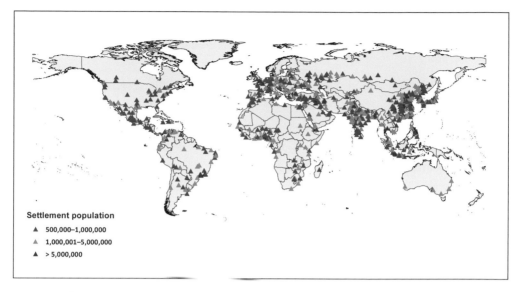

4.2 Settlements with a population of 500,000 and larger

Source: CIESIN–Columbia University, International Food Policy Research Institute (IFPRI), the World Bank, and Centro Internacional de Agricultura Tropical–CIAT. 2011. "Global Rural-Urban Mapping Project, Version 1 (GRUMPv1): Settlement Points." Palisades, NY: NASA Socioeconomic Data and Applications Center (SEDAC). http://dx.doi.org/10.7927/H4M906KR.

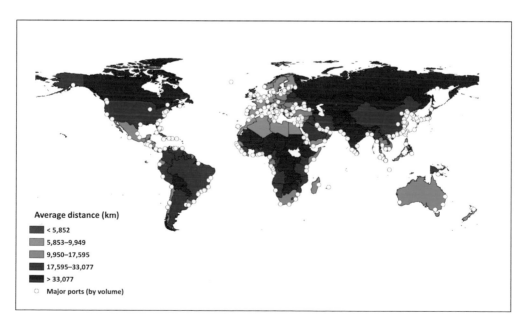

4.3 Country's average distance to major port

Source: McCord, Gordon, and Jeffrey Sachs. 2013. "Development, Structure, and Transformation: Some Evidence on Comparative Economic Growth." NBER Working Paper No. 19512. Washington, DC: National Bureau of Economic Research.

Large continental countries, such as Russia, have a big disadvantage. Most Russian cities and industrial zones are far in the interior of the country and face huge overland transport conditions to get to seaports (or to import international inputs for these industries). Much of tropical Africa is landlocked; in fact, with sixteen countries, Africa has the largest number of landlocked countries in the world.* In those sixteen countries, the populations are both physically and politically far from the ports. Goods not only need to be brought inland; they need to pass a political border as well. Coastal countries that have proximate access to international trade have tended to grow better and faster and to take off earlier. And the landlocked countries, or countries like Russia that are not landlocked but where most of the population and economic activity is far from seaports, have tended (with some important exceptions) to be laggards in the process of economic development.

The map in figure 4.4 shows yet another crucial aspect of physical geography: coal reserves. Energy is at the core of every economic activity, whether farming, industry, services, or transport. I have emphasized how the steam engine unleashed modern economic growth by dramatically expanding humanity's ability to concentrate energy on economic activities. Coal, followed later by oil and gas, gave a crucial and indispensable impetus to global economic development. Countries with plentiful fossil fuel resources have had an easier time of achieving economic growth. Countries that lack these fossil fuels can still achieve economic growth by exporting goods and services to pay for energy imports, or possibly by tapping other domestic energy resources such as hydropower where available, but development is considerably harder for countries where most primary energy must be imported compared with places that have the energy resources in the first place.

The geological distribution of fossil fuels is highly varied across the globe. Some parts of the world are blessed with massive fossil fuel reserves, while other places have almost none. In the nineteenth century, coal was "king," particularly for powering the steam engine. Figure 4.4 shows clearly that England, Western

---

* The list is: Botswana, Burkina Faso, Burundi, Central African Republic, Chad, Ethiopia, Lesotho, Malawi, Mali, Niger, Rwanda, South Sudan, Swaziland, Uganda, Zambia, and Zimbabwe.

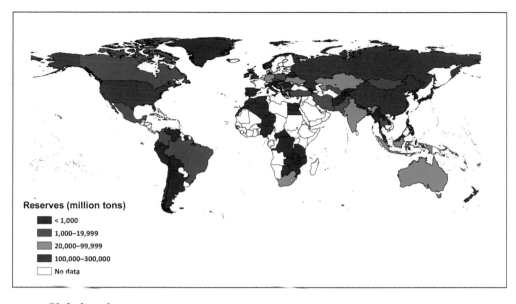

Reserves (million tons)
- < 1,000
- 1,000–19,999
- 20,000–99,999
- 100,000–300,000
- No data

4.4 Global coal reserves

Source: U.S. Energy Information Administration.

Europe, and the United States have lots of coal and that tropical Africa has almost none! That is no result of politics, imperialism, or culture. It is a matter of geology. The locations with accessible coal (that is, coal that can be mined at low cost and transported to population and industrial zones) were highly favored in economic takeoff, especially in the nineteenth century.

Figure 4.5 is another fossil fuel map, this one for oil. It does not depict the shape of Earth as we know it, but the shape of the planet if each country's size is drawn as proportional to its oil reserves. Saudi Arabia, with its massive reserves of petroleum, is right at the center of the map, and other large countries include Iraq, Kuwait, Iran, and Venezuela. Africa barely shows up on the map, because only a few places in Africa have petroleum. The differences of oil holdings are even more dramatic when measuring reserves relative to the national populations. Both Nigeria and Kuwait export 2–3 million barrels a day. Yet Nigeria has 160 million people, while Kuwait has around 1 million.

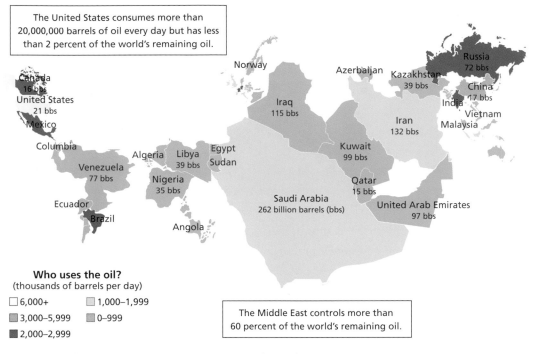

4.5 World map, country size proportional to oil reserves

Source: Environmental Action.

For tropical Africa in particular, the lack of fossil fuels need not consign these countries to a lack of economic development in the twenty-first century. The combination of modern technology and lots of sunshine has a lot of potential! The cost of solar photovoltaic (PV) power has fallen by a factor of around 100 since 1977. PV and other forms of solar power (such as concentrated solar thermal energy) could now offer Africa a great energy breakthrough, one that is especially important for countries that, through no fault of their own, simply lack the coal, oil, and gas reserves that have benefited other parts of the world.

Before leaving this discussion of energy resources, I should add that while having primary energy resources has generally been favorable for long-term development, there is an important exception known as "the resource curse." Some resource-rich (often petroleum-rich) countries have so badly mismanaged their resource wealth that this natural wealth has come to be seen as a curse rather than a blessing. This can happen, for example, when the large cash flows related to oil exports lead to massive corruption or perhaps to political instability as rival factions vie to grab the oil earnings. Foreign companies too have a long record of engaging in bribery and shadowy accounting practices to evade taxation. The misuse of oil revenues by Nigerian governments over the course of decades, compounded by rampant environmental degradation caused by oil spills of the producing companies without legal accountability, is probably the world's most notorious example of the resource curse in operation.

Yet another aspect to geography that makes a huge difference to long-term growth and sustainable development is climate. The importance of climate is obvious: all human beings need food, fresh water, and other ecosystem services (fiber, timber, protection from hazards) to survive and to thrive. Climate has a huge effect on crop productivity, disease, water scarcity or availability, and vulnerability to hazards.

Figure 4.6 shows a map of the Köppen-Geiger (K-G) climate classification system, which is one popular system for classifying climates widely used by geographers. The pink and the red areas in the K-G map are the tropical areas, characterized by year-round warmth. The tropical eco-zones have very distinctive challenges in food production and in disease burden, with many diseases such as malaria thriving in this ecology. (Places with winters typically have a seasonal break in the transmission of many warm-temperature diseases, making the disease burden lighter and also making it much easier to control or eliminate the diseases locally.)

The beige areas are the drylands, where total precipitation is low and the ability to grow crops is either small or entirely nonexistent. This is the region where pastoralism (maintaining livestock such as camels, cattle, goats, and sheep) tends to dominate and where many millions of people still live nomadic lives, moving from place to place to graze animals on seasonal grasslands. Since growing food in dry regions is so difficult, it is not surprising that these regions are particularly vulnerable to extreme poverty. There are a few wealthy dryland areas, but

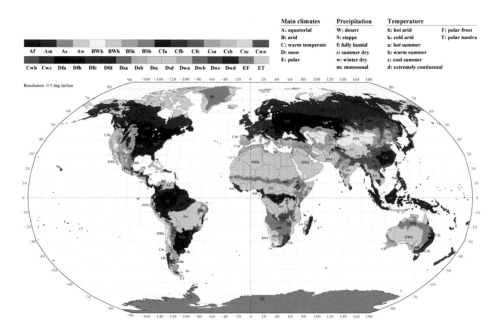

4.6 Global Köppen-Geiger climate classification

Kottek, M., J. Grieser, C. Beck, B. Rudolf, and F. Rubel. 2006. "World Map of the Köppen-Geiger Climate Classification Updated." Meteorol. Z. 15(3): 259–263. doi: 10.1127/0941-2948/2006/0130.

those cases are typically associated with vast mineral wealth under the ground, for example, the diamonds in the Kalahari Desert of Botswana and the hydrocarbons in the Arabian Peninsula.

The light green and the dark green areas are the world's temperate zones, which have both winters and summers. Most of these temperate areas have reasonably plentiful water throughout the year. The light green areas are particularly interesting. They are on the western side of the major landmasses: Eurasia, North America, South America, Australia, and a small part of South Africa (around Cape Town). This climate is called the "Mediterranean climate," since it prevails in the Mediterranean region of southern Europe, North Africa, and the Levant (the eastern Mediterranean). It is characterized by wet, fairly mild winters and warm, sunny, dry summers. You guessed it: this is just the perfect environment to grow the world's best wines.

The dark green areas, such as my own hometown of New York City and also most of western Europe (other than southern Europe), parts of China, Japan, Australia and New Zealand, and Argentina and Uruguay, are the temperate zones that are wet all year round. These are outstanding locations for grain production, especially the kind of mixed-grain and livestock farming we associate with Europe and much of the United States. Without doubt, this dark green part of the world has hosted the highest average incomes in the world, other than the special cases of rich zones that sit atop great mineral or hydrocarbon wealth. The temperate zones are, well, temperate: they are moderate, not too hot, not too cold, with a long growing season, and the absence or near absence of tropical infectious diseases such as malaria, yellow fever, dengue, and various tropical worm infections.

Modern economic growth began in the dark green temperate climate of England, and quickly spread to similar locations in North America, Australia, New Zealand, and the Southern Cone of South America (Argentina, Chile, and Uruguay). All of these locations exhibited high-productivity agriculture, mixed-grain and animal husbandry, strong forestry sectors with plentiful timber and other forest services, and healthful environments, notably for European settlers. We see that modern economic growth diffused not only according to geographical proximity (distance from London) but also what we might call "climate proximity," the similarity of a location to that of England.

The light purple and dark purple zones are the cold zones. The closer to the North Pole, the colder the climate and the shorter the potential growing season. Near the Arctic Circle, farming becomes impossible (with small groups engaged in hunting, fishing, or reindeer herding). Of course, there are many mining activities in the far northern regions, but almost always with small resident populations and very capital-intensive operations.

Let's consider in a bit more detail the single most important case of a climate-dependent disease: malaria. Malaria is a mosquito-transmitted disease. The malaria pathogen is a one-celled organism called *Plasmodium*. A mosquito becomes infected with *Plasmodium* when, in order to take a blood meal, it bites a human being who is already infected with malaria. It gets more than the blood; it also gets the *Plasmodium* in the individual's bloodstream. The mosquito infects another person, roughly two weeks later, when it once again takes a blood meal, this time from a human being not yet infected with malaria. In the process of drawing blood from the new victim, the mosquito also transfers the *Plasmodium* to the unsuspecting individual. Many days later, that person will develop a severe, life-threatening case of malaria.

Malaria is only transmitted by a certain kind of mosquito (certain species of the genus *Anopheles*) and only when the air temperatures are above around 18° Celsius/65° Fahrenheit. The warm air helps to make the mosquito infective when it bites the second time and transmits the disease to the uninfected individual. If the outside air is too cool, below 18°C, the *Plasmodium* living inside the mosquito probably will not have passed through its own life cycle rapidly enough in the gut of the mosquito for the next bite to pass the infection to another person.

Cooler climates, therefore, do not have malaria transmission, but places that are warm all year round, such as tropical Africa, generally have very high year-round malaria transmission. Africa seems to be uniquely burdened on three counts: high temperatures year-round; enough rainfall year-round to provide breeding sites for new generations of *Anopheles* mosquitoes; and strangely enough, the most deadly kind of *Anopheles* mosquito as well, called *Anopheles gambiae*. This kind of mosquito simply loves to bite people, as opposed to cattle and other animals. The result is that Africans are really in the malarial "line of fire."

Figure 4.7 is a map that I put together with colleagues around a decade ago, in order to combine the three key factors (temperature, moisture, and kind of *Anopheles* mosquito) to make a worldwide measure of which regions are most susceptible to malaria transmission, infection, and alas, deaths. Hands down, Africa is truly the most vulnerable part of the world (Kiszewski et al. 2004, 491). It puts together every ecological condition that contributes to year-round, intensive malaria transmission. The result is that perhaps 90 percent of all malaria deaths in the world today take place in tropical Africa.

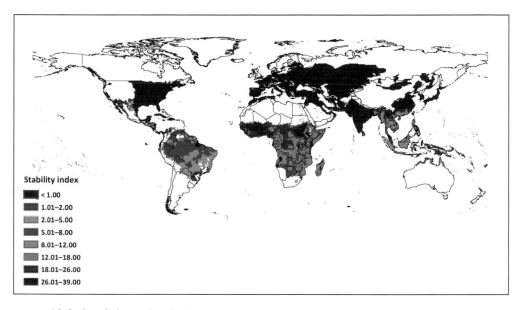

Stability index

- < 1.00
- 1.01–2.00
- 2.01–5.00
- 5.01–8.00
- 8.01–12.00
- 12.01–18.00
- 18.01–26.00
- 26.01–39.00

4.7 Global stability of malaria transmission

Source: Kiszewski, Anthony, Andrew Mellinger, Andrew Spielman, Pia Malaney, Sonia Ehrlich Sachs, and Jeffrey Sachs. 2004. "A Global Index Representing the Stability of Malaria Transmission." American Journal of Tropical Medicine and Hygiene *70(5): 486–498.*

Repeated bouts of malaria not only claim vast numbers of African lives, hundreds of thousands every year, but they debilitate African societies and economies. Children are repeatedly absent from school, and often never finish. Many kids die; others grow up with long-term physical and cognitive difficulties. Husbands and wives have large numbers of children out of fear that many will die, yet the consequence is that they often have more children than they can support with decent nutrition, health care, and education.

In this way, malaria has subtle and insidious ways of holding back overall economic development. Yet while the burden of malaria depends on climate, places with high malaria burdens need not just sit back and suffer. Geography is not deterministic. Geography points to measures that should be undertaken to reverse or offset the burdens caused by the physical environment. African nations with high malaria burdens now can use advanced methods of malaria control to

Table 4.1 Geography and Policy Implications	
Geographic condition	Implication for public policy
Landlocked	Build good roads, rail to the port; maintain good relations with coastal neighbors; emphasize Internet-based export activities to "defeat" location
Water stressed	Emphasize irrigation, e.g., using new solar-powered irrigation pumps for smallholder farmers; specialize in crops that do not require huge amounts of water
Heavy disease burden	Scale up public health interventions to control climate-related diseases
Natural hazards	Understand the changing probabilities of events like floods, droughts, cyclones, extreme storm, and others, and prepare for them with public awareness and physical and social infrastructure
Lack of fossil fuels	Examine and develop alternative options for domestic energy sources, such as geothermal, hydro, wind, and solar power; emphasize energy efficiency

reduce the number of infections and dramatically reduce the number of deaths from malaria.

The moral of the story is that a difficult geography does not prevent development, but it does signal the kinds of investments needed to overcome geographical obstacles. Table 4.1 shows a quick, short, illustrative list.

The point is: be aware of geography, don't give up ("geography is not destiny"), and come up with meaningful alternatives when underlying geographical conditions are difficult for one reason or another!

III. The Role of Culture—Demography, Education, and Gender

In the differential diagnosis of why some places are ahead and others behind, people very often turn first to the sixth category on the list: culture. Culture can be a glib and rather insulting explanation of somebody else's poverty. Rich people often like to think of themselves as rich because of their superior culture, for example, that they pray to the right god. They often have a hard time understanding the natural advantages that might have helped to propel their country forward. In short, the rich often like to blame the poor for their problems, attributing poverty to factors like laziness or the wrong set of religious beliefs.

The situation is often more humorous. When a place is poor, it has the reputation of being lazy. If and when it becomes rich, the reputation is turned on its head. This happened with Japan in the late nineteenth century. When Japan was still poor (around 1870), European observers condemned the Japanese for their alleged laziness. When Japan boomed, Europeans and Americans complained that Japanese culture led the Japanese to work too hard. This kind of reversal probably means that culture was not the key to understanding either the initial poverty or the subsequent wealth. Other factors were at play.

We should also remember that culture is not monolithic. Cultural attitudes, like economic structures, change over time. Think of the changing attitudes to women, African Americans, Jews, the Irish, and other groups who once faced terrible discrimination in the United States. With lots of struggle, attitudes (and laws) changed. The culture changed. The attitudes of today's youth in the United States regarding racism, religious minorities, gender roles, and other aspects of "culture" have changed markedly and relatively rapidly, with significant attitudinal changes in the course of one generation. I do not want to be misunderstood. Culture matters for economic development. But like geography, culture is not destiny. Attitudes evolve, and can evolve in ways that support sustainable development.

When it comes to the most important cultural beliefs that affect sustainable development, we should turn our attention to cultural attitudes toward family size, educational attainment, and the role of women.

When we look at the population challenge, a good place to start is the world map of the fertility rate shown in figure 4.8. The total fertility rate (TFR) in a country

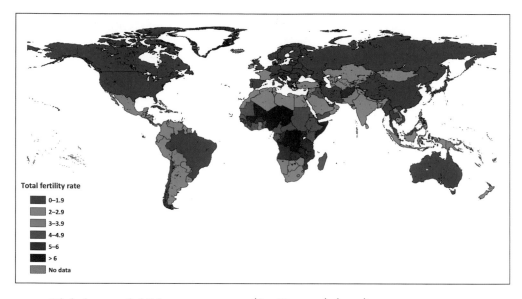

Total fertility rate
- 0–1.9
- 2–2.9
- 3–3.9
- 4–4.9
- 5–6
- > 6
- No data

4.8 Global map of children per women (fertility rate) (2011)

Source: World Bank. 2014. "World Development Indicators."

measures the average number of children a woman in the society will have during her lifetime. The world map shows the tremendous variation in TFRs in our world today. In many parts of the world, especially in the high-income countries, TFRs are below two. That means that each woman, on average, is having fewer than two children. When fertility rates are below two, so that on average each mother is not replacing herself in the next generation with a daughter (statistically speaking), the population will decline over the long term. When the TFR is above two, the population will tend to increase over the long term. In the high-income world, TFRs today are generally below two; in some of the world's poorest countries, notably in tropical Africa and in parts of South Asia, TFRs are still above four; and in many rural areas in low-income tropical Africa, TFRs are above six. This means that each woman is having three or more daughters. One can immediately see the potential for a dramatic rise in the overall population over the course of just a few decades.

High fertility rates tremendously affect economic development, because with very large populations of young children, poor families have a very difficult time providing the basics for all of their children. Perhaps only the eldest son is able to go to school, and the younger girls are married at a very young age without having attained a proper education. In the next generation, those young girls will grow up without the literacy and the skills they need to improve their own lives, and their own children are also likely to grow up in poverty. Countries that have made a transition from high fertility rates to low fertility rates have tended to have an advantage in economic development, while countries that have very high fertility rates tend to have much lower economic growth.

Over time the fertility rate shapes the population dynamics: whether the population is rising or declining in overall size, as well as the age structure of the population. A population's age distribution is illustrated by what is called the age-population pyramid—an example of Japan's changing age-population pyramid is shown in figure 4.9. These pyramids show the numbers of boys and girls, or men and women, at various ages. In the 1950s, TFRs in Japan were above replacement rate. (The TFR in 1950–1955 was around 3.0.) As a result, there were more young children than parents, and more parents than grandparents (because the TFR was above 2 in earlier generations as well). The age-population profile therefore looks like a pyramid, with a wide base (many young children), a smaller midsection of parents, and a narrow segment of grandparents at the top. By 2015, the shape of that age-population profile changes considerably. Japan has reduced its fertility rates, partly as a result of changing culture, partly as a result of economic development, and partly as a result of public policy and access to modern contraceptives. The number of children is actually much fewer than the number of parents because the TFR comes down to 1.3. In the projections for 2055, in the middle of the century, the continuing low TFR will lead to an age-population structure that is an inverted pyramid. Every generation will be smaller in number than the parents' generation, an astounding turn of events, but one that is implied by a TFR that stays well below 2.0.

These three cases provide a wonderful illustration of the link of TFR and the age structure of the population. With a high TFR, the population is young. With a moderate TFR around 2.0, the population is middle-aged. With a very low TFR, well below 2.0, the population is aged.

Most of today's very poor countries still have the pyramid shape, with a high TFR and a very wide base of young people. This means that populations are

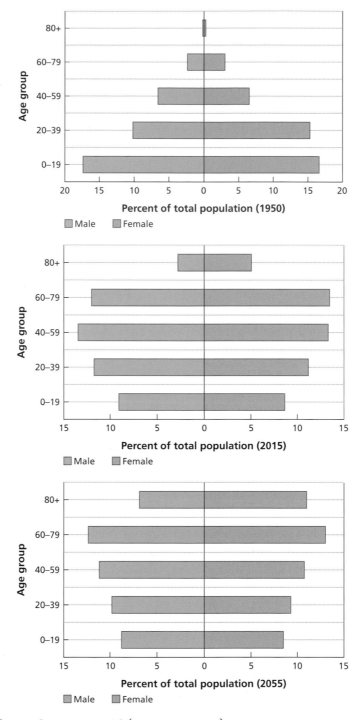

4.9 Japan's population pyramid (1950, 2015, 2055)

Source: United Nations Department of Economic and Social Affairs Population Division (DESA Population Division). 2013. "World Population Prospects: The 2012 Revision." New York.

continuing to soar. Each new generation is much more numerous than the parents' generation. The population of sub-Saharan Africa, for example, would reach almost 4 billion people by 2100, compared with around 950 million today, if the fertility rate declines only gradually. On UN data, Africa's population would have grown from around 180 million in 1950 to 3.8 billion at the end of the century, a greater than 20 times increase in 150 years—an unprecedented increase. With all of the difficulties of shrinking farm sizes, climate change, and depleting energy resources, this kind of dramatic and unprecedented rise in population would almost surely put prosperity out of reach.

The problems are as serious within each family. If a poor father and mother are raising six or eight children, how can they hope to provide each child with the human capital (health, nutrition, and education) that the child will need for lifetime success? More generally, how will all of today's very poor children obtain the education, health care, and nutrition that they need, and how can societies hope to keep up if the population continues to soar? Success will depend on today's high-fertility countries reducing the high fertility rates on a voluntary basis through public policy and changing cultural attitudes. The most important single step, it seems, is to help young girls stay in school. They will marry later, have fewer children, and be more oriented toward the workforce. They and their husbands will choose voluntarily to have many fewer children, a voluntary reduction of high TFR that has already occurred in most other places in the world.

Another attitude or cultural phenomenon, of course, one that is also shaped by politics and economics, involves a society's attitudes toward education. Some societies, even while in a state of great poverty, focused a huge amount of government and family effort and attention on literacy and education. South Korea is one such country—even when it was impoverished in the middle of the twentieth century, it had a very high literacy rate and a remarkably strong drive to raise educational attainment. This commitment to education helped South Korea to achieve some of the fastest and most successful economic development ever attained, and with widespread prosperity. A huge part of Korea's remarkable economic advance has been facilitated by its deep commitment to broad-based, high-quality education for all. This commitment to excellence in education shows up in international test scores. Figure 4.10 shows the 2012 Program for International

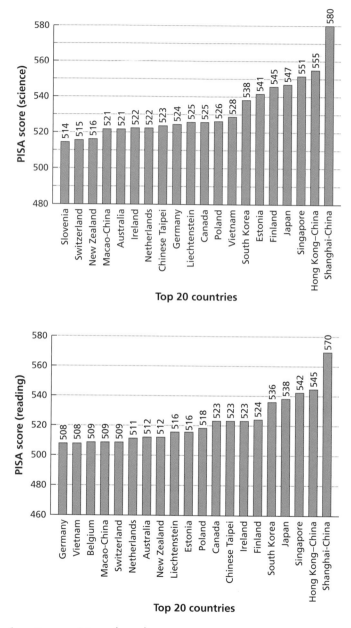

4.10 PISA education rankings (2012)

Source: Organization for Economic Co-operation and Development. 2014. PISA 2012 Results: What Students Know and Can Do—Student Performance in Mathematics, Reading and Science (1). PISA: OECD Publishing. http://dx.doi.org/10.1787/9789264201118-en.

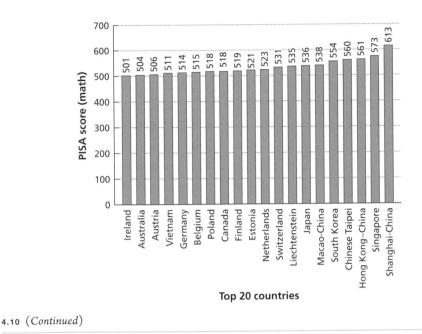

4.10 *(Continued)*

Student Assessment (PISA) rankings on international testing in math, science and reading, with Korea near the top in all three categories. This remarkable performance reflects not only public investments in education but also strong parental support in Korea for the educational attainment of their children. Note indeed the preponderance of East Asian excellence at the top of the charts, reflecting a cultural commitment to educational attainment that is widespread in East Asia.

Yet another cultural attitude that deeply influences patterns of economic development is the cultural attitude toward women. Do women have legal rights? Are women participating on an equal basis in the labor force? Do women continue to face massive discrimination? Gender equality, of course, also has political aspects (e.g., when women are denied the right to vote), but culture plays a very significant role. Once again, as with fertility and as with education, there are big differences around the world in attitudes toward gender equality and attitudes toward women, and this influences economic development in many ways.

There probably is not a society in the world where women still do not face at least some discrimination, because the long history of the world is a history of discrimination against women. It took exceptional political effort, social mobilization, and a lot of courage on the part of women to break through this discrimination, even in places that today we view as close to attaining gender equality. But there are many parts of the world where women still face profound barriers to their economic and political participation. What are the consequences of that? Most obviously and evidently, a society that tries to run on half its brain power and talents; that disenfranchises half the population; and that blocks women from problem solving, economic leadership, and politics, is bound to fall behind countries that are empowering all of their citizens, female and male.

This is an area where there have been huge positive changes over the last thirty years, although the progress is by no means uniform. For example, Rwanda's parliament is not only 64 percent female, but it also has the highest proportion of women in any parliament in the world. Female participation in politics has soared in Rwanda and is rising in other parts of the world, although there are still huge inequalities and

4.11 Female members of Rwanda's Parliament with President Paul Kagame

Parliament of Rwanda.

male domination of political power. In Rwanda, female empowerment has extended beyond parliament. Rwanda has made astounding progress in reducing child mortality and in improving education and social conditions. (The under-5 mortality rate has declined by roughly half in just one decade, from 145 deaths/1,000 births during 2000–2005, to 74/1,000 births during 2010–2015 [World Bank 2014c].) While there are many factors contributing to Rwanda's ongoing escape from the poverty trap, I believe the role of women in politics has played a significant role. Rwanda's success is a very powerful message for countries that are still lagging behind: for success in the twenty-first century, don't try to develop with only half of your citizenry, but take the lesson from a country that is mobilizing all of its citizens.

IV. The Role of Politics

In addition to geography, poverty trap, and culture, we must include politics (and governance more generally) in our poverty checklist. Politics can fail in many ways. I include four such types of failures on the checklist: bad policies (item 2); financial insolvency (item 3); poor governance (item 5); and adverse geopolitics (item 7).

Governance is so important because the role of government in economic development is absolutely crucial. The government is vital for building the infrastructure—the roads, rail, power transmission, port services, connectivity, water, sewerage, and the rest—that is necessary for any economy to develop. Government is essential for human capital development: the health, education, and nutrition of the population, especially of the children. If the government is not performing, public schools will be miserable. If the government is not performing, health conditions will be poor. Infectious diseases such as malaria will run rampant.

An effective government is necessary to ensure economic opportunity for all, including the poorest of the poor. What happens to a very poor family that cannot make ends meet? The children in such a family, unless supported through government programs, will be unable to obtain proper health, nutrition, education, and the ability to develop skills, all of which are essential to enabling a poor child to escape from poverty. Government therefore makes possible the intergenerational mobility out of poverty by helping poor kids to get an effective start in life and to receive a high-quality education.

Government is vital for the rule of law. Of course, without government, there can be anarchy and violence. If the government itself is massively corrupt, other institutions such as the banking sector will operate in a lawless environment and are doomed to fail, or at least fail to contribute to broad-based economic development. If contracts cannot be enforced, if courts are not working, then who can do business? When we see a country in crisis, in addition to the checklist of culture, poverty, geography, we want to look at the political situation across several dimensions.

Think simply about creating an effective road system in a country. This depends on effective policies (to design the road system, for example, and to hire the companies to implement it). This depends on adequate public finance; to be able to raise the needed funds, out of the budget or through bond issues or public-private partnerships. This depends on honesty. Many road projects never produce actual roads because of the high burden of corruption. And this requires decent geopolitics. The country must be at peace. It perhaps needs one or more international partners to get the job done. Obviously roads provide just one example of how good governance combines policy, politics, finance, and foreign affairs. One can say the same about education, health care, and countless other sectors of the economy.

China, one of the fastest-growing economies, has excelled in developing the capacity of government at all levels to undertake large-scale infrastructure investments. Rapid intercity rail now provides a tremendous national transport system. Major cities have urban metro systems. Mass electrification has enabled rapid industrialization. China is an example of a country where the government has played an essential role in enabling very rapid growth. On the other side, there are many poor countries where governments have not yet had the capacity, focus, or interest to undertake the large-scale infrastructure investments needed for effective development.

Government also has to regulate key sectors of the economy. Finance is one of them. Unregulated banking systems tend to get into crisis. The whole world experienced this in 2008 when the deregulation on Wall Street, at the very epicenter of the world's financial system, caused a massive financial crisis that spread throughout all of the arteries and veins of international finance. Government dropped the ball because powerful financial lobbies in Washington had successfully urged the deregulation of Wall Street. Wall Street gained massive profits; the rest of society suffered massive losses. When government fails or when governments encourage or allow illegality or fraud in the banking systems, financial panics often ensue.

Corporate lobbying can result in a massive amount of corruption and massive failures of the regulatory process. Corruption, of course, is not measured very well: corrupt companies and governments do not exactly go out of their way to document what they are doing! One leading NGO, Transparency International, provides a useful global public service by publishing the Corruption Perceptions Index, which uses expert opinions to measure the perceived level of public sector corruption in countries and territories around the world. A map of the Transparency International results for 2013 is shown in figure 4.12 (Transparency International 2013). Countries in dark blue (Canada and Scandinavia) are considered to be the least corrupt countries. Those in red are considered to be highly corrupt.

Governments have a major role in ensuring that even children from poor families have a chance of social and economic mobility. That requires public help for poor families, particularly to help poor children gain access to quality day care,

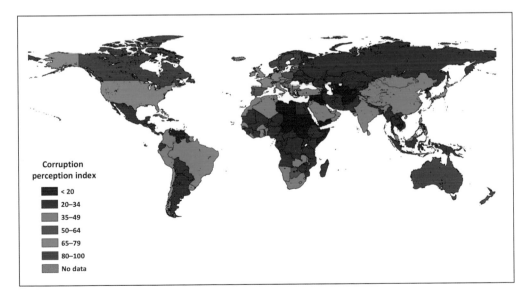

4.12 Global perceptions of corruption

Source: Transparency International. 2013. "Corruption Perceptions Index 2013."

preschool, nutrition, health care, and a safe environment. Such investments for poor children have all been demonstrated time and again to be tremendously powerful in enabling children, especially from poor households, to get that necessary added boost so they have a good shot at prosperity.

Governments differ significantly in their readiness to help children in poor families, and indeed to fight poverty more generally. Within the high-income countries of the Organization for Economic Co-operation and Development (OECD), there is tremendous variation between governments in their investments in social areas, such as protection against poverty and provision of social services. Figure 4.13 shows a graph of public social expenditures as a share of national income for the OECD countries. At the high end of such investments (measured as a share of GDP) are the Scandinavian social democracies; at the opposite end of the scale are countries, including the United States, Japan, and Ireland, where public help for poor families and poor children is very low. The consequences of this are that

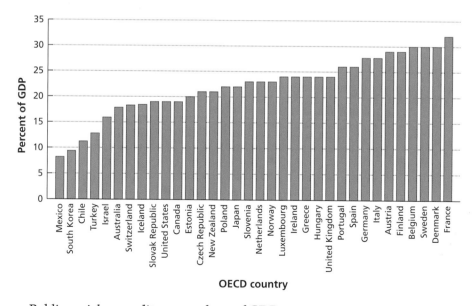

4.13 Public social expenditure as a share of GDP

Source: Organization for Economic Co-operation and Development. 2013. "Government Social Spending." PISA: OECD Publishing. http://dx.doi.org/10.1787/socxp-gov-table-2013-1-en.

poor children in these countries have a much higher risk of growing up in poverty and a much greater chance of ending up poor as adults.

Not surprisingly, poverty rates for children are lowest in the Scandinavian social democracies, where the social spending is highest. Child poverty rates are highest in the United States, Mexico, Italy, and Turkey, countries that invest much less in social programs as a share of national income. This pattern is clearly evident in figure 4.14, which plots social expenditures as a share of gross national product or GNP (on the horizontal axis) compared with the child poverty rate (on the vertical axis). Even among high-income market economies, we see there is huge variation. Some countries attend to the needs of the poor and create conditions for high social and economic mobility. Other countries, alas including my own (the United States), more or less leave the poor to their fate, resulting in a lack of intergenerational mobility and a replication of poverty across generations.

In order to reduce child poverty through funding higher levels of social expenditure as a share of national income, governments must collect higher taxation and with clearer explanations to their citizens of how their money is used.

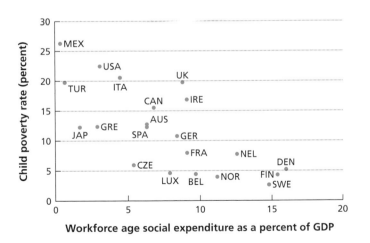

4.14 Social expenditures vs. child poverty

Source: UNICEF Innocenti Research Centre. 2000. "A League Table of Child Poverty in Rich Nations." Innocenti Report Card No. 1, June 2000. Florence: Italy.

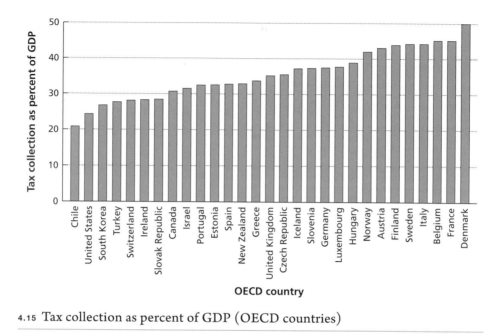

4.15 Tax collection as percent of GDP (OECD countries)

Source: Organization for Economic Co-operation and Development. 2014. "Total Tax Revenue." Taxation:
Key Tables from OECD, No. 2. PISA: OECD Publishing. http://dx.doi.org/10.1787/taxrev-table-2013-1-en.

As demonstrated by the graph in figure 4.15 of taxes as a share of GDP, once again the Scandinavian countries willingly incur higher taxes in order to reduce child poverty and to create conditions of social equality more generally. It is the countries at the other end of the scale, such as the United States, that have a low tax-to-national income ratio and that as a result invest much lower shares of national income in the social areas. The consequence for the United States is a much higher inequality of income, a much higher rate of child poverty, and a much lower rate of social and income mobility across generations. While the United States has long viewed itself as the land of opportunity and social mobility, unfortunately this largely is no longer true.

It is a kind of scourge for societies to become so divided between the rich and poor. Highly unequal societies are both unfair and inefficient, squandering the potential of the poor by a failure to help the poor invest in their skills and health in order to achieve high lifetime productivity. As a result, the economic pie is both smaller and unfairly divided. One of the dimensions of sustainable development is social inclusion: that everybody should have a chance, including children born into poor families. To achieve social inclusion, we will need to focus on the proactive, positive role of government. Is it doing its job to make sure a poor child has the chance to get ahead? When governments carry out that role, they provide a major boost for sustainable development.

V. Which Countries Are Still Stuck in Poverty?

In our quest to understand sustainable development and economic development, we have focused on the modern era of economic growth and have undertaken a differential diagnosis to understand why the moving ripples of economic development have reached some places in the world and failed to reach others. We can now put the pieces together by focusing on the areas in the world that are still stuck below the threshold of self-sustaining growth. Using the $2,000 dollars per person per year threshold as a marker, the map of global development in figure 3.3 shows that those countries shown in red are still below that takeoff level today. These remaining regions are mainly tropical Africa; a number of landlocked countries, including Afghanistan, Nepal, Mongolia, and Laos; and a few other parts of the world. Without question we must regard sub-Saharan Africa as the greatest challenge of development, as it is still the place in the world with the highest poverty rates and with the biggest challenges in meeting basic needs.

The good news is that in recent years, especially since the year 2000, economic growth in sub-Saharan Africa has picked up. There have been major advances in some key areas of disease control, improved access to education, and building infrastructure. However, there is still not self-sustaining, rapid, and dynamic growth, though this prospect now feels very much within reach. We can do a differential diagnosis on sub-Saharan Africa and ask what we learn about sustainable development by taking that multidimensional view of a region.

The African tropics have many distinctive features that are relevant for economic development. The disease burden is very heavily concentrated in the tropics, where malaria, vector-borne diseases such as dengue fever, and worm infections prevail. Agriculture can be very difficult. There is often water scarcity with the very high temperatures, a vulnerability to drought, and a high instability of rainfall. The depletion of soil nutrients can be extremely pernicious in the tropical context. There is nothing impossible about these challenges, because diseases like malaria are fully controllable, but they need to be controlled. They are serious issues and have their highest burden in these geographic areas, and so special attention needs to be given to those challenges.

We have noted that Africa has the distinctive feature of the most landlocked countries of any continent. Roughly one-third of African countries are landlocked: sixteen of the forty-nine countries. This is a big problem, but it is worth pausing to ask why this is the case. One key reason is the colonial legacy. Nature does not draw national boundaries; politicians do. When the politicians divided up Africa, notably at the Berlin Conference of 1884–1885, they divided it up into little parcels and often cut across natural ecological areas or artificially divided ethnic groups. This difficult legacy made it hard for many populations to even reach the coasts. The distance from the ports may also have to do with history in another sense. Some historians have argued that as a defense against the slave trade, some African populations moved from the coasts to the interior, where they would be safer from capture.

There is also a physical geographic aspect to these distances as well. In many parts of Africa the coastal physical environment is rather hostile. In East Africa, around Somalia and Kenya, the coast tends to be very dry. The easterly winds do not bring precipitation to the coast, but to the highlands in the interior. This means that the high population densities in East Africa are not at the coast but are in landlocked interior countries like Rwanda or Uganda, where there is much more rainfall than in the port areas like Mombasa, Kenya, which is in a much drier region.

The colonial legacy played other adverse roles as well. At the time of Africa's independence in the late 1950s and into the 1960s and onward, there were very few Africans with higher educations. On the eve of independence in 1960, the Democratic Republic of Congo had fewer than twenty university graduates in a population of 13 million. The European powers did not provide education, which they saw as a political risk. When African nations achieved independence, many countries had just a tiny fraction of the population with a high school degree, much less a university education.

The European powers also left behind a deficient physical infrastructure. Figure 4.16 compares the map of Africa's railway system today with the Indian railway system, built largely during the British colonial period. In India, there is a full railway grid. This is because the single ruling colonial power, Great Britain, created a unified infrastructure in part to facilitate the extraction of India's natural resources, such as India's cotton for use in England's cotton mills. In Africa, where there were more difficult topographical and geographical conditions and many political divisions, the European colonial powers did not sit down together to construct a railway network. Each imperial power typically constructed its own rail, perhaps from a port to a mine or plantation. The rail system left by colonial powers therefore was not a grid but just individual lines running from ports to interior locations that were important to the colonial powers. The lack of an effective rail grid has been an enormous burden for Africa. When India had its agricultural Green Revolution of the 1960s, its railways played a crucial role in bringing fertilizer into the interior and bringing grain from the agricultural regions to the cities. But in Africa, the rail cannot serve that purpose, since it does not exist. Even in the twenty-first century, Africa's rail network has yet to be built.

The legacies of colonial rule in Africa, in short, have been extremely pernicious. This is not an explanation of everything, as there is never a single-factor explanation. A differential diagnosis does not necessarily yield a simple answer. We do not need more simplistic answers, but we need accurate answers. A differential diagnosis helps with accuracy by identifying the challenges that need to be addressed.

Our conclusions must not be occasions for pessimism, however. I have emphasized that historical or geographical burdens are not fate; they are not destiny; they are reasons for action. The problems of extreme poverty in Africa and elsewhere can be solved. The tools for such solutions are more powerful than ever, in education, health care, agriculture, power, transport, finance, and many other areas. There are proven methods of public policy to scale up these solutions. We will explore many of those practical solutions in the chapters ahead.

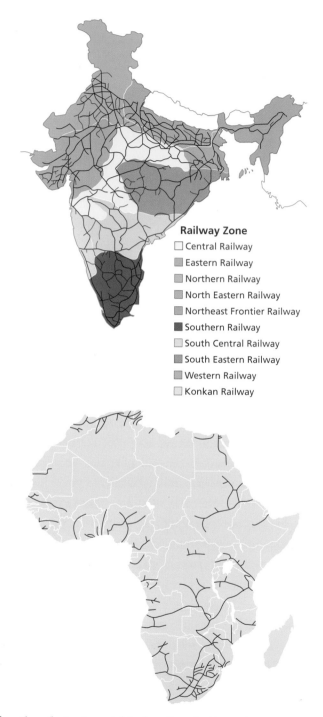

Railway Zone
- ☐ Central Railway
- ☐ Eastern Railway
- ☐ Northern Railway
- ☐ North Eastern Railway
- ☐ Northeast Frontier Railway
- ☐ Southern Railway
- ☐ South Central Railway
- ☐ South Eastern Railway
- ☐ Western Railway
- ☐ Konkan Railway

4.16 India's railroads (*top*) vs. Africa's railroads (*bottom*)

Source: India's railroads: Copyright © Compare Infobase Ltd. Africa's Railroads: African Studies Center. "Africa's Railroads." Michigan State University. http://exploringafrica.matrix.msu.edu.

5

ENDING EXTREME POVERTY

I. The Reasons to Believe That Extreme Poverty Can Be Ended

We have studied the process of modern economic growth and have seen how a persistent rise in gross domestic product (GDP) per person has occurred and spread throughout almost all the world, while a few remaining regions of the world have not yet taken off. Where growth has occurred, extreme poverty has declined, often to negligible levels. There is reason to believe that sustained economic growth can spread to the remaining regions— especially to tropical Africa—and thereby eliminate the remaining pockets of extreme poverty. Yet for reasons that I will discuss, the favorable prospect is by no means guaranteed. It will have to be achieved through deliberate local, national, and global effort and will not simply happen on its own.

We will analyze the possible pathways to ending extreme poverty in the next fifteen to twenty years. Yet to do so, we first need a working definition of extreme poverty. The World Bank's poverty line is certainly the most widely used. The World Bank defines extreme poverty as an income below a poverty line of $1.25 per day, measured in U.S. dollars at international prices of 2005. By this measure, there were an estimated 1.2 billion people below the extreme poverty line as of 2010, the year of the most recent data.

The World Bank's definition is surely too narrow. It would be better to define the extreme poverty line according to the ability of individuals to meet basic material needs (SDSN 2012b). These material needs include: food, clean water, sanitation, shelter, clothing, access to health care, access to basic education, and access to essential services such as transport, energy, and connectivity. These basic needs are the minimum needed for survival and human dignity. We could define those living in poverty as individuals that by lack of household income or public services are unable to meet their basic needs. By this broader definition, the number of those in poverty would surely rise above 1 billion, and perhaps could reach 2 billion people. Unfortunately, as of now, there is no comprehensive, worldwide data on this broader sense of extreme poverty. We therefore tend to fall back on the more limited World Bank definition. Perhaps in the period of the Sustainable Development Goals (SDGs), a broader and sounder definition will be practicable and measured globally.

The World Bank also measures other threshold lines above the $1.25 per day mark. Another common line is drawn at $2 per day, also at international 2005 prices. Naturally, a higher proportion of the world falls under the $2 per day mark, an estimated 2.4 billion people as of 2010.

The *headcount poverty rate* measures the share of the population under a given poverty line. The recent trend from 1981 to 2010 is shown in figure 5.1. Note how steeply it has come down: from 52 percent of the developing world population in 1981 to 43 percent in 1990, 34 percent in 1999, and 21 percent in 2010. Note that the poverty rate has declined by half between 1990 and 2010. The first Millennium Development Goal (halve, between 1990 and 2015, the proportion of people whose income is less than $1.25 a day) has thereby been achieved, if we consider the developing countries as a single entity. This gives us hope that extreme poverty can be reduced in those places where it remains high until today, most importantly in sub-Saharan Africa.

The headcount poverty rate by major region is shown in figure 5.2, for the years 1981, 1990, 1999, and 2010. We see that China has achieved the most remarkable poverty reduction in history, with extreme poverty falling from 84 percent in 1981 to just 12 percent in 2010. This remarkable progress has, of course, been accompanied by an equally remarkable rate of economic growth, roughly 10 percent per year during these three decades. On the other extreme is sub-Saharan Africa. The poverty rate actually rose between 1981 and 1999, from 51 percent to 58 percent.

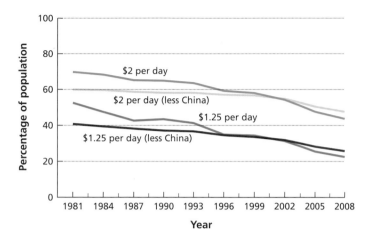

5.1 Poverty rates for the developing world (1981–2008)

Source: PovcalNet/World Bank.

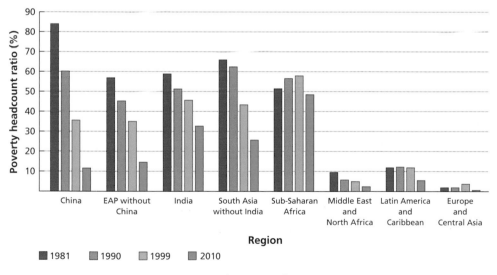

5.2 Extreme poverty rates by region (1981–2010)

Source: PovcalNet/World Bank.

It was only after the adoption of the Millennium Development Goals in 2000 (discussed later in the chapter), that the rate of extreme poverty began to fall. South Asia is the region in between. In India, the poverty rate declined from 60 percent in 1981 to 33 percent in 2010. In the rest of South Asia, the poverty rate went from 66 percent to 26 percent in that time interval.

By adopting the methods of differential diagnosis, we can help regions still stuck in poverty to overcome the chronic low growth that has kept poverty rates high throughout modern history. Indeed, with sound policies in today's high-poverty regions, it is possible to realistically foresee the end of extreme poverty on the planet within this generation, perhaps by 2030 or 2035. The idea that humanity could actually put behind it the ancient scourge of extreme poverty is a thrilling idea. It may seem fanciful or utopian, but it is actually very practical. It is based on strong evidence and the experiences of recent years.

Those parts of the world still stuck in extreme poverty can get out of the poverty trap if they pursue policies aimed at overcoming the specific barriers to growth that now hold them back. Indeed, sub-Saharan Africa has already embarked on that effort, and growth rates have recently risen to around 6 percent per annum. They can go even higher. Success, though, will require not only sound domestic policies but also the partnership of other parts of the world.

It is our job to understand how the end of poverty can be achieved and then act to make it happen. It is extraordinarily important to take note of and appreciate the progress that has already been made, as well as recognize that the setting of a global goal to end extreme poverty is, by itself, one of the most important tools that we have.

The great British economist John Maynard Keynes raised the idea of ending poverty back in 1930, though he was certainly referring at that stage to the industrialized countries, not to the entire world. In his famous essay, "Economic Possibilities for Our Grandchildren," Keynes begins by noting that from the time of the Roman Empire until the early eighteenth century, the rate of technological progress remained extraordinarily low (Keynes 1930). It was so low, Keynes notes, that a peasant from the Roman Empire would have felt relatively at home in rural England in the early 1700s. Keynes then goes on to describe the explosion of technological advances from the Industrial Revolution onward, and draws the lesson that soon enough, productivity would rise to the point where poverty in Britain and other high-income countries would actually be brought to zero. Here is how he puts it:

> I would predict that the standard of life in progressive countries one hundred years hence will be between four and eight times as high as it is to-day. There would be nothing surprising in this even in the light of our present knowledge. It would not be foolish to contemplate the possibility of a far greater progress still . . . I draw the conclusion that, assuming no important wars and no important increase in population, the economic problem may be solved, or be at least within sight of solution, within a hundred years. This means that the economic problem is not—if we look into the future—the permanent problem of the human race. (Keynes 1930, 3)

When Keynes refers to "the economic problem," he means poverty; thus, he is stating that poverty could be a thing of the past within one century, by 2030. In fact, Keynes was proved right roughly a half-century from the writing of his essay. By around 1980, extreme poverty was a thing of the past in the high-income world, consigned to the "dustbin of history."

What is interesting is that Keynes' 100-year forecast might prove to be correct for the entire world, not just the "progressive countries," as he termed the industrialized countries of his day. Even more remarkable, perhaps, is that when Keynes made his forecast, the world population was just 2 billion. The world population is now 7.2 billion, more than three times the population of 1930, and by the middle of this century it will be likely be more than 9 billion. Keynes also added another condition that barred further world wars. However, there was another major war—World War II. Despite both of these conditions, the massive increase of world population and the continuing tragedies and destruction of war, Keynes's basic insight that technological progress can bring about the end of poverty remains true and prescient and is now within reach of the entire world.

The Millennium Development Goals

> We will spare no effort to free our fellow men, women and children from the abject and dehumanizing conditions of extreme poverty, to which more than a billion of them are currently subjected. We are committed to making the right to development a reality for everyone and to freeing the entire human race from want. We resolve therefore to create an environment—at the national and global levels alike—which is conducive to development and to the elimination of poverty.
>
> —UN General Assembly, "United Nations Millennium Declaration"

In September 2000, a remarkable thing happened. More than 160 heads of state and government gathered at the United Nations to usher in and convey the world's hope for the new millennium. The secretary-general of the United Nations at that time, Kofi Annan, put forward to the world leaders a pathbreaking "Millennium Declaration." The declaration called on the world to honor the new millennium by committing to great global goals: universal human rights, peace and security, economic development, environmental sustainability, and the drastic reduction of extreme poverty. As part of the "Millennium Declaration," the world leaders adopted eight specific development goals that quickly became known as the Millennium Development Goals (MDGs), shown in figure 5.3.

Why a cartoon drawing for each goal? The goals are meant for the average person in the street, not for high theorists. This is important to appreciate. The goals are phrased in a way that they can be understood in the villages, the slums, the places where poor people live and work and fight for their survival. The goals serve to orient humanity around a great moral challenge: to improve the life conditions of the most vulnerable people on the planet. They exist to spur action across the society: by governments, businesses, communities, families, faith groups, academicians, and individuals. They are meant to spur broad social change, not just a few technical fixes here and there.

Goal number 1 calls for eradicating extreme poverty and hunger. Goal number 2 is to achieve universal primary education. Goal number 3 is to promote gender

5.3 The eight Millennium Development Goals

UNDP Brazil.

equality so that women, like men, have rights and access for economic progress. Goal number 4 is to sharply reduce child mortality. Goal number 5 is to sharply reduce maternal mortality and ensure a healthy childbirth process for mothers and their children. Goal number 6 is to fight the raging pandemic diseases of AIDS, malaria, and other mass killers. Goal number 7 is to promote environmental sustainability. Finally, goal number 8 is to promote a global partnership whereby rich countries help poor countries to achieve the first seven goals.

Beneath this general description are some specific quantitative targets and many dozen indicators; the example for MDG 1 is described in table 5.1. For the eight MDGs there are twenty-one specific quantified targets, as well as approximately sixty detailed indicators to measure the progress. It has been my great honor to serve as special advisor to the UN secretary-general on the MDGs, first for UN Secretary-General Kofi Annan (during 2001–2006), and now for UN Secretary-General Ban Ki-moon (2007 to the present). My responsibility has been to help analyze and design strategies to support countries in achieving the MDGs and to work with the UN agencies and donor governments as well to help implement those strategies.

Table 5.1 Targets and Indicators for MDG 1

Goal 1: Eradicate extreme poverty and hunger

Target 1A: Halve, between 1990 and 2015, the proportion of people whose income is less than $1 a day	1.1 Proportion of population below $1 (PPP) per day 1.2 Poverty gap ratio 1.3 Share of poorest quintile in national consumption
Target 1B: Achieve full and productive employment and decent work for all, including women and young people	1.4 Growth rate of GDP per person employed 1.5 Employment-to-population ratio 1.6 Proportion of employed people living below $1 (PPP) per day 1.7 Proportion of own-account and contributing family workers in total employment
Target 1C: Halve, between 1990 and 2015, the proportion of people who suffer from hunger	1.8 Prevalence of underweight children under five years of age 1.9 Proportion of population below minimum level of dietary energy consumption

Source: United Nations Statistics Division 2008.

It's been wondrous to see how the MDG goal setting has energized civil society and helped to orient governments that otherwise might have neglected the challenges of extreme poverty. The MDGs have drawn global attention to the plight of the poor and also have helped to motivate problem solving around the world to overcome the remaining extreme poverty. Of course, as economic history shows, and as Keynes emphasized, the long-term fundamental forces driving poverty down are technological. Yet the MDGs have been important in encouraging governments, experts, and civil society to undertake the "differential diagnoses" necessary to overcome remaining obstacles.

Progress has been quite notable, and breakthroughs have occurred in some of the poorest countries and regions of the world. The overall decline of the rate of extreme poverty, as we have noted, has been dramatic: down by more than half since 1990. Of course the MDGs were not the main factor in the biggest success of all: China. However, in Africa, the MDGs have played a far more significant role in helping to end a long period of stagnation and rising poverty and to begin a period of falling poverty, improving public health, and more rapid economic growth.

The gains have been made not only in reducing poverty but also in many of the other MDGs. Consider the fight against disease for example. Figure 5.4 shows the rapid increase in the number of HIV-infected people kept alive by antiretroviral medicines (ARVs), shown by the blue curve. If not controlled by ARVs, the HIV virus causes AIDS and almost-certain death. Now, with the spur of the MDGs and the health programs they have promoted, millions of people in low-income countries are alive today, because they have been given free access to lifesaving ARVs.

Figure 5.5 demonstrates another public health triumph that I would ascribe to the public awareness and problem solving promoted by the MDGs: the reduction

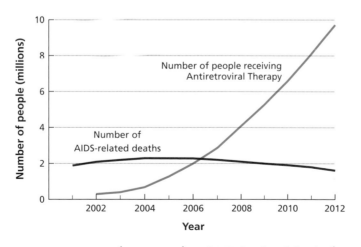

5.4 AIDS treatment recipients (2002–2010) and AIDS-related deaths (2000–2010)

Source: Joint United Nations Programme on HIV/AIDS (UNAIDS).

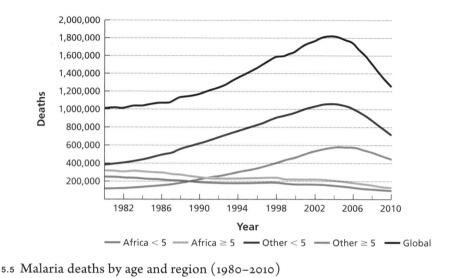

5.5 Malaria deaths by age and region (1980–2010)

Source: Murray, Christopher J. L. et al. 2012. "Global Malaria Mortality Between 1980 and 2010: A Systematic Analysis." The Lancet 379: 413–431.

of the malaria burden and malaria deaths. Note how malaria deaths in Africa peaked around 2005 and then began to decline markedly. This was achieved through the scaling up of malaria-control programs based on a number of cutting-edge technologies, including long-lasting insecticide-treated bed nets; a new generation of antimalaria medicines; and various other advances (such as rapid diagnostic tests) enabled by scientific progress. The MDGs encouraged the creation of several special programs to fight malaria, and these programs have by now led to a remarkable decline of malaria prevalence and malaria deaths, especially in sub-Saharan Africa.

The combination of continued rapid technological change and a good "differential diagnosis" to identify priority needs of each low-income region can thereby direct investments toward high-return antipoverty programs, whether for infrastructure (such as roads, rail, power, connectivity, and ports), health care, safe drinking water and sanitation, or improved access to schooling. Just as public health has improved with the scaling up of programs to fight AIDS and malaria, similar breakthroughs can be made in other areas: more productive farming, new industrial development, and greatly improved educational attainment.

II. Strategies to End Extreme Poverty

The end of extreme poverty is within reach. As we've noted, there are roughly 1.2 billion people still living below the World Bank's current poverty line of $1.25 per person per day. Thankfully, this number has been sharply reduced from 1.9 billion people in 1990. So where are those remaining areas of extreme poverty?

There are two main regions of the world still stuck in a poverty trap (World Bank 2014d). The most poverty-stricken region of the world is tropical sub-Saharan Africa. In 2010, an estimated 48.5 percent of the population of tropical sub-Saharan Africa remained below the poverty line. Fortunately that rate is declining now and has been declining since the start of the new millennium. Some estimates put the poverty rate even lower today, though the data are much debated. The other place where extreme poverty in large numbers remains is South Asia, where the poverty rate in 2010 was estimated to be 31 percent of the population. In raw numbers, in 2010, around 413 million people lived in extreme poverty in tropical sub-Saharan Africa, and 507 million people lived in extreme poverty in South Asia. Just these two regions constitute around 76 percent of all of the world's extreme poverty.

In East Asia, around 20 percent of the total population, or 250 million people, are still in extreme poverty, even though East Asia has enjoyed by far the fastest decline of extreme poverty of any region, in conjunction with its remarkably high rate of economic growth. In the Middle East and North Africa, around 10 percent of the total population lives in extreme poverty, around 100 million people. The remaining 100 million or so of the world's poor are scattered in the other regions of the developing world (Latin America and the Caribbean, Europe, central Asia, small island states).

The two big regions needing future breakthroughs are therefore sub-Saharan Africa and South Asia. Let us first make a differential diagnosis for sub-Saharan

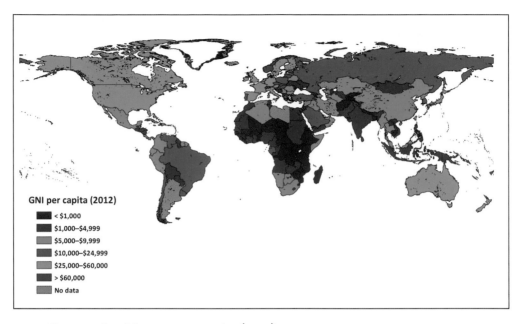

5.6 Gross national income per capita (2012)

Source: World Bank. 2014. "World Development Indicators."

Africa, to see what can be done to accelerate Africa's economic growth and poverty reduction. We will then turn to South Asia.

Ending Extreme Poverty in Sub-Saharan Africa

There is definitely good news in Africa. Figure 5.7 shows the year-to-year growth
rates of the world economy and of sub-Saharan Africa (International Monetary
Fund [IMF] 2014). The average growth rate in sub-Saharan Africa picked up significantly after the year 2000. Indeed, sub-Saharan Africa has been growing faster
than the average of the world economies, at around 5 percent per year and even
faster in certain years. For 2014, the IMF forecasts annual growth of around 6 percent. This growth rate implies a doubling time of around 12 years (= 70/6). With
population growth at around 2.5 percent per annum, however, the growth of GDP

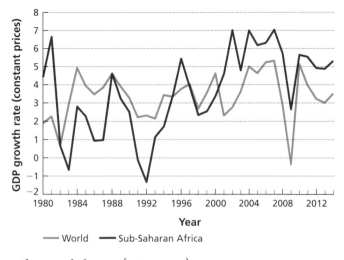

5.7 GDP growth annual change (1980–2012)

Source: International Monetary Fund. 2014. World Economic Outlook Database.

per capita is considerably lower, around 3.5 percent per year, with a doubling time therefore of around 20 years (= 70/3.5).

Something is beginning to go right, and it is possible for sub-Saharan Africa to achieve even faster progress. A differential diagnosis of Africa's problems shows that there are challenges in nearly all of the seven big categories: poverty trap, economic policy framework, fiscal framework, physical geography, governance patterns and failures, cultural barriers, and geopolitics. To organize a complex discussion, I will focus on four particular areas where Africa can achieve rapid breakthroughs: farm productivity, urban productivity, national infrastructure, and human capital investment.

Figure 5.8 shows the crop productivity (tons of grain per hectare) in different parts of the world. Africa is almost solid blue, which in this map means that it achieves very low farm yields. On average, smallholder farmers in sub-Saharan Africa have produced a yield between half a ton and one ton of grain per hectare. This is very poor in international comparative terms. Many other parts of the developing world achieve four or five times that yield. In the most productive

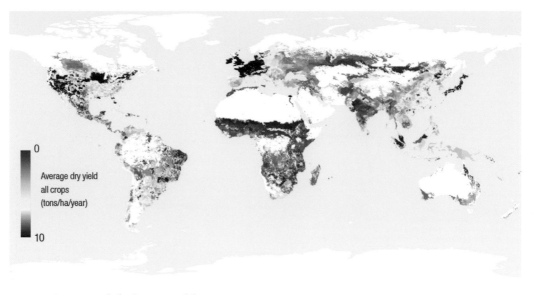

5.8 Average global crop yields

West, Paul C., Holly K. Gibbs, Chad Monfreda, John Wagner, Carol C. Barford, Stephen R. Carpenter, and Jonathan A. Foley. "Trading Carbon for Food: Global Comparison of Carbon Stocks vs. Crop Yields on Agricultural Land." Proceedings of the National Academy of Sciences. *November 1, 2010. doi:10.1073/pnas.1011078107.*

grain belts of the world, for example, in the United States, western Europe, and Japan, yields often rise to ten times Africa's yield.

What is the cause of Africa's low farm yields? In fact, African farms face many obstacles. One key challenge is soil-nutrient depletion. Africa's farmers have generally been too poor to keep their farms' soils replenished with the nitrogen, potassium, and phosphorous necessary for decent crop yields. Just as with undernourished human beings, undernourished crops also fail to grow and thrive. In Africa, with farmers unable to use fertilizers to replenish their soils, the farmlands have been exhausted of the nutrients needed for high yields. The map in figure 5.9 shows the details.

Farmers in almost all other parts of the world use extensive fertilizers, both organic and chemical, to replenish the key nutrients that are removed with each harvest. When a crop is harvested, the nitrogen and other nutrients leave with it.

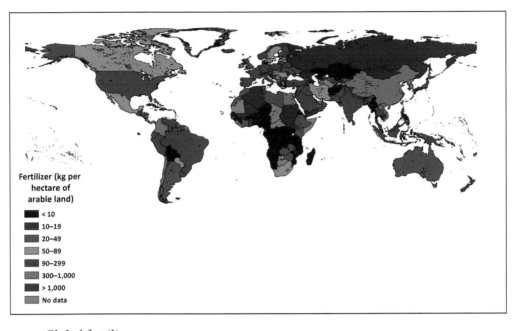

Fertilizer (kg per hectare of arable land)
- < 10
- 10–19
- 20–49
- 50–89
- 90–299
- 300–1,000
- > 1,000
- No data

5.9 Global fertilizer use

Source: AAAS Atlas of Population and Environment.

Somehow those nutrients have to be put back in the soil, whether through green manures, chemical fertilizers such as diammonium phosphate and urea, or long fallow periods, in which nitrogen is replenished through natural processes.

Yet most of Africa's peasant farmers have been so poor that they have been farming without the advantages of those added nutrients, and the resultant low yields trap the farmers in a poverty trap (SDSN 2013c, 6). They get low yields year after year. Since the farmers are too poor to buy the fertilizers that they need, their soils continue to be depleted of key nutrients. The yields remain low, and every year the farmers get a very meager income that does not help their families' struggles with hunger or give them the income necessary to buy inputs that would enable higher production.

In addition to fertilizer, other inputs are also necessary for high yields, such as good water management and irrigation where possible. This typically requires wells and pumps. Additionally, good seed varieties are needed to contribute to high yields. All of these improved inputs are beyond the means of Africa's peasant farmers. In the same way that African farmers have lacked the means to replenish the soil nutrients, they have also lacked the means to invest in irrigation and high-yield seeds.

The problem adds up to an agricultural poverty trap. A high priority for Africa is to invest in its smallholder farmers, with government programs that enable even the poorest farmers to get the inputs that they need, whether on credit or as a grant, so they can enjoy higher farm yields and higher incomes and thereby start investing in these crucial inputs on their own. Over time, these farm households will build up their capital and their creditworthiness. The government programs needed to help them at the start can then gradually be withdrawn, allowing banks rather than government aid to do the same job of financing inputs.

To end extreme poverty in Africa will also require a major buildup of infrastructure, including roads, rail, power, ports, and communication networks. As in many other areas, Africa's colonial rulers left the newly independent African nations with a poor start in infrastructure upon independence. This can be illustrated by comparing India's rail grid with Africa's rail grid, as shown in figure 4.16. Remember that India had just one imperial ruler, Great Britain. Britain constructed a full rail network in part to be able to bring India's cash crops such as cotton to the coastal ports. In Africa, by contrast, there were several European imperial powers (Great Britain, France, Italy briefly, Spain, Portugal, Germany until World War I). They did not connect their separate investments and indeed never built much of a rail system at all. Africa's rail system was mainly single lines running from ports to particular mines and plantations. Modern-day Africa faces extremely high overland transport costs, in part because of the weakness of the rail network, combined with a wholly insufficient highway system. And the fact that the African continent hosts fifty-four countries, including forty-nine in the sub-Saharan region, makes the creation of a modern, continent-wide transport network a continuing unmet challenge.

Other aspects of infrastructure are also especially important in the twenty-first century (SDSN 2013a, 21). There can be no economic development on a sustained basis without mass electrification. Figure 5.10 is a well-known satellite photo of

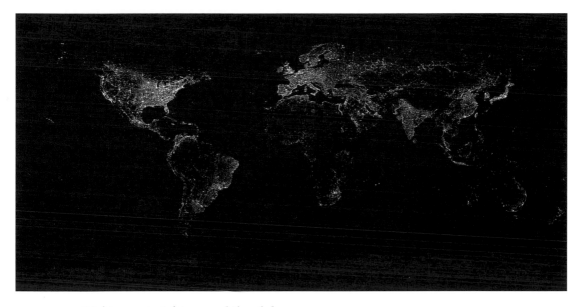

5.10 Lights on at night around the globe

Data courtesy Marc Imhoff of NASA Satellite GSFC and Christopher Elvidge of NOAA NGDC. Image by Craig Mayhew and Robert Simmon, NASA GSFC.

the Earth at night, indicating the places with nighttime electrification and illumination. It shows the bright lights of the United States (especially in the populous eastern half of the country), Europe, coastal China, Japan, coastal South America, India, Southeast Asia, and the Arabian Peninsula. But in sub-Saharan Africa, the lights are out at night. To this day, a large part of rural Africa still lacks access to electricity. In addition to not having access to lights at night or electricity for home activities, there is a critical lack of power for pumping water for irrigation; for refrigeration; for preservation of agricultural outputs; for industrial processing of food, textiles, and apparel; and for every other kind of industrial activity.

The absence of electrification has been a huge chronic barrier to Africa's development and another aspect of Africa's poverty trap (SE4All 2012). Without electricity, productivity is very low. Low productivity means very low output per person, which in turn means low income and thus poverty. Poverty means low tax collections by government, and therefore the inability of the government to invest in the electricity needed to lift the region out of poverty. Once again, we see the vicious circle of poverty. African governments know very well they need to build the power capacity; yet they lack the resources to do so out of their own revenues and the creditworthiness to do so through borrowing. They are stuck, trapped, and in need of a temporary boost of grants and low-interest loans to move out of the rut.

Another critical dimension of infrastructure in the twenty-first century is information technology (IT). The good news is that because these technologies are so powerful and their costs have fallen so far, Africa is already on its way to mass coverage by mobile telephony, which already reaches even the most remote villages. Private investors have already laid, or will soon lay, submarine fiber-optic cables that will slash Internet prices and facilitate the spread of broadband throughout the continent. Since these investments are being made by the private sector, with

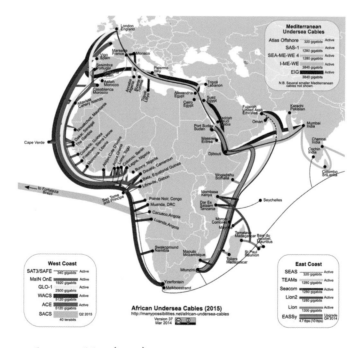

5.11 African undersea cables (2013)

Map courtesy of Steve Song (http://manypossiblities.net).

favorable profitability and lower fixed costs than power generation, the Internet grid and mobile telephony are spreading without the need for public financing or foreign aid. ICT has already given a huge boost to Africa's development, and the boost will be even larger in the years ahead when mobile broadband dramatically improves access to health care, education, banking, and other services.

The moral of the story is that Africa, like the rest of the world, is now poised for a breakthrough—if it can mobilize targeted investments in agricultural productivity, in health care, and in continent-wide infrastructure. In my view, Africa will be able to make the breakthrough in long-term economic growth that has so far eluded it throughout its modern history.

However, there is one final challenge that Africa must surmount. Africa still has a very high fertility rate, meaning that family sizes are very large on average, and the population is growing remarkably rapidly (UN Department of Economic and Social Affairs Population Division [DESA Population Division] 2013). The fertility rate for 2010–2015 is estimated to be 5.1 children, meaning that each woman on average is having more than two daughters to replace her in the next generation. Not surprisingly, the population is growing rapidly.

Note in figure 5.12 that in 1950, sub-Saharan Africa's population was a mere 180 million people. Just sixty years later, sub-Saharan Africa's population is now around 900 million, an increase of five times. And the United Nations projects further rapid growth of the population during the twenty-first century, unless Africa is able to make a transition to a lower fertility rate faster than on the current trend. Based on a *moderately rapid* decline in the high fertility rate, Africa's population is projected to reach an astounding 3.8 billion people as of 2100, roughly four times larger than now. (This is the so-called medium-fertility variant of the UN Population Division.) If the decline in the fertility rate is slower than in the medium-variant, the UN high-fertility variant finds that Africa's population would be even larger, roughly 5.3 billion people. On the other hand, if the fertility rate falls more rapidly than the UN now projects as likely, the population in 2100 in the low-fertility variant would be 2.6 billion people, lower by more than 1 billion persons compared with the medium-fertility variant (DESA Population Division 2013).

Africa will reap many development benefits if it keeps the population on the low-fertility variant. First, there would be a lower population, and therefore more

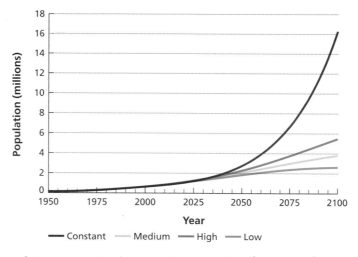

5.12 UN population scenarios for sub-Saharan Africa (1950–2100)

Source: United Nations Department of Economic and Social Affairs Population Division (DESA Population Division). 2013. "World Population Prospects: The 2012 Revision." New York.

land, oil, timber, water, and other natural resources per person. Second, and crucially, families would be smaller, so that each family could invest more per child in education, health, and nutrition. Third, the population age would be higher on average, as there would be a better balance between parents and children in each generation. Fourth, the population would grow less rapidly, so a smaller fraction of savings and investment would be used simply to keep up with the growing population. More of the savings and investment could be used to raise the amount of capital (such as roads, infrastructure, vehicles, and machinery) available to each person. In short, there are great benefits for Africa in fostering through voluntary means a lower fertility rate and thereby faster economic development.

It is worth emphasizing the huge gains in African educational attainment that would likely ensue with lower fertility rates. With a smaller youth cohort, each family would be able to ensure the health, education, and nutrition of every child. Now, poor families choose among their children—perhaps educating the first son but not the rest—while with a smaller cohort of children, all could be educated. Moreover, the government would not be facing an endless race against time and against a rising population, always needing to build more schools and train more teachers, at great expense, not to improve education but simply to keep pace if possible with the rising population.

In summary, in addition to the vital investments in agriculture, health, education, physical infrastructure, fiber optics, and electrification, Africa will benefit by investing more in the rapid voluntary reduction of today's high fertility rates. How does a government "invest" in voluntary fertility reduction? First, the government ensures that girls are enabled to stay in school at least through the high school diploma level, in order to discourage child marriages. Second, the government should invest in child survival, to convince each family that having fewer children is "safe" in terms of their ultimate survival. Families do not need to be large simply to ensure the survival of a few of the children. Third, the government should make sure that family planning and modern contraceptives are available for free or low cost for those households that voluntarily decide to reduce their fertility rate.

III. South Asia—the Continuing Challenge of Food Supply

We have seen that there are two main regions in the world where there is still extensive extreme poverty: sub-Saharan Africa and South Asia. With the necessary investments, Africa can break free of extreme poverty. So too can South Asia, comprising India, Bangladesh, Bhutan, India, Nepal, and Pakistan. South Asia is already making notable progress in poverty reduction, but still has around 500 million poor people out of the region's total population of around 1.6 billion people. There are still major challenges of poverty in both the rural and the urban areas of South Asia.

What distinguishes South Asia from other regions? There are, of course, various aspects of wondrous culture, traditions, and the many remarkable dimensions of the physical environment. But the one distinguishing aspect to underscore is the extraordinary population density of South Asia.

Consider India with its 1.2 billion people, equaling roughly 16 percent of the world's population. Yet India has just 2.5 percent of the world's land area, and many

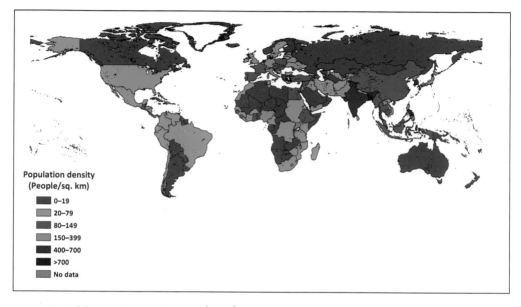

5.13 World population density (2013)

Source: World Bank. 2014. "World Development Indicators."

parts of that landmass of India are very dry or even desert. Figure 5.13 shows countries shaded according to their population density, and India and its next-door neighbor, Bangladesh, are shaded as two of the most densely populated parts of the world. The numbers indeed are quite staggering. Bangladesh has on average 1,200 people per square kilometer. India has about 410 people per square kilometer, but many of those square kilometers are nearly empty desert regions, making the populated regions even more densely packed. The United States, by contrast, has about 32 people per square kilometer. The population density in India is more than 10 times higher than in the United States.

The implications of this very high population density throughout India's history have been adverse. Indian farms are very small, and farmers traditionally have been able to grow only a small amount of food. The cities are extraordinarily crowded, and India's and South Asia's cities more generally have dramatically increased in population in recent decades.

Many people in the 1950s and 1960s thought the situation was hopeless for South Asia. They thought that the population was so large (and increasing so rapidly) that India and its neighbors would not be able to feed themselves. These observers forecast mass deaths from famine. When Bangladesh gained its independence from Pakistan in the early 1970s, Henry Kissinger notoriously called it a "basket case." Thankfully, the forecasts of mass famine have proven to be off the mark. Indeed, India not only has avoided famine, but it has grown reasonably rapidly over the past twenty years. It takes pride in being one of the world leaders of the IT revolution, with wonderful engineering and innovation in using IT for economic development. Through IT, India has become integrated into the world economy, often in cutting-edge industries, using creative programming and IT systems developed by India's top-notch engineers.

How did India avoid the fate that was so widely predicted for it? India's success begins naturally with agriculture, because India was overwhelmingly a smallholder peasant society deeply challenged with food insecurity. It was a great breakthrough in agricultural technology in the 1950s and 1960s that enabled India to overcome chronic famines of the past and to begin the liftoff into sustained economic growth. That breakthrough in agricultural technology has been famously dubbed the "Green Revolution."

What is the Green Revolution? It started with the individual pictured on the left in figure 5.14, Norman Borlaug. Borlaug was a highly skilled agronomist who used his great ingenuity and determination to develop high-yield seed varieties for wheat while working in Mexico in the 1940s and the 1950s. (He later won the Nobel Peace Prize for these accomplishments.) Borlaug was invited to India in the early 1960s to see whether his high-yield seed varieties might help India to raise farm yields. His counterpart was yet another great agronomist pictured on the right in figure 5.14, M. S. Swaminathan.

Borlaug and Swaminathan took the special seeds that Borlaug had developed for Mexican conditions and planted them in Indian soils in Indian conditions. The first year did not work out well. They reviewed the situation and decided on a different approach. The second year proved that, lo and behold, the varieties developed by Borlaug for Mexican conditions worked beautifully in the Indian conditions if planted in the right way. Borlaug and Swaminathan quickly realized

5.14 (*top left*) Norman Borlaug *Photo courtesy of the Norman Borlaug Institute for International Agriculture at Texas A&M University.*

(*top right*) M.S. Swaminathan *"Monkombu Sambasivan Swaminathan—Kolkata 2013-01-07." Biswarup Ganguly, Wikimedia Commons, CC BY 3.0*

(*bottom left*) Chidambaram Subramaniam *Age Fotostock/DINODIA* that a Green Revolution for India was within technological reach. To make it happen, they had to add a third partner to form a historic triumvirate. He is seen at the bottom in figure 5.14, Chidambaram Subramaniam, who was the dynamic minister of agriculture of India in the early to mid-1960s. The core idea of the new Green Revolution was to multiply Borlaug's Mexico seeds for use in India, and then to plant them with added fertilizer, irrigation, and transport facilitation in order to jumpstart a major takeoff of crop yields.

The results were spectacular. India's yields soared, and then the concept of high-yield varieties began to spread around the world. A true Green Revolution began to unfold worldwide by the end of the 1960s. Figure 5.15 shows the impacts on yields for the developing countries as a whole. Up to the mid-1960s, average yields were still less than 1,000 kilograms per hectare of arable land, that is, less than 1 ton per hectare. But then, with improved seed varieties and greater use of fertilizers and irrigation, yields began to rise significantly. By 1980, yields averaged 1.5 tons per hectare. By the year 2000, they were above 2.5 tons per hectare. In many parts of the developing world, yields routinely exceed 3 tons per hectare,

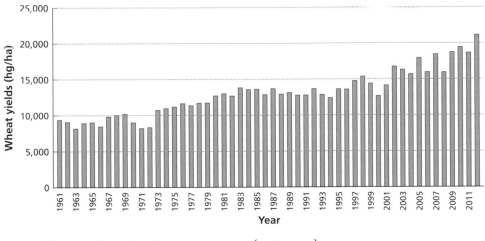

5.15 Wheat yields in developing countries (1961–2012)

Source: Food and Agriculture Organization of the United Nations. 2014. "Crops." Latest update: 7/18/2014.
http://faostat3.fao.org/faostat-gateway/go/to/download/Q/QC/E.

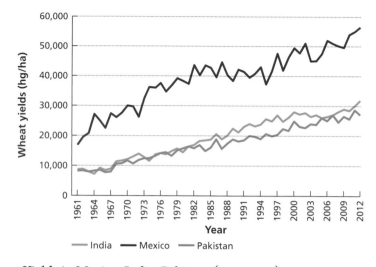

5.16 Wheat Yields in Mexico, India, Pakistan (1961–2012)

Source: Food and Agriculture Organization of the United Nations. 2014. "Crops." Latest update: 7/18/2014.
http://faostat3.fao.org/faostat-gateway/go/to/download/Q/QC/E.

one example being Mexican wheat, as shown in figure 5.16. India and Pakistan have not reached Mexican productivity levels, but they have increased their yields by three to four times since the mid-1960s.

However, there is still a problem: India's population growth remained rapid as well after 1965. The population did not grow so rapidly that it literally and figuratively ate up all the gains in grain yields, but it did grow rapidly enough so that many of the agricultural gains, when measured in per capita terms, eventually diminished and by now have created a renewed food crisis in some parts of India and South Asia.

Figure 5.17 shows India's population growth (UNFPA 2013). In 1950, India's population was about 400 million; India was already a huge and densely populated country. Yet by 2014, that population has roughly tripled. So while grain production has roughly increased fourfold, population has tripled, unfortunately undoing most of the gains in production per person.

Figure 5.18 shows the feed grains per capita from the beginning of the 1950s to today. The curve was rather significantly rising up until around 1990. The spikes in the curve come from the fact that some years were favorable monsoon years,

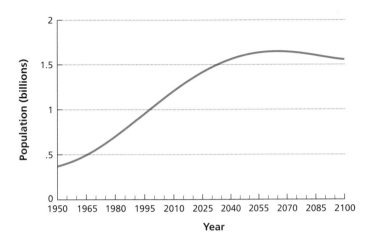

5.17 India's population (medium-variant after 2010)

Source. United Nations Department of Economic and Social Affairs Population Division (DESA Population Division). 2013. "World Population Prospects: The 2012 Revision." New York.

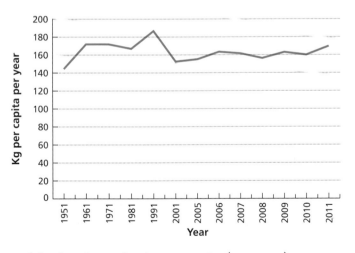

5.18 India's total food grain production per capita (1951–2011)

Source: Food and Agriculture Organization of the United Nations. 2014. "Crops." Latest update: 7/18/2014. http://faostat3.fao.org/faostat-gateway/go/to/download/Q/QC/E.

while others were bad monsoon years, which strongly impact the yields. Yet from the early 1990s onward, India's continued population increase meant that the increase of grain output per Indian essentially stopped. India is now actually producing less feed grain per person than it did twenty years ago.

The stagnation, even decline, of grain output per person has created a new round of troubling hunger issues and stresses in the Indian countryside. India's recent rapid economic development, while very real, is still burdened and held back by problems of hunger and poverty in the countryside. One stark condition, childhood stunting, exemplifies the problem. Childhood stunting is an indication of chronic undernutrition of young children. When young children do not get the nutrients they need, they do not achieve their potential height for age. Stunting signifies a significant reduction of height for age relative to the potential of the population at each age. Figure 5.19 shows where childhood stunting is

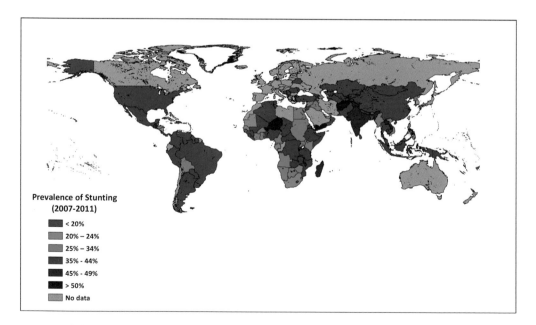

Prevalence of Stunting
(2007-2011)

■ < 20%
■ 20% – 24%
■ 25% – 34%
■ 35% - 44%
■ 45% - 49%
■ > 50%
■ No data

5.19 Global prevalence of stunting

Source: UN data.

highest in the world today. As with extreme poverty, stunting is highest in tropical Africa and South Asia. India is the country with the largest number of stunted children. While there are many wondrous aspects of India's development—its rapid growth in IT and manufacturing, its leadership in global engineering, and its potential to grow in the future—there remains the worry over food security and decent nutrition, especially among poor farmers.

It is also true to say, as M.S. Swaminathan has emphasized repeatedly for decades, that India needs a *second Green Revolution*, or what he has called an *Evergreen Revolution*. This second Green Revolution or Evergreen Revolution would not be exactly like the first one. In view of the rising environmental threats faced by India and the world, the second Green Revolution must emphasize not only crop yields (tons per hectare) but increased crop efficiency in the low use of water, fertilizers, and other inputs. The first Green Revolution used massive amounts of groundwater, but that groundwater is now close to depletion in many sites. The first Green Revolution called for a massive increase in fertilizer use, and some of that fertilizer has polluted India's rivers and coastlines. The first Green Revolution did not pay heed to long-term climate change, which was not yet recognized. The second Green Revolution will need to develop crop varieties that are resilient to heat waves, droughts, floods, and other shocks that will rise in the future as part of the consequences of human-induced climate change.

India and South Asia also face the continuing challenge represented by MDG 3 on gender equality. In many traditional South Asian cultures, women face massive burdens. Many are not allowed to be in the labor force or to own or inherit property. They may not be allowed to manage money. Girls are often left with insufficient nutrition, health care, and access to even basic education. The burdens of gender inequality are passed from mother to daughter. One of the recent breakthroughs in South Asia has been the empowerment of women and girls, but there are still major areas of discrimination to overcome.

One of the noteworthy ways that rural women have been empowered in recent decades has been through microfinance institutions, a new method of small-scale lending that is well adapted to the needs of impoverished rural women. The key innovations were pioneered in Bangladesh by two NGOs that are now rightly world renowned, Grameen Bank (founded by Nobel Peace Prize laureate

Muhammad Yunus) and BRAC (founded by social entrepreneur and innovator Sir Fazle Hasan Abed).

Both NGOs pioneered women's empowerment in the villages through self-help groups and undertook a massive expansion of microfinance through a group lending process. In group lending, seen in figure 5.20, an entire group of women jointly guarantees the repayment of loans made to a single member of the group, thereby lowering the risk of default and enabling the loan to take place. The members allocate the funds to the other group members and manage the loan repayments. Each borrower might receive a few dozen dollars in a month, which provides working capital such as the inventory for a small retail shop or the inputs for a bakery. The repayment rates of Grameen and BRAC and other such microfinance providers have generally been very strong, except when the national economy has been hit by macro-scale shocks. Because of these successful results, both in managing small loans and in empowering rural women, microfinance has

5.20 Grameen women's microfinance group

Grameen Foundation.

spread throughout the world as a powerful tool for grassroots empowerment, gender equality, and income generation.

One of the notable features of female empowerment, sometimes in the context of the self-help groups, is that it has given young women the incentive to marry later and reduce their total fertility. A mother in the labor force who earns her own income knows through experience and through knowledge from her peers that having fewer children will not only enable her to spend more time at work to earn a higher income but will also enable the household to invest more in each of her children so that he or she will have a chance for a better life.

Bangladesh has seen a significant decline of the fertility rate, as shown in figure 5.21 (DESA Population Division 2013). Back at the time of independence in 1971, Bangladesh's total fertility rate (TFR) was around seven. For every 1,000 women there would be 7,000 children, of whom 3,500 would be girls. In one generation, therefore, every 1,000 mothers would be raising 3,500 future

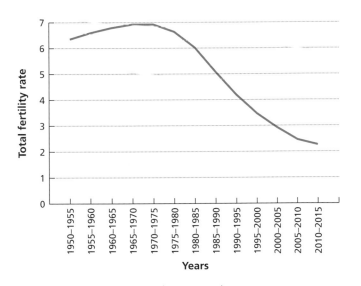

5.21 Total fertility rate in Bangladesh (1950–2015)

Source: UN data.

mothers, signifying a dramatic expansion in population from one generation to the next. Yet through the movement for women's empowerment—backed by microfinance, expanded educational opportunities, and less onerous cultural and legal barriers for women—the TFR began to decline very rapidly on a voluntary basis. By now, the TFR is at the so-called replacement level of two. Each woman, on average, has two children, and therefore one daughter. Each woman, on average, replaces herself with a daughter in the next generation. Over time, the population will tend to stabilize, thereby improving the overall prospects for economic development.

Another feature of South Asia's poverty has been the chronic undernutrition of children, which in turn has hindered their survival, health, and capacity to learn. Disturbingly high proportions of young children in South Asia are stunted, that is, very short for their age. There are at least three factors in this stunting: inadequate dietary intake, chronic infections by worms and other disease-causing agents, and the lack of access to safe water and latrines to prevent a rapid reinfection after each bout of illness. The result is that children eat too little and lose too many of the nutrients they receive to parasites and frequent bouts of disease. Stunting could be overcome through a three-pronged effort: better diets, deworming (and other disease control), and provision of safe water and sanitation (latrines). By overcoming stunting, the South Asian region would also be positioned to make much greater strides in primary education. Of course the education systems of the South Asian countries would benefit from higher budgets and the introduction of innovative twenty-first-century approaches, such as the effective use of new information technologies in the classroom and home study.

South Asia, like sub-Saharan Africa, has the end of extreme poverty within reach. But South Asia will require major efforts to achieve the Second Green Revolution, as well as focused investments in infrastructure, water and sanitation, health care, education, and in the empowerment of girls and women to complete the demographic transition and to raise the skill levels of the population. By mobilizing its great skills in IT and other areas of high-tech knowledge, India is especially well poised to achieve a sustainable development breakthrough. It will need effective leadership and good governance for success.

IV. A Closer Look at Official Development Assistance

Our differential diagnoses of sub-Saharan Africa and South Asia have shown how targeted investments in agriculture, health, education, infrastructure, and women's empowerment can help these regions to free themselves from the ancient scourges of extreme poverty. I have often described these targeted investments as getting onto the "first rung of the development ladder" (Sachs 2005). By that I mean that key investments in basic education, health, infrastructure, and farming can enable a poor household, or indeed a poor region, to earn enough added income and garner enough wealth to be able to finance the next stage of development. By getting on the first rung of the ladder, the household (or region) is able to ascend to the second rung, then the third rung, and so forth, thereby enjoying self-sustaining growth that eventually will lead to the end of extreme poverty.

The problem with the poverty trap, however, is that a country may be too poor to get on the ladder by itself. The country's leaders may be visionary; they may have an excellent idea of how to carry out the needed investments. Yet they simply lack the cash flow—whether out of government revenues or new borrowing—to do so. In short, the impoverished country (and the individual impoverished households within it) needs a "hand up" to get onto the development ladder. This is the main argument for foreign development assistance.

The idea of official development assistance (ODA), meaning development aid from governments or international agencies, has been around since just after World War II, when the United States launched the famous Marshall Plan to help postwar Europe rebuild and recover after the devastation of the war. The Marshall Plan offered a temporary injection of funds, given mostly not as a loan but as a grant, to help jump-start a renewal of economic life and self-sustaining growth. The Marshall Plan lasted for about four years, from 1948 to 1952, and did wonders in helping western Europe get back on its feet. It provided an inspiration for a growing system of grants and low-interest loans not only for postwar reconstruction but for jump-starting long-term economic growth, for example, in the poor, newly independent countries of Africa and Asia.

It is important to understand that from the very start, few people advocated the use of ODA as a long-term way of life. Advocates of foreign assistance, including myself, believe that aid is a *temporary* measure to help a poor country make the crucial early investments needed so that the economy can soon stand on its own and begin climbing the development ladder. Aid is not a permanent need or solution. Countries that receive aid can reach a level of income through economic growth whereby they soon "graduate" from the need for aid entirely. China and Korea are two examples of countries that relied on aid when they were poor and then graduated from aid and indeed more recently became significant donor countries. Roughly speaking, graduation from aid can occur when a country passes from low-income to middle-income status. This occurs at a GDP per capita of around $1,200 per year (measured at the market exchange rate), or roughly $3,000 per person per year when GDP is measured at international (purchasing power parity, or PPP) prices.

Official development aid became a basic pillar of the global community around 1970 (OECD 2010). A commission on international development headed by a former prime minister of Canada and Nobel Peace Prize laureate, Lester Pearson, recommended a global commitment to ODA. The commission's report, *Partners in Development,* called on high-income countries to become donors to poor countries. The report suggested that the high-income countries donate around 1 percent of their GDP to help the low-income countries to overcome poverty. Of that 1 percent of national income, around two-thirds, specifically 0.7 percent of national income, should come through official channels, mainly government-to-government grants and low-interest loans. The remaining 0.3 percent of GDP should come through private contributions, mainly from corporations, foundations, individual philanthropists, and charitable organizations. Based on this commission report, the UN General Assembly in 1970 formally adopted the goal that high-income countries should contribute 0.7 percent of their national income to ODA.

Consider the United States, today a $16 trillion economy. The 0.7 percent of GDP standard would lead to ODA equal to $112 billion dollars of ODA each year. Alas, the United States is not close to that standard. The ODA given by the United States is around $30 billion dollars per year, closer to 0.18 of 1 percent of the U.S. national income and therefore less than one-third of the international standard.

Figure 5.22 shows the ODA given by high-income countries (OECD 2014). Only five countries among the donors typically reach the targeted threshold of 0.7

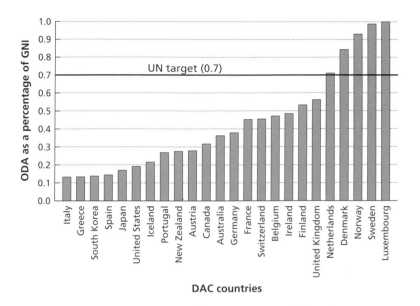

5.22(A) Official development assistance (percent of GNI) (2012)

Source: Organization for Economic Co-operation and Development. 2013. "Compare your country—Official Development Assistance 2013." Paris: OECD. http://www.oecd.org/statistics/datalab/oda2012.htm.

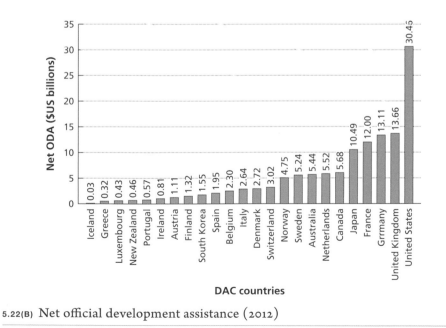

5.22(B) Net official development assistance (2012)

Source: Organization for Economic Co-operation and Development. 2013. "Compare your country—Official Development Assistance 2013." Paris: OECD. http://www.oecd.org/statistics/datalab/oda2012.htm.

percent of national income: Sweden, Norway, Denmark, Luxembourg, and the Netherlands. Sweden and Luxembourg indeed are at 1 percent of national income. At the other end of the spectrum are countries that give quite low proportions of national income, including the United States. Since the United States is such a large economy, it still gives a lot of money in absolute terms and indeed is the largest single donor. The combined income of the donor countries is around $40 trillion per year. At the official aid target level of 0.7 percent of national income, total donor aid for the poor countries would be about $280 billion dollars per year of aid. In fact, the aid is around $120 billion per year, or just 0.3 percent of gross national income.

What kind of spending does the ODA support? Official development assistance has to fit the following criteria. First, the money must go to poor countries. Second, the money must be provided by an official agency of the donor country. Third, the money must be used for economic development in the recipient country. It cannot be used, for example, for arms purchases or to support troops, sports games, or cultural events.

There is another important distinction to make between types of aid. Aid given as emergency relief, for instance, food aid in the middle of a famine, is called humanitarian relief. Similarly, emergency help after a natural disaster is also counted as aid, but it typically will save lives rather than promote long-term development. The kind of development aid that can help a country make a break-through out of poverty is something quite different. The most effective kinds of development assistance build capital—such as paved roads, an expanded power grid, and more clinics and schools—or capacity, such as training and salaries for teachers and health workers, or social investments such as health care delivery.

There is a lot of confusion about whether aid works or not, because not all aid is the same. If a donor rather cynically gives boxes of cash to warlords because it thinks that such bribes will be good for a war effort, or it gives money to governments for corrupt reasons (such as to secure an arms deal), then such "donations" may be called aid but will do nothing to foster economic development. The kind of ODA that works for long-term development and poverty reduction is used to support investments in the critical areas I have discussed in this chapter. When *that* kind of aid is given, the evidence is very strong that it can have a large and important effect. Make no mistake about it—aid that is poorly directed or misused can be wasted. But aid that is well targeted to urgent needs can be crucial to help countries achieve the MDGs and to get onto the ladder of economic development.

During the MDG period, the most effective scale-up of ODA was in the area of public health. After the year 2000, there was a major increase of ODA for health. That increased assistance played an enormous role in helping poor countries to control AIDS, malaria, and tuberculosis and in helping to ensure that mothers are safe in childbirth and newborns can survive the difficult first days of life. That kind of aid helps to ensure that young children get adequate nutrition and are protected against childhood scourges for which vaccines exist. That kind of aid can help to ensure that children can attend school and thereby reach their full individual potential. We have already noted the big breakthroughs that have come after the year 2000 in lower mortality rates of children and of mothers during pregnancy and childbirth. We have already noted the large gains in fighting AIDS and malaria. We have seen the increased economic growth of sub-Saharan Africa. In all of those cases, ODA played a positive role alongside other factors.

Official development assistance, in other words, can make a huge difference when it is operated for the real purpose of development and on a professional basis grounded in an accurate differential diagnosis of the needs of a low-income country. ODA can be the difference between success and failure in breaking free of the poverty trap. It comes at very low cost, less than 1 percent of the national income of the donor countries. If the rich world makes that effort, and if the funds are well used, ODA indeed can help to ensure that we are the generation that ends extreme poverty.

V. Designing Practical Interventions—the Case of the Millennium Villages

After you as a clinical development specialist have made the correct differential diagnosis, mobilized the development aid, and understood the key concepts of targeted investments in basic needs, then the real-life problem of implementation of development programs becomes the key. Real-life implementation of targeted investments is a major operational challenge. When the MDGs were first enunciated and I was asked by then UN Secretary-General Kofi Annan to advise the UN system on how the MDGs could be achieved, I called on colleagues and professionals from around the world to suggest the most effective approaches for implementing the needed investments, in a four-year project called the UN Millennium Project (2002–2006).

The expert advice came from many different disciplines: agronomy, education, public health, urban and rural engineering, and community development, among others. In 2005, the UN Millennium Project presented a long synthesis report (UNMP 2005b) and many supporting volumes of detailed information to the member states of the United Nations. In a special session of the General Assembly in the fall of 2005, the UN member states adopted a number of the key ideas on how to proceed in a practical way to achieve the MDGs.

My colleagues and I then undertook to implement these ideas in a few selected places in rural Africa, to learn how our recommendations could best work on the ground. That is how the ten-year Millennium Villages Project (MVP) got started. The recommendations from the UN Millennium Project became the basis for village-level work in ten countries across sub-Saharan Africa. The goal was to demonstrate pathways to achieve the MDGs (MVP 2011).

The map in figure 5.23 shows the locations of the Millennium Villages. It also shows Africa in a brightly colored depiction, based on the distinctive farm systems in Africa. The yellow areas along the east coast of Africa, for example, covering parts of Kenya, Tanzania, and Malawi, are maize-producing regions. The beige area of northern Ethiopia is a highland region where the main staple crop is a grain called teff, which is used for wonderful bread traditional in the Ethiopian diet but not widely known in the rest of the world. The brown-shaded region stretching east to west from West Africa are regions of cereal crop production in dry regions. And the orange area just above it, known as the Sahel, is an even drier region of crops mixed with pastoralist livestock management.

We wanted to see how the MDGs could be approached in each of these distinctive "agro-ecological zones," because each eco-zone poses specific challenges. How can farmers best grow each type of crop? How can pastoralists best manage

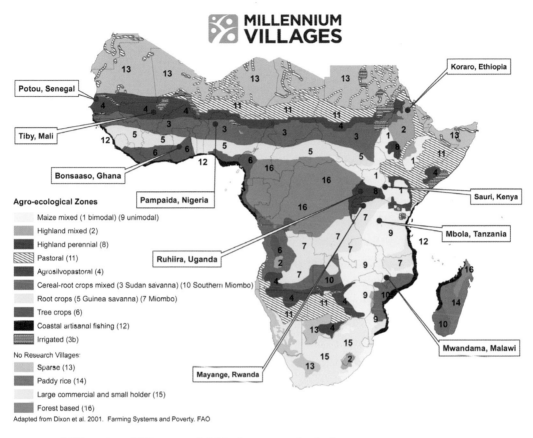

5.23 Millennium Villages and Africa's agro-ecological zones

From the Millennium Villages Project.

their livestock, especially in the face of repeated droughts? The disease burdens are also quite different across these eco-zones. In the highlands, for example, malaria is not a crushing problem, while in the tropical lowlands, malaria can be holoendemic, meaning that it infects just about everybody year-round unless it is brought under control. With the help of the host governments, in 2005–2006, the MVP identified ten very poor rural villages as the base for the project. Each of these Millennium Villages was initially a "hunger hot-spot," meaning that there was chronic undernourishment of at least 20 percent of the population. In other words, not only were the villages in poor countries, they were also in very poor parts of these poor countries. The idea was to use all eight MDGs as the guiding principles to promote the long-term development of these villages.

Applying the MDGs meant designing programs to achieve all eight of the MDG goals. There are two big reasons for that holistic approach. One is that each of the eight MDGs is meritorious in its own right. But another reason for the holistic approach is that the goals are synergistic. Providing safe water in a community can not only rid the community of part of the disease burden but also can help the children be healthy enough to go to school. Similarly, fighting malaria not only protects the lives of the community but also helps protect its productivity. Malaria control helps to ensure that people are not sick when it is planting or harvest season and that the children are not too sick to go to school. Not only do we want to achieve the eight MDGs because each is important, but because achieving one of the MDGs helps to achieve the others as well.

The MVP used development assistance of $60 per person per year during the first five years of the project (roughly 2006–2010). The local government and local NGO partners provided an additional $60 or so. Total spending therefore amounted to around $120 dollars per villager per year to address the MDG challenges. This development assistance made it possible to build schools, clinics, water points, road, power grids, and other infrastructure. The project has shown that even a very small amount of money, if properly directed and based on a proper differential diagnosis, can have a big impact in improving health, education, and infrastructure. The holistic approach seems to be working, though the final evaluation of the project will take place in 2015 and 2016, at the conclusion of the MDG period.

One of the most exciting developments in the Millennium Villages has been the development of the local health system. We are witnessing a major improvement in public health, including sharp reductions in child and maternal mortality. The project has helped to spur innovations in health care delivery, for example, by empowering Community Health Workers (CHWs) to reach even the poorest households in the villages (One Million Community Health Workers Campaign 2013).

The new CHW system is one of my own favorite developments of the Millennium Villages Project. People from the poor communities are becoming effective guardians of their own good health. A CHW is typically a young woman from the community, maybe with ten to twelve years of schooling in total. She has no medical degree or nursing degree. Yet with a little bit of training over a few months, the local worker with a backpack with the right kinds of medical supplies can transform, improve, and save lives in her community.

Each CHW carries in her backpack the tools to fight malaria. First, she will do a rapid diagnostic test for malaria with just a drop of blood from a child. There is no longer the need for the test to be done at a laboratory in a clinic many kilometers away. Second, the CHW will have the necessary medicines to fight malaria if the diagnostic test is positive. Again, the parent does not have to carry a very sick and feverish child to a clinic; the CHW can effectively treat the child at home. Third, the CHW will have a mobile phone. It will be possible to call an ambulance or

5.24 CHW with a backpack of supplies

Credit: Yombo Tankoano.

to call the clinic for advice from the nurse or doctor on duty. More and more, these smartphones are also being programmed with expert information systems to receive needed advice and information automatically by phone and to track information about patients.

I am happy to say that the Millennium Villages have already inspired many of the host governments to scale up large national programs in malaria control, AIDS treatment, help for smallholder farmers, and electrification with off-grid solar-based systems. Many other innovations have been tested, demonstrated, and pioneered in the Millennium Villages. The successful projects are now spreading. The project began in ten countries, but has already expanded to more than twenty countries. Many of the ideas tested in the individual villages are now applied across entire countries. It is very exciting to see this kind of progress on the ground. It is especially thrilling to see what is now possible through improved technologies: information systems that work at very low cost for better health, better education, and improved access to infrastructure. These are the technological and systems advances that encourage us to envision the end of extreme poverty in this generation.

6

PLANETARY BOUNDARIES

Long-term economic growth stretching over centuries would seem to present a puzzle. How can the world economy and population continue to expand if Earth itself is finite? Are there limits to growth? Have we hit them already? Is there still "room" on Earth for poor countries to raise their living standards? Does Earth have adequate resources—water, land, air, and ecosystem services such as the harvests of forests and fisheries—to sustain a growing world economy? In short, can economic growth be reconciled with environmental sustainability?

These questions bring us to the heart of sustainable development. We aim for a world that is prosperous, socially inclusive, and environmentally sustainable. Yet can all of these objectives really be reached? Since the late eighteenth century, great thinkers have pondered this question. They have wondered whether gains in living standards would prove to be illusory as the world ran short of primary resources. Would scarcity doom humanity to poverty in the long term? Are our gains in living standards merely a temporary overshooting, with humanity to get its comeuppance in the form of a future environmental crisis? These worries are increasingly heard, as the multiple crises of climate change, land degradation, water scarcity, and loss of biodiversity continue to deepen.

Yet I will argue that by very careful and science-based attention to the real and growing environmental threats, we can indeed find ways to harmonize growth—in the sense of material improvement over time—with environmental sustainability. That confidence, indeed, lies at the heart of sustainable development as a normative idea. By taking precautions, respecting resource constraints, recognizing the dangerous environmental destruction we are wantonly committing, and changing course, humanity has the option to achieve its objectives of ending poverty; raising living standards; ensuring social inclusion; and protecting the environment for ourselves, other species, and future generations. To do so, we need to understand the real natural boundaries—the planetary boundaries—that we must observe as responsible stewards of the planet.

I. Planetary Boundaries

Economic growth is complicated, but sustainable development is even more complicated. To achieve sustainable development, countries need to achieve three goals simultaneously: economic growth, broad-based social inclusion, and environmental sustainability. While many countries have "solved" the growth puzzles, few have succeeded in achieving all three aspects of sustainable development.

Indeed, we can go further. Since many of the environmental challenges—such as climate change, ocean acidification, and the extinction of species—are global-scale crises, and since all countries are feeling the effects of these crises, we can say that no country is actually on a path of sustainable development. Even when individual countries are making maximal efforts on their own part, they are still vulnerable to a world economy that has failed to take adequate actions to head off environmental calamities.

The problems are getting harder, not easier. The main problem is one of scale. The world economy has become very large relative to the finite planetary resources. Humanity is pushing against the limits of the environment. In the words of world-leading ecologists, humanity is exceeding the planetary boundaries in several critical areas.

6.1 Thomas Robert Malthus

Portrait of Thomas Robert Malthus by John Linnell.

Let's review the global circumstances very briefly. Back in 1798, Thomas Robert Malthus wrote the seminal work *An Essay on the Principle of Population,* warning humanity that population pressures would tend to undermine improvements in living standards. If humanity were able to raise its living standards, wrote Malthus, then the population would expand accordingly, until the rise of population would put strains on the food supply and thereby reverse the gain in living standards. Malthus's vision was decidedly pessimistic about sustainable development!

We now know that Malthus was too quick to assume that population pressures would automatically reverse the gains of economic development. Certainly Malthus could have had no idea about the dynamism of science-based technological advances that would occur after his essay. Certainly he could not foresee the Green Revolution in particular, which would dramatically expand the capacity to grow more food to feed a larger global population. Nor could Malthus have foreseen the demographic transition, by which richer households would choose to have fewer children, so much so that populations are already stabilizing or even declining in some of the world's richest places.

Yet Malthus had many things right. When he wrote, the world's population was around 900 million people. It has since risen more than seven times. Population has indeed increased sharply alongside the long-term rise of productivity. And there is more to come: perhaps up to 10.9 billion people by 2100 (according to the medium-fertility variant of the UN Population Division).

To gauge the scope of human impact on the environment—the pressures that humanity is putting on Earth's ecosystems—we need to combine the sheer numbers of people with the increased resource use per person. For that, we can look at rough estimates of the world output per person. In 1800, the gross world product (GWP) per capita was around $330 in 2013 prices. Now it is around $12,600 per capita. That means that per capita income has increased by around 38 times.

Since total world output (GWP) is the product of population and GWP/population, we find that the total world product has increased by around 275 times, roughly from $330 billion for the entire world in 1800 to around $91 trillion. Of course these are very rough estimates, but they do give us a sense of order of magnitude of the increase of global production. Alas, that production has also translated into an increase in the adverse human impact on the physical environment.

Humanity has become so numerous and so productive that we can say that we are "trespassers" on our own planet. By that I mean we are crossing boundaries of Earth's carrying capacity, thereby threatening nature and even our own species' survival in the future. The concept of planetary boundaries is an extremely useful one. When the world-leading environmental scientist Johan Rockström brought together other leading Earth systems scientists, they asked: What are the major challenges stemming from humanity's unprecedented impact on the physical environment? Can we quantify them? Can we identify what would be safe operating limits for human activity, so we can urgently begin to redesign our technologies and our economic growth dynamics to achieve development within the planet's limits?

They came up with a list of planetary boundaries across nine areas, shown in figure 1.16 (Rockström et al. 2009).

The first and most important of the planetary boundaries relates to human-induced climate change. We will study human-induced climate change in detail in chapter 12. For now we should note that human-induced climate change is the result of the rising levels of greenhouse gases (GHGs) in the atmosphere. The GHGs include carbon dioxide, methane, nitrous oxide, and a few other industrial chemicals. These GHGs have a shared property: they warm the planet. The greater the concentration of GHGs in the atmosphere, the warmer on average is planet Earth. Because of industrial activity, the GHG concentrations have risen dramatically in the past century, and the Earth has already warmed by around 0.9° Celsius compared with the temperatures before the Industrial Revolution. On current trends, the Earth will warm by several degrees Celsius by the end of the twenty-first century.

Greenhouse gases allow incoming solar radiation, in the form of ultraviolet radiation, to pass through the atmosphere to Earth, thereby warming the planet. The Earth, in turn, reradiates that heat as infrared radiation. The Earth warms to the point that the incoming (ultraviolet) radiation is exactly balanced by the outgoing infrared radiation. The GHGs, however, trap some of that outgoing infrared radiation, thereby making the Earth warmer than it would be without an atmosphere. (Without the GHGs in the atmosphere, the Earth would be like the moon, considerably colder.) So far, so good. The problem is that with rising concentrations of GHGs, the Earth is becoming warmer than it was before industrialization began. And those rising temperatures are pushing the planet to a new climate, one that is different from the climate that has supported human life during the entire period of civilization. This change of climate is deeply threatening (as we will discuss in more detail later). It threatens the global food supply; it threatens the survival of other species; it threatens to cause much more intense storms; and it threatens a significant rise in ocean levels, which could disrupt life in many parts of the world.

The most important of the GHGs is carbon dioxide (CO_2). The main source of human-induced CO_2 comes from burning coal, oil, and gas. (The other major source we will study is land-use change, such as deforestation.) The release of energy in the fossil fuels results from the combustion of carbon in these energy sources. The carbon atoms combine with oxygen to release energy plus CO_2.

In this way, CO_2 is the inevitable by-product of burning fossil fuels. Fossil fuels have created the modern economy. Without them, the world would be poor, as it was for the millennia until the Industrial Revolution. Yet now the CO_2 emissions from fossil fuels pose an unprecedented threat. We need to find new ways to produce and use energy, so we can enjoy the benefits of the modern economy without the dire threats of human-induced climate change.

The second of the planetary boundaries, ocean acidification, is closely related to the first. The oceans are becoming more acidic as the atmospheric concentrations of CO_2 increase. The CO_2 in the atmosphere dissolves in the ocean, producing carbonic acid (H_2CO_3). Carbonic acid dissociates to a hydrogen ion (H^+) and bicarbonate (HCO_3^-). The rise of H^+ signifies the increased acidity of the oceans. This rising acidity threatens various kinds of marine life, including corals, shellfish, lobsters, and very small plankton, by making it hard for these species to form their protective shells.

The pH of the ocean has already decreased by 0.1 unit on the pH scale, which runs from 0 (most acidic) to 14 (least acidic). A change of 0.1 in the pH of the oceans might not seem like all that much, but the scale is logarithmic, so a decline of 0.1 signifies an increase of protons in the ocean of 10 to the power 0.1, or about 0.26 ($= 10^{0.1}$), a 26 percent increase of acidity in the oceans, with a lot more acidification to come as the atmospheric concentration of CO_2 continues to rise. The map of the ocean in figure 6.2 shows that the changes in the pH scale are already being noticed in different parts of the world. The oceans are not uniformly becoming more acidic; the local effects depend on ocean dynamics and on regional economic activities. Yet the pH map in figure 6.2 shows that we are already on a trajectory of dangerously rising ocean acidity.

The third planetary boundary is ozone depletion. Brilliant atmospheric scientists in the late 1970s discovered that particular industrial chemicals called chlorofluorocarbons (CFCs), which were used mainly for refrigeration and aerosols at the time, tended to rise into the upper atmosphere and dissociate (that is, split up into smaller molecules). The chlorine in the CFCs, when dissociated from the rest of the molecule, attacked the ozone (O_3) in the upper atmosphere (the stratosphere). By chance, a new NASA satellite was in place to take pictures from space of the ozone layer, and shockingly, the pictures (shown in figure 6.3) in the

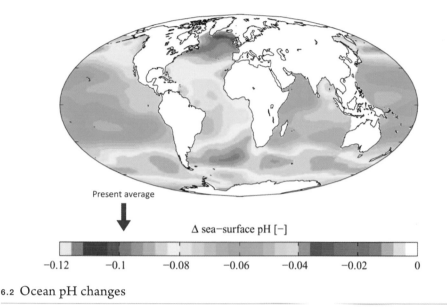

Present average

Δ sea–surface pH [−]

| −0.12 | −0.1 | −0.08 | −0.06 | −0.04 | −0.02 | 0 |

6.2 Ocean pH changes

"Estimated change in annual mean sea surface pH between the pre-industrial period (1700s) and the present day (1990s)." Plumbago. Wikimedia Commons, CC BY-SA 3.0.

mid-1980s demonstrated a huge ozone hole (site of ozone depletion) over the South Pole.

This was a dramatic discovery. The ozone level in the upper atmosphere protects human beings from receiving too much ultraviolet radiation from the sun. Ozone depletion was suddenly a newly recognized, very dire threat to human survival. The real fear was that skin cancers and other disorders would soar as the ozone level faced depletion.

Fortunately, because of great science and technology, humanity was spared the worst. The public was informed in the nick of time that industrial chemicals that were thought to be harmless were in fact a grave threat to public health. The CFCs needed to be eliminated before they caused a catastrophe. The good news is that the world has acted on this one, introducing a new treaty to phase out CFCs from industrial systems and to replace CFCs with safer chemicals. This is now occurring step-by-step. Without the scientific discoveries, technological insights, and global agreements, ozone depletion would be a grave threat to humanity. Yet we

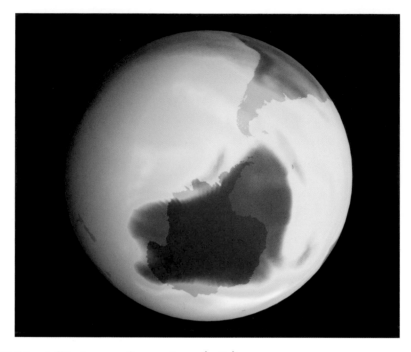

6.3 NASA satellite image of ozone layer (1985)

NASA/Goddard Space Flight Center Scientific Visualization Studio.

are not yet fully past the threat: we still need to fully eliminate CFCs and to ensure the replacement chemicals are indeed fully safe.

The next planetary boundary shown in figure 1.16 (moving clockwise around the circle) is pollution caused by excessive flows of nitrogen and phosphorous, especially as the result of the heavy use of chemical fertilizers by the world's farmers. Here too, something of profound benefit for humanity—chemical fertilizers—turns out to have a hidden and serious danger. Farmers must put nitrogen, phosphorus, and other nutrients into farm soils to ensure decent yields on their crops. Without fertilizers, yields would still be around 500 kilograms to 1 ton per hectare, rather than the 3–5 tons or more that farmers can achieve on their grain production. Without chemical fertilizers, it would not be possible to feed the 7.2 billion people on the planet. It has been estimated that perhaps 4 billion people today are fed as a result of chemical fertilizers.

The problem is that much of the nitrogen and phosphorous is not taken up by the crops. Much of it actually returns to the air and is carried downwind to other locations. Much of it enters the groundwater and rivers, with heavy concentrations of nitrogen and phosphorous reaching the estuaries where rivers meet the oceans. In turn, the heavy influx of nitrogen and phosphorous leads to dangerous ecological changes in the estuaries. The nutrients give rise to "algal blooms," which are massive increases in algae in the estuaries that grow as a result of the high availability of the nitrogen and phosphorous nutrients. When these algae die, they are consumed by bacteria, which in turn deplete the oxygen in the water, giving rise to hypoxic (low-oxygen) dead zones and killing the fish and other marine life. This process of "eutrophication" (high nutrient concentrations leading to algal blooms and then hypoxia) is already occurring in more than 100 estuaries around the world. Figure 6.4 shows a young boy swimming in an algal bloom off the coast of Shandong, China.

The fifth planetary boundary arises from the overuse of freshwater resources. Humans and other species need fresh water every day to stay alive. Of the total amount of freshwater that humanity uses, about 70 percent is used for agricultural production; about 20 percent is used by industry; and the remaining 10 percent is for household use, meaning cooking, hygiene, and other household uses. Humanity is now using so much water, especially for food production, that in many parts of the world societies are depleting their most critical sources of freshwater. Farmers around the world are tapping into groundwater aquifers, taking water out of the ground faster than it is being recharged by rainfall. The result is that these aquifers are being depleted. When they are depleted, the farmers depending on this groundwater will suffer massive losses of production, and food scarcity will result. Groundwater depletion is now a worldwide phenomenon, affecting areas including the U.S. Midwest, northern China, and the Indo-Gangetic plains of Northern India and Pakistan.

Freshwater scarcity will be exacerbated by countless other problems: growing populations, industrial use of water (e.g., for mining and power plants), changing rainfall and soil moisture conditions due to human-induced climate change, and the loss of meltwater from glaciers as glaciers retreat and eventually are eliminated as a result of global warming. All in all, the planetary boundary of freshwater will pose a major crisis for many regions of the world in the decades to come.

6.4 Young boy swimming in algal bloom in Shandong, China

Photo: Reuters/China Daily.

The sixth planetary boundary is land use. Humanity uses a massive amount of land to grow food, graze animals, and produce timber and other forest products (e.g., palm oil), and for our expanding cities. Humanity has been converting natural lands such as forests to farmlands and pasturelands for thousands of years. Many regions of the world that were once dense forests are now farmlands or cities. The resulting deforestation not only adds CO_2 to the atmosphere (as the carbon in the plants and trees returns to the atmosphere), thus adding to human-made climate change, but it also destroys the habitats of other species. Human land-use change, whether for farms, pastures, or cities, is causing a massive disruption to ecosystems and species survival in many parts of the world.

The seventh planetary boundary is biodiversity (biological diversity). The evolution of life on Earth has created a remarkable diversity of life, somewhere between 10 million and 100 million distinct species, most of which have not yet been catalogued. That biodiversity not only defines life on the planet, but also contributes in fundamental ways to the functions of ecosystems, the productivity

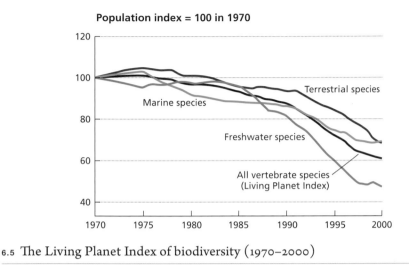

Population index = 100 in 1970

6.5 The Living Planet Index of biodiversity (1970–2000)

Source: World Wildlife Fund. 2012. "Living Planet Report 2012." Gland, Switzerland: WWF International.

of crops, and ultimately to the health and survival of humanity. We depend on biodiversity for our food supply, our safety from many natural hazards (e.g., coastal flooding), countless construction and industrial materials, our freshwater, and our ability to resist pests and pathogens. When biodiversity is disrupted, ecosystem functions change markedly, usually in an adverse way (e.g., the productivity of crops diminishes).

Humanity is massively disrupting biodiversity. We are doing so in countless ways, including through pollution, land-use change such as deforestation, human-induced climate change, freshwater depletion, ocean acidification, and nitrogen and phosphorus flux. Many species are declining in numbers, genetic diversity, and resilience. Figure 6.5 gives some broad sense of the decline of populations of major groups of species. Indeed, countless species face the risk of complete extinction, and prevailing science holds that humanity is now causing the Earth's sixth great extinction wave. As summarized in table 6.1, the other five extinctions in Earth's history resulted from natural processes, such as volcanoes and meteorites, as well as the internal dynamics of Earth itself. This sixth megaextinction is not natural. It is the result of one species—humans—damaging the planet so severely that we are putting millions or even tens of millions of other

Table 6.1 The First Five Great Extinctions	
1. End of the Ordovician, 440 millions of years ago (M.Y.A.)	Enormous glaciation and lowering of sea levels; 60 percent of species disappeared.
2. End of the Devonian, 365 M.Y.A.	Glaciation and falling sea levels; possibly caused by a meteorite impact; 70 percent of species wiped out.
3. End of the Permian, 225 M.Y.A.	Huge volcanic eruptions; Earth became winter; 90–95 percent of all species extinct.
4. End of the Triassic, 210 M.Y.A.	Possibly caused by a comet shower; most ocean reptiles extinct; many amphibians extinct.
5. End of the Cretaceous (called the KT extinction), 65 M.Y.A.	Meteorite struck Earth; dinosaurs, marine reptiles, ammonoids, and many species of plants were wiped out; mammals, early birds, turtles, crocodiles, and amphibians less affected.

species at risk. Since humanity depends on those other species, we are of course gravely endangering humanity as well.

The eighth planetary boundary is called *aerosol loading*. When we burn coal, biomass, diesel fuels, and other sources of pollution, small particles called aerosols are put into the air. A tremendous amount of air pollution is created that is very damaging to the lungs, claims many lives per year, and has a significant impact on changing climate dynamics. Very fine particles less than 2.5 micrometers in diameter (written as 2.5 μm) can cause life-threatening lung disease. China's major cities have been experiencing catastrophic levels of aerosol pollution, leading to urban smog so thick that on some days it is dangerous to venture outside.

The next (very broad) category is chemical pollution. Industries such as petrochemicals, steel production, and mining not only use a huge amount of land and water for their processing but also add a tremendous load of pollutants back into the environment, many of which accumulate. They can be very deadly for humans as well as for other species. China, the world leader of economic growth over the past thirty years, has also become the leader of polluted waterways of its major cities because of the extent of its heavy industrial processing, a major environmental problem it will have to deal with.

When humanity trespasses on these planetary boundaries, meaning that human pressures on the environment become greater than the ability of the Earth's natural systems to absorb those human pressures, the result is a major change in the function of the Earth's ecosystems. Those changes, in turn, threaten human wellbeing and even human survival when the shocks occur in places where populations are very poor and do not have the buffers of wealth and infrastructure to protect them. When fisheries die, fishing communities die with them. When groundwater is depleted, farming collapses. When the climate changes, regions can be thrown in turmoil and even war, as has increasingly occurred in the dryland regions of Africa, the Middle East, and western Asia.

Human-induced climate change is already having such dire impacts in many parts of the world. The most direct manifestation of human-induced climate change has been the rise in temperatures. Consider, as an illustration, a world map prepared by NASA of the average temperature in 2013 in each location on Earth compared with the average temperature in that location during 1951–1980, as shown in figure 6.6. We see that almost all the world was warmer in 2013 than in the base period. Only very tiny spots in the ocean (e.g., off the coast of Peru) were actually colder on average than during the base period. The same would apply to just about any period in recent years: warming is pervasive and covers nearly all of the world's land and sea surface.

Along with the rise in the average global temperature has come the rising frequency of extreme heat waves. World-leading climate scientist James Hansen has analyzed the extreme heat events on the planet from the 1950s till now, with the results shown in figure 6.7 (Hansen, Sato, and Ruedy 2012, E2417). The red spots on the world map indicate occurrences of extreme heat waves. Note the years for the nine maps, starting in 1955 and ending in 2011. We see clearly that the numbers of red blotches on the map—signifying extreme heat waves—have increased dramatically between 1955 and 2011. Indeed, events that only occurred one or two

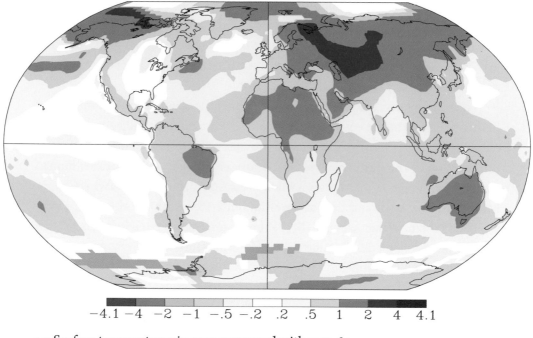

$-4.1 \quad -4 \quad -2 \quad -1 \quad -.5 \quad -.2 \quad .2 \quad .5 \quad 1 \quad 2 \quad 4 \quad 4.1$

6.6 Surface temperatures in 2013 compared with 1951–80

Source: Calculated at http://data.giss.nasa.gov/cgi-bin/gistemp/nmaps.cgi?sat=4&sst=3&type=anoms&mean_gen=0112&year1=2013&year2=2013&base1=1951&base2=1980&radius=1200&pol=rob.

times per 1,000 days in the 1950s are now occurring at a frequency of 50–100 times per 1,000 days in our time.

II. Growth Dynamics

It is a stunning reality that humanity is already pushing against the limits of Earth's planetary boundaries. Yet the environmental pressures are likely to increase in the future, not decrease. That is because the world population and gross domestic product (GDP) per capita both continue to grow. And indeed, we are interested in the success of poor countries in raising their living standards. We are therefore

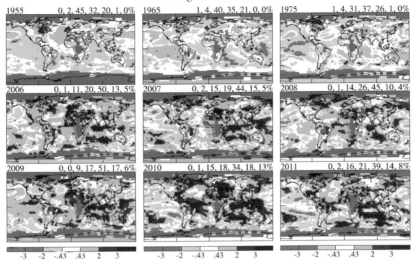

6.7 Changes in global extreme temperatures (1955–2011)

From Hansen, James, Makiko Sato, and Reto Ruedy. 2012. "Perception of Climate Change." Proceedings of the National Academy of Sciences *109(37): E2415–E2423.*

faced with the most important challenge of sustainable development: how to reconcile the continued growth of the world economy and the sustainability of the Earth's ecosystems and biodiversity.

This challenge is profoundly significant and profoundly challenging. We want economic development, and we need environmental sustainability. The two seem contradictory, though I will argue that they are in fact compatible if we follow smart policies. Still, making growth and environmental sustainability compatible will be no easy feat. To put in another way, we need to learn to achieve economic growth that remains within planetary boundaries.

To get a quantitative sense of the extent of this challenge, let us first consider the amount of "pent-up growth" that is now in the world economic system. By pent-up growth, I mean the amount of economic growth that we might expect as the result of poorer countries catching up with richer countries, even if the richer countries do not grow rapidly in the future.

We are now a world of around $91 trillion a year (the International Monetary Fund's estimate for 2014) when measured in U.S. dollars at international (purchasing power parity) prices. There are 7.2 billion people with an average output per person of approximately $12,000. The average income in the high-income countries is roughly three times the world average, meaning that high-income countries have an average per capita income of around $36,000. The average income of the developing countries (low-income and middle-income countries) is roughly $7,000. Suppose the poor countries successfully catch up with the rich world. That catching-up process would raise the income of the developing countries to $36,000 and would raise the world average income to that level as well. Since the average income would rise three times, total world output similarly would increase by three times, from around $91 trillion to around $275 trillion.

That is a stunning increase for a world economy that has already trespassed major planetary boundaries. Yet is understates the potential impact, since the three time increase is what would happen with today's population. Now let us factor in future population growth. Today's world population of 7.2 billion people is projected to rise to around 9.6 billion by midcentury, and 10.8 billion by the end of the century. Just the rise by 2050 is an increase of 33 percent by midcentury. With full catching up, the world economy would therefore grow to 9.6 billion people at $36,000 per person, or a total world income of $346 trillion, nearly four times today's GWP.

It is true that convergence of income levels is not likely to occur by 2050. Today's developing countries are not likely to *entirely* close the per capita income gap with the high-income countries by midcentury. Yet our calculations also assumed that the rich countries would stay in place at $36,000. But they are likely to achieve continued economic growth. So our calculations must adjust for two factors: incomplete catching up and continued economic growth in the high-income countries. We need a statistical model of future growth in order to make an educated assessment about possible outcomes.

Here is one simple rule of thumb. Compare the growth rates of the United States and countries with lower per capita incomes. Generally speaking, a country at half of the per capita income of the United States (i.e., $25,000 per person) will tend to grow roughly 1.4 percentage points per year faster than the United States

Table 6.2		
Country	Per capita income (PPP) ($)	Growth rate (tendency per year) (%)
Least-developed	1,613	8.0
Low-income	3,125	6.6
Lower-middle-income	6,250	5.2
Upper-middle-income	12,500	3.8
Lower-high-income	25,000	$2.4 \, (= 1 + 1.4)$
United States	50,000	1

in per capita GDP. If the United States grows at 1 percent per year in per capita terms, the country at $25,000 per capita would tend to grow at around 2.4 percent per year. A country at half the level of $25,000 (i.e., $12,500 per capita) would tend to grow *another* 1.4 percent per year faster or at a rate of 3.8 percent per year $(= 1\% + 1.4\% + 1.4\%)$. Using this principle, we find the typical growth rates shown in table 6.2.

The poorer a country's starting point (assuming no poverty trap or other fundamental barriers to growth), the greater the headroom for rapid catching up. Over time, the poorer countries narrow the gap with the richer countries by growing faster. As the income gap narrows, so too does the growth rate of the poorer country. There is a gradual convergence of living standards over several decades, as well as a convergence of growth rates to the long-term growth rate of the technological "leader" (in our example, to the 1 percent growth of the United States). The poor country starts out growing very fast, and then as it becomes richer and closer to the technological leader, its growth rate also slows down and eventually converges with that of the technological leader.

The convergence theory helps us understand why the developing countries are indeed achieving faster economic growth than the high-income countries. If we trace this out for the next forty years from 2010 to midcentury, assuming that the high-income world averages 1 percent per year and the poorer regions catch up gradually with the high-income region along the lines of the convergence formula,

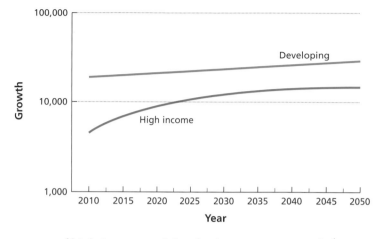

6.8 Convergence of high-income and developing country growth (2010–2050)

the result is the kind of graph shown in figure 6.8 (shown with a logarithmic scale for the vertical axis). While the high-income and developing countries start out quite far apart, basically with a fivefold advantage for the high-income countries, the gap between the two groups narrows significantly to the point where the high-income countries are only two times, not five times, larger than the developing world by the middle of the century.

What would this gradual convergence imply for total world production and the resulting planetary pressures? To answer this question, we need to now add in the population dynamics as well. As noted, today's population of 7.2 billion people will reach 8 billion people around 2024 and 9 billion by around 2040 (DESA Population Division 2013). By the end of the twenty-first century, in the medium-fertility variant of the UN Population Division, the world will reach almost 11 billion people. This is shown in figure 6.9, again using a logarithmic scale for the vertical axis. With the logarithmic scale *the slope of the curve tells us the growth rate of the world population*, so when we see the curve leveling off by the end of the century, it also means that the growth rate of the world population is slowing to a low number. By the end of the century the population is projected to stabilize,

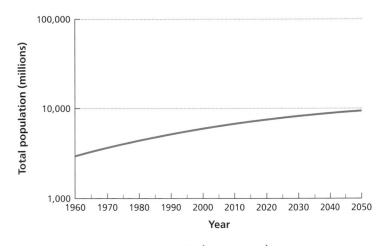

6.9 Global population on a semi-log scale (1960–2050)

Source: United Nations Department of Economic and Social Affairs Population Division (DESA Population Division). 2013. "World Population Prospects: The 2012 Revision." New York.

as signified by the flattening curve. Combining population forecasts with the convergence theory, and assuming the scale of the planetary boundary challenge can be met so that convergent growth can continue, the world economy would rise from around $82 trillion in 2010 to around $272 trillion by the middle of the century, a more than a threefold increase, but slightly lower than our previous calculation based on the full convergence of the developing countries.

We live in a world already bursting at the seams, with humanity pushing against planetary boundaries. We also live in a world where the developing countries seek to close the income gap with the rich world and have the technological means to do so over time. Yet if this continued economic growth is pursued using today's technologies and business models, humanity will completely burst through the planetary boundaries, wreaking havoc on the climate system and the freshwater supplies, increasing the oceans' acidity, and negatively impacting the survival of other species. In order to reconcile the growth that we would like to see with the ecological realities of the planet Earth, we are going to need the world economy to develop in a fundamentally different way in the future.

III. The Case of Energy

Of all of the problems of reconciling growth with planetary boundaries, probably none is more urgent and yet more complicated than the challenge of the world's energy system. The world economy has developed (one could say "grown up") on the basis of fossil fuels, starting with the eighteenth-century steam engine, and then the nineteenth-century internal combustion engine, and then the twentieth-century gas turbine. Indeed, until James Watt invented the improved steam engine in 1776, there was no realistic way to achieve sustained economic progress. Fossil fuels allowed the breakthrough to the era of modern economic growth, and that history reminds us of how deep the challenge is of moving away from fossil fuels in the twenty-first century. The energy sources that have been central to global economic development for more than two centuries are now a clear and present danger to the world, because of the CO_2 they emit.

A simple solution might seem to be simply to use less energy. But this is not actually so simple, because energy means the ability to do work. Any useful work in an economy depends on access to high-quality energy. Yes, energy efficiency must clearly be part of any solution for sustainable development, as we waste a lot of energy in the form of driving unnecessarily large cars, living and working in poorly insulated buildings, and so forth. Yet the world needs energy resources, and the use of energy, even with a substantial gain of efficiency, is likely to increase in total as the world economy grows. So we have a basic problem. More energy will be needed in the future, but the traditional forms of fossil fuel energy cannot do it for us, because they would create a massive intensification of human-induced climate change.

The graph in figure 6.10 shows on a logarithmic scale the income of different countries and their primary energy use. Total energy use combines fossil fuels, wood burning, hydroelectric power, geothermal energy, wind and solar power, nuclear power, and biofuels (other than wood). This graph shows the total output compared with the total primary energy use. The graph of the GDP per capita of an economy versus its energy consumption is close to a straight line, signifying that a doubling of the size of an economy tends to be associated with a doubling of primary energy use. As the economy grows, the energy use

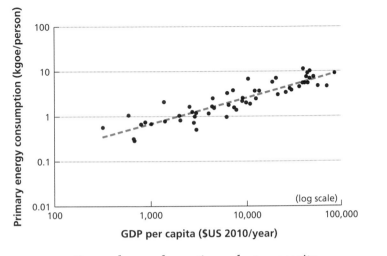

6.10 Energy consumption and gross domestic product per capita

Source: U.S. Energy Information Administration, the World Bank.

tends to grow alongside it, though of course with energy-saving efficiency gains over time as well.

It is useful to quantify how much energy we use, how much CO_2 we therefore emit into the atmosphere, and what that implies for how much climate change we are causing. On average, for every $1,000 of total production (expressed in 2005 dollars) in the economy, the economy use (expressed in metric tons of oil-equivalent energy) tends to be 0.19 tons of oil-equivalent energy. One metric ton is 1,000 kilograms, so 0.19 of a metric ton is 190 kilograms; therefore, for each $1,000 of production we use approximately 190 kilograms of oil or its equivalent in energy content.

Every ton of oil-equivalent energy used in the world releases 2.4 tons of CO_2 emissions. The exact amount of CO_2 depends on the energy source. Since nuclear power is not a fossil fuel, for example, it does not by itself create CO_2 emissions. On the other hand, coal is almost all carbon (with some impurities), so it creates the most CO_2 emissions per unit of energy of any fuel, about 4 tons of CO_2 for an amount of coal equal in energy units to 1 ton of oil. One ton of oil itself creates about 3.1 tons of CO_2 emissions. The amount of natural gas equivalent to 1 ton of oil creates about 2.4 tons of CO_2. And hydroelectric power, solar power, and wind power all release zero CO_2 and are clearly highly desirable from the point of view of heading off climate change.

Let us now put the pieces together. The world economy in 2010 (measured in 2005 dollars) was about $68 trillion (the 2014 world economy is $91 trillion measured in 2014 dollars). Multiplying $68 trillion by 0.19 tons of oil equivalent per $1,000 and then by 2.4 tons of CO_2 per ton of oil-equivalent energy (please do the calculation!), results in the 31 billion tons of CO_2 that the world released into the atmosphere in 2010. Humans also put CO_2 into the atmosphere in some other ways, such as by chopping down trees and releasing the biologically sequestered carbon previously stored in the trees. Approximately 46 percent of every ton of CO_2 released stays in the air. The other 54 percent is typically stored in what are called "natural sinks," the oceans, land, and vegetation. That means if we put 31 billion tons into the air in one year, a little over 14 billion of those tons stayed in the air.

Now comes the next question. How much is 14 billion tons of CO_2 compared with the entire atmosphere? To answer that we can look at the total volume of the atmosphere (how many molecules are in the atmosphere) and how many molecules of CO_2 are in those 14 billion tons. Doing the calculations, we find that for every 7.8 billion tons of CO_2 released into the atmosphere, the CO_2 in the atmosphere rises by one molecule per million molecules. This gives us a translation factor: each 7.8 billion tons of CO_2 in the atmosphere raises the CO_2 concentration by 1 molecule per million. Scientists speak of "parts per million" instead of molecules per million, and use the abbreviation ppm. In 2010, the 14 billion tons of CO_2 in the atmosphere therefore raised the CO_2 concentration by around 1.8 ppm (parts per million).

Is that a big increase for one year? Yes. Should we be frightened by it? Yes. Figure 1.15 shows a graph of the concentration of CO_2 in the atmosphere measured over hundreds of thousands of years (Scripps 2014). The concentration of CO_2 fluctuates over geological times (thousands of years) as a result of normal Earth processes such as changes in the Earth's orbital cycle. The graph shows the peaks and declines of CO_2 in the Earth's geological history over the past 800,000 years, driven mainly by natural changes of the Earth's orbital cycle until the most recent 200 years.

Consider the graph in figure 1.15; all the way to the right is the present age. During the past 200 years, and especially the past 100 years, the CO_2 concentration has shot straight up, breaking out of the natural range of the past 800,000 years. This is the result of humanity discovering how to use fossil fuels in huge quantities. For 800,000 years, the concentration of CO_2 fluctuated between roughly 150 and 280 parts per million. Then suddenly, in the blink of an eye in geological time, humanity has caused the CO_2 to soar way above 280 parts per million. Within just 150 years, the CO_2 concentration has soared from 280 ppm to 400 ppm. We have reached a level of CO_2 in the atmosphere not seen for the past 3 million years!

What the climate scientists tell us is that this kind of change is consistent with a significant rise of temperatures on the planet. Indeed, if we reach 450–500 ppm of CO_2, as we soon will likely do, humanity will very likely be living on a planet that is on average 2°C warmer than before the Industrial Revolution. A 2°C rise in the global average temperature might not sound like much, but it implies even larger increases of temperature in the higher latitudes and also massive changes of the Earth's climate system, including patterns of rainfall, droughts, floods, and extreme storms. Moreover, the sea level will rise significantly, perhaps by 1 meter during the century, and with bad luck (such as the disintegration of part of the Antarctic or Greenland ice sheets) by much more than 1 meter. We are talking about changes in CO_2 concentrations that, when translated into climate change and environmental change more generally, are unprecedented in human history—large, dangerous, and happening now.

How fast are these changes occurring? If we are at 400 ppm today, and the CO_2 concentration is rising by around 2 ppm per year, we will reach 450 ppm just 25 years from now and 500 pm in 50 years. If economic growth leads to an even faster rate of CO_2 change, we might reach the range of 450–500 ppm even earlier. Indeed, if the world economy were to triple, and energy use were to triple alongside it, then CO_2 would be rising around 6 ppm each year rather than 2 ppm.

In other words, if we do not dramatically change course quickly, we are on a path of extraordinary peril. Because of our fossil fuel reliance, we would be seeing a great increase in frequency of the heat waves already evident in the maps by James Hansen (figure 6.7). We would mostly likely be seeing mega-droughts, mega-floods, more extreme storms, more species extinction, more crop failures, a massive sea level rise over time, and a massive acidification of the oceans as that CO_2 dissolves into the ocean and produces carbonic acid. Some regions will be more vulnerable than others. Not every place on the planet will experience each kind of disruption. But in a world that is 3°C warmer (or even more) in temperature than now, the disruptions will be widespread. And we could well be on our way to 4°C warmer or even more by the end of the twenty-first century according to the best evidence.

The solutions, which we will study later in this book, involve a "deep decarbonization" of the energy system, meaning a way to produce and use energy with far lower emissions of CO_2 than now. There will be at least three main "pillars" of deep decarbonization. The first is energy efficiency, using much less energy per unit of GDP than now. The second is low-carbon electricity, meaning that we produce electricity with wind, solar, nuclear, or carbon capture and storage technologies, so that emissions of CO_2 per megawatt of electricity are drastically reduced. The third is to shift from burning fossil fuels to using electricity generated by a low-carbon source, a process called "fuel switching" or "electrification." For example, automobiles can shift from internal combustion engines powered by petroleum to electric motors powered by batteries charged by low-carbon electricity (e.g., a grid running on solar power). Instead of homes heated by oil furnaces, homes can be heated with electric heat pumps, run on electricity generated by a low-carbon source. Every part of the world will need to join in this three-part process.

We must indeed change course on energy and we must do it quickly—far more quickly than what the politicians are telling us. But there is some good news. There are powerful low-carbon technologies available at sharply declining prices for solar power, wind power, energy efficiency, electric vehicles, and more. These technologies will be crucial to a low-carbon future.

IV. The Case of Food

Intuitively, fossil fuel use (and the mining that goes along with it) would seem to be the dominant means by which humanity impacts the physical planet. Energy use is everywhere, in transport systems, power generation, industry, offices, and homes. Yet there is actually an economic sector with comparable or even greater environmental impact than the energy sector: agriculture.

Perhaps this is not entirely surprising. Agriculture is, of course, key to our very survival. We must eat. And since the beginning of civilization, most of humanity has been engaged in farm life. Even now, in the early twenty-first century, half of the world's population resides in rural areas, with some fairly direct connection with agriculture. Yet the extent of agriculture's impact on the environment is even bigger than it appears. Think of the planetary boundaries—almost every one of them is related to agriculture.

Consider each of the planetary boundaries in turn from the point of view of agriculture (SDSN 2013c).

- Climate change. When land is cleared for farmland and pastureland, the resulting CO_2 emissions contribute to climate change. So too does the energy use on farms and in the transport and preparation of foods; the methane released in rice production and by livestock; and the nitrous oxide that results in part from the volatilization of nitrogen-based fertilizers.
- Ocean acidification. Agriculture contributes to the CO_2 emissions that in turn are the main culprit in ocean acidification.
- Ozone depletion. CFCs used in food production and storage (e.g., refrigerants) are the drivers of ozone depletion.
- Nitrogen and phosphorous fluxes. The use of chemical fertilizers is the main source of anthropogenic nitrogen and phosphorous fluxes.
- Freshwater depletion. Agriculture, we have seen, is by far the greatest user—and therefore cause of depletion—of freshwater resources.
- Biodiversity. The grand tradition of agriculture, unfortunately, is to "simplify" the biodiversity of a given landscape. A complex natural ecology is replaced by a human-managed ecology that often involves a single genetic variant of a single crop such as rice, wheat, or maize. Monoculture farming can cause a sharp decline in biodiversity that eventually reduces crop productivity as well as other ecosystem functions. Agriculture can reduce biodiversity in other ways as well, for example, through the application of pesticides and herbicides that end up poisoning the local environment or through the introduction of nonnative species that disrupt local ecosystems.

- Aerosols. Agriculture can contribute to aerosols through many pathways: dust, burning of crop residues, combustion of diesel and other fossil fuels, and so forth.
- Chemical pollution. Agriculture in high-income settings is often highly chemical intensive, involving chemical fertilizers, pesticides, herbicides, and other soil treatments. Pollution may also arise from food processing, waste management, use of antibiotics in animal feeds, and so on.

In addition to crossing these planetary boundaries, the global agriculture system has other important adverse impacts. One issue is that the food system is also giving rise to new pathogens. For example, the industrial breeding of poultry causes recombination of genes of bacteria and viruses. When livestock and poultry mix with wild species, there are further viral recombinations. The interaction of the food industry with wild-type pathogens has probably given rise to several emerging infectious diseases, most likely including the frightening outbreak of the SARS virus in 2003.

All of these huge, and unsustainable, environmental consequences of farming are deeply ironic. They recall Malthus's warning about the physical limitations of growing food on the planet. Malthus noted that population tends to increase geometrically (at a given growth rate), while the ability to grow food, he believed, increases only arithmetically (that is, by a given quantum, not a given growth rate, per year). He noted that geometric growth would necessarily overtake arithmetic growth, so the growth of the human population would necessarily overtake the ability to grow food. At some future point, warned Malthus, there would be so many people that hunger would necessarily ensue, with devastating feedbacks, such as war, famine, disease, and other scourges that would push the population back down. Malthus argued that in the long run, humanity would therefore not break free of the physical constraints on the ability to grow food.

Malthus did not anticipate the scientific advances of the nineteenth and twentieth centuries. He did not anticipate the science of soil nutrients, founded by the great German scientist Justus von Liebig in the 1840s. He did not anticipate the science of seed breeding made possible by the science of modern genetics, which has its roots in the discoveries of the Silesian monk Gregor Mendel in the 1860s. He did not anticipate the invention in the early decades of the 1900s of human-made nitrogen fertilizers in the Haber-Bosch process. And he did not anticipate the great synthesis of these advances in the Green Revolution that occurred from the 1950s to the 1980s. For these reasons, most economists and others have long scorned Malthus. Modern science indeed allowed a geometric growth of food production in line with a geometric rise of the world's population.

I am going to make a different point, though. Malthus really had a stronger case than we recognize, and we should thank Malthus wholeheartedly for pointing out a deep conundrum that continues to this day. First, when Malthus wrote his famous text, the world population was one-eighth of what it is now. Malthus was correct to worry. Second, when economists claim that Malthus neglected the potential for technological advance, we can note that economists on their part neglect the environmental damage caused by modern farming. Yes, the global farm system feeds the planet (though not necessarily very well, as I emphasize later in the book), but it does not do so in an environmentally sustainable way. Until global farming itself is a sustainable activity, we should not be too quick to brush Malthus aside. We don't want Malthus to have the "last laugh" (that indeed would be a tragedy for humanity), but we do want to correct the farm system before it does irreversible damage to the global environment.

Just as we are going to need to find a new energy pathway based on energy efficiency and low-carbon energy supplies, we are also going to need to find new farm systems, adapted to local ecological conditions and causing much less ecological damage. What is common to nearly all of the world's major farm regions is that the farm systems are still not sustainable. We have yet to prove Malthus wrong! His specter will loom large until the world population is stabilized (or declining) and our production methods are environmentally sound. The challenge of a sustainable global food supply is therefore a fundamental part of any twenty-first century agenda for achieving sustainable development.

V. Population Dynamics and Sustainable Development

A major part of our ability to achieve sustainable development will depend on the future dynamics of the world's population. The more people there are on the planet, the more challenging it will be to reconcile the economic objectives of raising living standards per person with the planetary boundaries. The more rapidly population is growing in a particular country, the more difficult it will be to combine economic growth, social inclusion, and environmental sustainability in that place.

Poor countries with high fertility rates (with more than three children per woman, and in some countries reaching six or seven children per woman) are often stuck in a "demographic trap." Because households are poor, they have many children. Yet because they have many children, each child is more likely to grow up poor. These societies end up in a vicious circle in which high fertility and poverty are mutually reinforcing.

Facing the question of high fertility (and the rapid population growth that accompanies it) is therefore crucial for breaking free of poverty. When poor families have large numbers of children, they are not able to provide the necessary investment for each child in the human capital—health, nutrition, education, and skills—the child needs to be healthy and productive as an adult. Moreover, governments are not able to keep building the infrastructure—roads, power, ports, and connectivity—needed to keep up with the growing population. And the country's fixed natural capital such as land and depleting natural capital such as hydrocarbons must be subdivided among an ever-growing population. Reducing the fertility rates voluntarily, while respecting human rights and family desires, is therefore essential to sustainable development and the end of poverty. The world's governments have enshrined sexual and reproductive rights as core human rights for women, yet often these rights are not realized because countries are too poor to implement programs for safe pregnancy and family planning, or sometimes because governments do not implement the programs they have been committed to provide.

The world's demographic future is still up for grabs, depending on the fertility choices that households (especially low-income households) make in the future

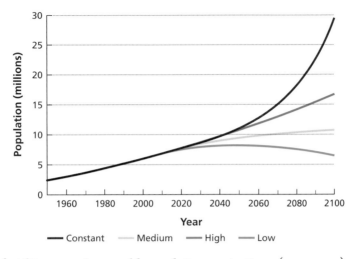

6.11 Four fertility scenarios, world population projections (1950–2100)

Source: United Nations Department of Economic and Social Affairs Population Division (DESA Population Division). 2013. "World Population Prospects: The 2012 Revision." New York.

and the support of public health programs to make those choices. Figure 6.11 shows the four fertility scenarios produced in 2012 by the UN Population Division (DESA Population Division 2013). The single line between 1950 and 2100 shows the actual change of population from 2.5 billion to 7.2 billion in those years. There are four scenarios after 2010 depending on alternative assumptions about fertility rates between 2010 and 2100.

The medium-fertility variant shown by the light blue line reaches about 10.8 billion people in the year 2100. This would signify a net increase of another 3.6 billion people by 2100, roughly half again of today's population. The medium scenario is the one that the United Nations regards as the most plausible continuation of current trends.

The red line at the top shows something unthinkable, but still very interesting. Suppose fertility rates do not change at all from their current levels. In each country and age group, the fertility rate would remain as it is currently. Simply running the clock forward based on the current fertility rates, the world population in 2100 would be 28.6 billion, four times higher than today! The Earth could not sustain this, so it will not happen. Yet this scenario does tell us that fertility rates must decline from their current levels.

The green line is called the high-fertility variant. It is a bit more plausible than the constant-fertility variant, and yet still pretty frightening. It says that if women were to have on average just *one-half child more* (as a statistical average, or five children more per ten women) than on the medium-fertility variant, the world would reach 16.6 billion. A small change in the fertility rate, of 0.5 children per woman, has an effect of nearly 6 billion more people on the planet by 2100. Fertility rates matter!

The low-fertility variant is the blue line below the other three. This last scenario is preferable to the others from a sustainable development standpoint. In this variant, each woman has on average 0.5 children fewer than in the medium-fertility variant (or to put it another way, every ten women have five children fewer than in the medium-fertility variant). The population would peak around 2050 at 8.3 billion and then gradually decline to 6.8 billion by 2100, fully 4 billion people fewer than in the medium-fertility variant! Such an outcome, with the population at the end of the century less than now, would make it much easier to meet the social, economic, and environmental needs and goals of humanity.

These scenarios show that small changes of fertility rates will have big effects on outcomes. They suggest that if steps are taken to help facilitate a faster reduction of fertility in today's high-fertility regions, for example, by helping girls to stay in school through age 18 rather than marrying young, the positive impacts from the household to the planetary scale could be huge.

Figure 6.12 shows the annual rate of change of population in the medium scenario for different groups of countries. The solid blue line is the world average, which shows the world's population growth peaked at about 2 percent around 1970. At that time the world population was about 4 billion people, so with a 2 percent growth rate the world was adding about 80 million people per year.

In the year 2010, the growth rate dropped to 1.1 percent to 1.2 percent per year, but now the base on which that percentage growth is occurring is twice as large as back in 1970. Multiply 1.1 percent by 7.2 billion people, and there is still the same 80 million increase as of forty years ago. This says that while the *proportionate*

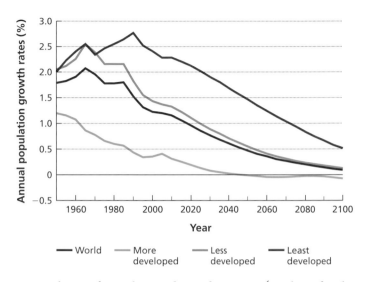

6.12 Average annual rate of population change by region (medium fertility scenario) (1950–2100)

Source: United Nations Department of Economic and Social Affairs Population Division (DESA Population Division). 2013. "World Population Prospects: The 2012 Revision." New York.

growth rate of population has slowed, the *arithmetic* increase each year remains around 75 to 80 million people.

In the medium-fertility variant, the world's population growth rate tends to decline to almost zero by the end of the century, because fertility rates basically come down to replacement. The replacement fertility rate means that each mother has two children, one daughter and one son, so each mother is replacing herself with a daughter who will become the mother of the next generation. This keeps the population stable in the long term. (The replacement rate, technically, is a little bit above 2.0 to take account of the early mortality of children who do not reach adulthood.)

Figure 6.12 shows clearly that the least-developed countries (LDCs) have the highest population growth rate. In the poorest places, there are many regions where family planning is not used; girls are pulled from school very young; and women face massive discrimination and are not in the labor market. In these

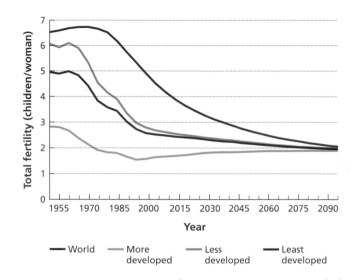

6.13 Total fertility trajectories by region (medium fertility scenario) (1950–2100)

Source: United Nations Department of Economic and Social Affairs Population Division (DESA Population Division). 2013. "World Population Prospects: The 2012 Revision." New York.

circumstances, fertility rates tend to be extremely high, for example more than six children per woman. It is these countries where a rapid, voluntary transition to the replacement rate is most important.

The graph in figure 6.13 shows the actual total fertility rates between 1950 and 2010, and then shows the medium-fertility variant projections by the United Nations to the year 2100. As of 2010, the more-developed countries, at the bottom of the curve, are already below replacement rate. If their fertility rates continue to be so low in the years ahead, their populations will decline. The highest fertility rates at the top of the graph are the LDCs. For the less-developed regions as a whole, and for the world on average, the fertility rates are a bit above replacement but not as high as in the LDCs.

What could lead to a faster transition to a replacement fertility rate in today's high-fertility regions? There are many determinants of the fertility rate. Age of marriage is key. In traditional societies, girls are often not schooled at all or leave school and marry very early, sometimes as young as age 12, perhaps for economic or cultural reasons. Childbearing starts very soon thereafter, and these young girls remain without economic, political, or social empowerment, and often end up giving birth to six to eight children or even more. A second determinant of fertility is the access (or lack of access) to modern contraception and family-planning services. Places where contraceptives are widely available, where clinical services work, and where there is culturally sensitive advising of households, tend to have lower fertility rates. Family-planning programs that are culturally sensitive and operating effectively in low-income countries can dramatically lower fertility rates on a wholly voluntary basis. A third determinant of the total fertility rate is women's role in the labor force. In some countries, women are not allowed to work or are restricted to working in the home or in just a few occupations. Fertility rates in these settings tend to be high. When women are working outside the home, the fertility rates are much lower. There is a direct "opportunity cost" of foregone earnings when women are home raising many children.

Another possible factor is the urban versus rural location of the household. In farm households, parents often view their children as "farm assets." Children do farm work, such as milking the cows, carrying fuel wood, and fetching water. In an urban setting, by contrast, children are much more likely to be in school and not working in a formal way (though there are of course painful exceptions). This means that on average, families in urban areas see the net cost of raising children to be higher than do families in rural areas. When families migrate from rural to urban areas, their fertility rates thus tend to come down.

Child survival is another key determinant of fertility. If most children survive to adulthood, families may choose to have few children; but if the parents worry that many children will die early, they will likely have more children to ensure the survival of at least some children. One of the keys to a quick voluntary reduction of fertility therefore is to lower the mortality rate of children, thereby giving confidence to parents to have fewer children as well. The legality of abortion also plays an evident role as well. Different societies have widely divergent views about abortion, but the data suggest that those countries with legalized abortion tend to have lower observed fertility rates than countries where abortion is illegal.

Public leadership also plays a big difference, because the choice of family size is also influenced by social norms. In most traditional societies, the cultural norm was to have as many children as possible. But when economic, social, and health conditions change, fertility rates also change. And public policy can speed or slow that change depending on the messages sent by leaders in the community and government. Role models also influence fertility rates. Sociologists have found that when television broadcasting arrives in a poor area, fertility rates tend to come down, often quickly. One hypothesis is that people watch role models with small families on television and therefore choose to emulate these examples.

Population dynamics are very important for sustainable development. The chances for sustainable development will be very different if the world population reaches 10.8 billion at the end of the century or instead peaks by 2050 and declines to 6.8 billion by 2100. The latter trajectory would be much easier from the point of view of achieving a higher quality of life, greater poverty reduction, higher income per capita, and environmental sustainability. There is also good reason to believe that lower fertility rates would be the truly preferred choice of most households if they have affordable and convenient access to family planning; education for their girls; child survival; and decent jobs and the absence of discrimination for women. When those conditions exist, it is most likely households would take the opportunity on a voluntary basis for a sharp reduction of fertility rates, helping to move the world more quickly to a peak and then gradual decline of the world population. This would enormously help to put the world on a sustainable development trajectory, where living standards can be raised while respecting the planetary boundaries.

VI. Economic Growth Within Planetary Boundaries

Many environmentalists alarmed by humanity's trespassing of planetary boundaries have concluded that economic growth must end now, that further economic growth and respect for planetary boundaries are a fundamental contradiction. They suggest indeed that rich countries should significantly lower their consumption levels to make room for higher living standards in poor countries. This attitude is understandable: the crisis of planetary boundaries is urgent and unaddressed after decades of alarm bells sounded by world-leading scientists. Perhaps the economic juggernaut itself is untamable and must be stopped in its tracks, with an urgent focus on redistribution rather than development.

I argue differently. Most importantly, choosing the right technologies, we can achieve continued economic growth and also honor the planetary boundaries. Consider the case of energy once again. Our energy crisis, to repeat, is not the overuse of energy per se, but the release of CO_2 through the burning of fossil fuels (in the absence of technology to capture and store the CO_2). By harnessing wind and solar power, for example, it would be possible to expand access to energy, support more economic activity, and avoid dangerous greenhouse gas emissions all at the same time. Similarly, through better agricultural techniques, it is possible to grow more crops with less water (more "crop per drop") and less, not more, application of fertilizers (with more precision in the use of the fertilizers). The goal of continued growth is a valid one, especially in low-income and middle-income countries, for which growth means more health, better education, more access to travel and leisure time, and more safety from various threats to wellbeing. It is even valid for high-income countries as long as they base their growth on resource-saving technologies so as not to violate planetary boundaries or to leave less space for the poorer countries aiming to catch up in living standards.

Why don't global markets by themselves ensure that economic growth is sustainable? There are two major reasons. The first is that most of the planetary damages are kinds of "externalities," meaning that those who impose the damages (e.g., more CO_2 emissions) don't pay the costs of the damages. They impose losses on others without those losses being controlled by market incentives. When a factory burns coal and causes pollution and climate change, there is nothing in the price of the coal that persuades the coal user to switch to a safer form of energy such as solar or wind power. When a farmer uses fertilizer that runs off the farm and creates eutrophication downstream, the farmer bears no penalty, and the price of the fertilizer does not include the costs that will be imposed on others. The result is overuse of fertilizer just like overburning of fossil fuels.

The second reason is intergenerational. Today's generations impose costs on future generations. Those alive today despoil the environment without having to bear responsibility to future generations. It is the role of government and our ethical standards—for example, religious teaching in many faiths to respect the creation—that must guide us to be good stewards on behalf of future generations. This is not to say that the present generation must bear all of the costs of environmental sustainability. Some outlays for a clean environment can be funded by public debt, for example, that will be paid by later generations. Even in that case, however, the current generation must think ahead—morally and practically—to ensure the wellbeing of generations not yet born.

Much of environmental economics studies the question of how to use various kinds of incentives—both market based and socially based—in order to reduce externalities. When such incentives are ignored, the externalities are rampant. We have, in the famous words of ecologist Garrett Hardin, a "tragedy of the commons," in which the commons of the oceans, rivers, and atmosphere are despoiled by overuse and overpollution. This tragedy of the commons can be controlled through a variety of "economic instruments" or policy tools, including:

1. Corrective taxation that puts a "price" on the pollutant, thereby causing businesses and individuals to use less of the polluting technology. A popular idea, for example, is to put a "carbon tax" on each ton of CO_2 emitted into the atmosphere in order to create incentives to shift to low-carbon energy.

2. Permit systems that limit the overall amount of polluting activity, such as a permit to emit CO_2. These permits may trade in the open market (in which case they are called tradable emissions rights), and the price of the permit acts like a corrective tax. By polluting less, a business can sell its emissions permit to another user, thereby pocketing a market profit.

3. Liability rules that allow those hurt by pollution (e.g., by downstream eutrophication) to sue those upstream causing the damage. This can cause potential polluters to reduce their harmful practices.

4. Social institutions that engage the community in prosocial practices, such as protecting scarce land, scarce forest products, endangered species, and threatened fish stocks. Nobel laureate Elinor Ostrom brilliantly emphasized the power of communities to "internalize" the externalities, that is, to stop the harms caused by externalities through social institutions that promoted cooperative behavior at the community scale.

5. Public financial support to discover more sustainable technologies through "directed" research and development aimed at particular breakthroughs. There is now considerable, yet still insufficient, public support for new discoveries in photovoltaics (solar power), advanced biofuels, safer nuclear power plants, carbon capture and storage, and other technologies to "decarbonize" the energy system.

In harnessing these various powerful policy instruments, the goal should be to eliminate externalities and achieve intergenerational fairness as well—in short, to achieve growth within planetary boundaries. The end result, if successful, would be to "decouple" growth and dangerous overuse of primary resources and ecosystems. Decoupling means that growth can continue while pressures on key resources (water, air, land, habitat of other species) and pollution are significantly reduced rather than increased. Such decoupling is technologically feasible, but surely requires the right policies and incentives to achieve it.

And yes, such decoupling will be much easier in a world with a stable or gently declining world population rather than a world with a still rapidly growing population. Remember that material wellbeing of each person on average depends not on output per se, but on output per person. In a world that is trespassing on planetary boundaries, a high level of output per person is much easier to achieve if the number of people is finally stabilized rather than continuing to grow at a rapid rate (now equal to 75–80 million net addition to the world's population each year). Thus, growth of material wellbeing per person is best protected if the astounding rise in global population is finally brought under control this century through a voluntary reduction of fertility rates to their replacement rate or below, thereby leading a peak and then gradual decline in the global population during the twenty-first century.

7

SOCIAL INCLUSION

I. The Ethics of Wealth, Poverty, and Inequality

Sustainable development targets three broad goals for society: economic development, social inclusion, and environmental sustainability. In most of the world, countries struggle with all three of these goals. Even the high-income countries, which have largely banished extreme poverty, struggle with high and rising inequalities of income, wealth, and power. And, of course, all countries are vulnerable to the violation of planetary boundaries. Even a country that is an environmental "saint," taking every measure on its own to protect the environment, will face the onslaught of global environmental crises, from climate change to the loss of biodiversity to ocean acidification.

The goal of social inclusion is unfinished business in almost all parts of the world. Traditional societies in most places developed strong cultural and legal barriers to the equitable participation of women in the economy, so that gender inequalities cast a long shadow still not overcome. Indigenous populations have faced shocking discrimination, sometimes verging on attempts at genocide. Native Americans in the United States, First Nations in Canada, Aboriginals in Australia, the Maori in New Zealand, the Orang Asli in Malaysia, the Scheduled Tribes in India, all share the distinction of combining indigenous status and massive poverty and exclusion.

Discrimination may be based on characteristics of ethnicity, religion, race, gender, caste, or sexual orientation.

In modern economies, class is another possible barrier to social inclusion. There are reasons why kids who grow up in poor families can easily find themselves stuck in poverty, in an intergenerational poverty trap. Overcoming poverty may require education, working capital, access to health care, and access to social networks (e.g., to land a good job). Those who are born into poor families may be unable to obtain the education, health care, and social networks they need to escape from poverty.

The Ethics of Equality and Rights

When we think about the questions of social exclusion, there are multiple dimensions of moral and ethical choices to face. One set of issues addresses income and wealth inequality. Should society as a whole, through government and social institutions, work to narrow income and wealth inequalities? Are there trade-offs between income redistribution and growth? For example, if government raises taxes on the rich in order to provide social services to the poor, will the tax-and-redistribution policy lead to lower economic growth, as is sometimes claimed?

A second related (but not identical) issue is the question of economic discrimination, both through legal and cultural channels. Laws in many parts of the world continue to discriminate against some groups in society: women, religious minorities, indigenous groups, LGBTs, and others. In most of human history until the nineteenth and twentieth centuries, the legal framework even allowed for slavery. It took a bloody civil war in the United States to break the back of slavery, and apartheid in South Africa did not end until 130 years after the U.S. Civil War. In some places human trafficking and slavery continue today, though generally in illegal processes that are carefully kept hidden from view.

A third dimension of social inclusion involves cultural norms. Sometimes practices are not strictly illegal, but the attitudes in society, such as discrimination against minority groups, are prolonged by cultural and social attitudes. What can and should be done about that? How should we think about the relationship of ethics, culture, and law?

There are many different aspects of values and value systems to explore in order to investigate and promote social inclusion. It is particularly insightful to understand the many schools of thought about these deep questions that have existed throughout history. There are six important ethical approaches to social inclusion to highlight (Helliwell, Layard, and Sachs 2013, chap. 5).

The first of these is virtue ethics. The Buddha, Confucius, and Aristotle are three important exemplars of virtue ethics. The Buddha's influence extends throughout South and East Asia. Confucius has a huge and lasting influence to this day on ethics in China and other parts of northeast Asia. Aristotle has a long and deep legacy of influence on Western thinking about values, with major influence in Christianity, Judaism, and Islam.

In all three of those great sages' thinking, there is a call on individuals to cultivate a set of attitudes and behaviors appropriate to life in society and to achieve wellbeing. For Aristotle, virtue was the key to *eudaimonia*, which is often translated as "a flourishing life." Man is a social animal, said Aristotle, and therefore man has to cultivate the characteristics, attitudes, habits, and behaviors to be a good citizen in the city-state (or *polis*), which for Aristotle was the ideal form of political organization. Such virtue requires moderation in all things. Individuals need not starve themselves of worldly goods, but they should not greedily covet them either; rather, individuals should aim for a middle course, in materialism and in other values as well (e.g., steering between cowardice and recklessness).

Confucius, like Aristotle, thought that individuals should undertake the self-cultivation of virtue. A stable society depends on the virtue of its members. Virtue includes altruism, humaneness toward others, and a disposition to behave properly. The family plays a central role in Confucian thought. Obedience to parents (filial piety) is one of the greatest virtues.

For Buddha, the aim of life was escape from suffering by ridding the mind of illusions. Buddha warned individuals to beware of their cravings for sensory delights and material possessions. These are traps that are bound to disappoint. True happiness is found by training the mind to reject such cravings and to seek happiness in other ways, especially through compassion toward others and meditation and mindfulness. For Buddha, as for Confucius and Aristotle, our material

7.1 (*left*) Buddha *Seated Buddha from Gandhara, Mike Peel, Wikimedia Commons, CC BY-SA 4.0.*

(*middle*) Aristotle *Bust of Aristotle, Marie-Lan Nguyen, Wikimedia Commons.*

(*right*) Confucius *Confucius circa 1770.*

desires are poor guides to our long-term happiness. Compassion, moderation, and mindfulness mark the true path to long-term wellbeing.

Virtue ethics therefore marks one vital approach to social inclusion. It is based on the idea that human beings have a responsibility to others and must cultivate their own attitudes and virtues in order to meet those responsibilities. Compassion is the common thread of all these philosophies: one must pay less attention to one's own cravings and more to one's responsibilities toward others. Through that path not only will society thrive, but the compassionate individual will thrive as well.

A second philosophical view arises from the great religions (we should recognize as well that the religious traditions also draw upon philosophical traditions). The three great monotheistic religions—Judaism, Christianity, and Islam—all champion the Golden Rule: "Do not do to others what you would not want them to do to you." (This is also a basic principle of Confucianism.) The underpinning of this doctrine is a principle of equality, in which all humanity is viewed as equal children of god. The religions preach humility before god, and the need for righteousness in one's behavior toward others as living up to god's plan. Unlike secular ethics, religious ethics are often infused with the promise of eternal happiness and salvation in return for good behavior on Earth.

The major religious traditions bid special attention to the poor. Biblical Judaism includes provisions for releasing the poor from their debts in special "Jubilee" years, and also freeing those held in debt bondage. Jesus's teachings, of course, pay particular attention to the needs of the poor, notably in his teaching that he who feeds the hungry, clothes the naked, and cares for the sick serves the Lord: "I tell you, whatever you did for one of the least of these brothers and sisters of mine, you did for me." (Matt. 25:40) In Islam, one of the five pillars of the faith is charity for the poor (*Zakat*).

A third approach to ethics is called deontological ethics, or "duty ethics." This approach, epitomized by the great Enlightenment philosopher Immanuel Kant, holds that ethics is a matter of duty to rational principles. Kant argued that ethics means adopting a universal standard of behavior. He famously described this as the "categorical imperative": that individuals should behave according to those maxims (rules) that can serve as universal laws. This is a secularized version of the Golden Rule, that individuals should act as they would wish others to act, but it is put more generally, that individuals should act according to those principles that can apply generally.

Philosopher John Rawls offered a modern variant of Kant's categorical imperative in an influential book called *A Theory of Justice* (1971). Rawls suggests that individuals can discover Kantian principles through a special thought experiment. Suppose you are asked to design the general principles of the society (e.g., for example as libertarian capitalism or as tax-and-transfer capitalism), but must do so behind a "veil of ignorance," meaning you do not know what role you will play in the society you are designing. Will you be the billionaire or the starving individual? According to Rawls, you would choose to hedge your bets by ensuring that every member of the society can have a decent life. Rawls goes further by claiming that you would *maximize the condition of the least well-off member of society*, since it may turn out that you will be in that unfortunate position. Other philosophers have disputed Rawls's conclusion, saying that individuals would not be so risk averse as to only consider the situation of the least fortunate member of the society.

A fourth approach to ethics, utilitarianism, is secular rather than religious. Utilitarianism arose near the end of the eighteenth century and has sustained a continuing impact on policy thinking till today. The founder of utilitarianism was the British political activist and philosopher Jeremy Bentham. Bentham wrote that the goal of society is happiness, and the goal of an ethical system, moral philosophy, and indeed practical politics should be the greatest happiness for the greatest number of people. Bentham said that society should maximize the "utility" of the people, where utility signifies a psychological state of wellbeing. In utilitarian doctrine, legislators are to investigate how proposed policies will affect the psychological wellbeing of the public and then to adopt those policies conducive to the greatest good for the greatest number.

Built into utilitarianism is some measure of support for the redistribution of goods across the members of the society. Utilitarianism builds on the idea of the "diminishing marginal utility of income," which holds that the added happiness of an extra $100 of income is very high for a poor person but very low or even negligible for a rich person. Thus, taxing $100 from a billionaire and giving the $100 to a hungry and impoverished individual would only negligibly diminish the utility of the billionaire (if at all) but would markedly raise the utility of the hungry person. Since the utilitarian's goal is to maximize overall wellbeing in the society, typically thought of as the *sum* of the utilities of the members of society, the followers of utilitarianism will generally favor such a tax-and-transfer scheme of income redistribution from rich to poor.

Economists have traditionally argued that such utilitarian redistribution comes at the cost of economic efficiency. As I noted earlier, the mainstream view holds that redistribution may divide the pie more equitably (fairly) but at the cost of baking a smaller pie. The act of redistribution, for example, through taxes of the rich and transfer payments to the poor, is alleged to cause distortions, inefficiencies, and waste, leading people to work less hard and to deploy capital in less productive ways. Yet I've also emphasized that such a view is far too pessimistic.

7.2 Jeremy Bentham

Jeremy Bentham by Joseph Wright.

Many kinds of redistribution to the poor are actually highly efficient investments in the health, skills, and productivity of the poor. These are high-return investments that the poor themselves would be eager to make if only they had the household income or borrowing capacity to make them! In this way, the utilitarian emphasis on redistribution can be doubly justified, not only as equitable but also as efficient, leading to high-return investments that otherwise would not be made because of the failure of market forces alone to provide adequate investment funds to the poor.

A fifth philosophical approach to social ethics also arose in Britain around the same time as utilitarianism and goes by the name of libertarianism. This philosophical approach has since become a favorite of part of the right wing of American and British politics. The libertarian position holds that the greatest moral precept is liberty. The meaning of life, in the libertarian view, is the freedom to choose one's own life course. The greatest harm, in this view, is when the state takes away the liberties of individuals. Limited government, which does little more than protect the national borders and provide for the rule of law internally, is viewed as the best form of government. "He who governs least governs best."

A libertarian rejects the utilitarian idea of redistribution through taxation. Taxing a billionaire to provide for a hungry individual, in libertarian thinking, is an unjustified intrusion into the freedom of the billionaire. The libertarian might urge the billionaire to give the $100 as charity, but would defend the billionaire's right to choose whether to give. Libertarians therefore view most taxation, other than to pay for the national defense and the justice system, as an illegitimate intrusion on the liberty of the taxpayers.

A sixth philosophical approach, one that is very much part of global culture today, is the human rights philosophy. Human rights offer another justification for social inclusion. The human rights approach, which also has its roots in some religious traditions, holds that every human being on the planet has basic human rights that must be protected by the society, including by government. There are five basic categories of such rights: political, civil, economic, social, and cultural rights.

The fundamental idea is that all human beings have inherent rights simply because they are human beings. That is true no matter what society they happen to be born into and no matter in which country they reside. These human rights include not only political and civil rights, but also economic rights, essentially the right to fulfill one's basic material needs. To realize these rights means that societies must organize themselves to protect the individual, whether through tax-and-transfer systems or other means. Note the fundamental difference, for example, compared with the libertarian idea. The libertarian says that government's only responsibility is to create a framework of law, order, and security; it should not redistribute income or property for the benefit of the poor. By contrast, the human rights approach says a poor person has basic rights to health, to education, to means of livelihood; and so society must be organized, perhaps through taxation and provision of public services, to help meet those basic needs.

This brief overview helps, I hope, to distinguish these six different ethical approaches. Throughout human history, ethical issues have been considered in these distinctive ways (and others). The great sages thought about our responsibility to be virtuous. The religions call upon us to honor the Golden Rule. Kant taught us to consider the ethical duties we can derive through rational thought. Rawls built on Kant by asking us to design social institutions behind a "veil of ignorance." Utilitarianism considers our collective responsibility to maximize happiness within the society. Libertarianism emphasizes the dangers of an over-reaching government. Finally the human rights paradigm emphasizes the global and legal environments needed to meet basic human standards for every person on the planet, irrespective of the specific political or social system he or she may live under.

The human rights approach is perhaps the dominant framework today of the international system of nations (SDSN 2013b). It is the one that the member states of the United Nations have signed on to, and it has certain very powerful and attractive aspects to it. It says that we should meet the basic needs of every-body on the planet or at least strive to meet them as soon as possible. If we are not able to meet all needs today, because of technological or other resource limita-tions, there should be a "progressive realization" of these human rights, meaning that even under current constraints, the world's governments should continue to take steps toward fully meeting all human rights.

Economists sometimes give another name to basic needs, calling them "merit goods." Merit goods are those goods and services that should be accessible by all individuals in society irrespective of their ability to pay (or their identity in terms of race, gender, class, religion, ethnicity, and so on). For example, health and edu-cation are both widely judged to be merit goods, because individuals need them and because their universal coverage benefits society at large. Adam Smith, who is generally thought of as a free-market economist, noted that education is a merit good, because universal education produces better societal outcomes than educa-tion only for a few. Smith wrote the following in Book V of *The Wealth of Nations*:

> The State derives no inconsiderable advantage from the people's instruction. The more they are instructed, the less liable they are to the delusions of enthusiasm and superstition, which, among ignorant nations frequently occasion the most dreadful disorders . . .
>
> The expense of the institutions for education and religious instruction, is likewise, no doubt, beneficial to the whole society, and may, therefore, without injustice, be defrayed by the general contribution of the whole society. (Smith 1776, 642)

Smith is saying that all of society is advantaged when the population is properly educated. There will be less tendency of the public to follow delusions and superstitions. The benefits in terms of social stability will be large. For this reason, mass education should be paid for by society at large, enough to ensure that every child can be educated, even the children of very poor families. This is one of the earliest and most influential defenses of mass public education.

In addition to education, health care is also widely considered to be a merit good, partly because most of us believe that individuals have a basic right to health, but also because helping people to remain healthy helps the rest of society as well. When a significant part of a population does not have access to medical care, communicable diseases are more likely to spread rapidly throughout the community, indeed throughout the world. Most observers therefore agree that governments should ensure universal health coverage (e.g., vaccines), in part to control the spread of infectious diseases.

This focus on meeting universal basic needs, especially for health and education, can be justified through the lens of human rights or through the lens of utilitarianism. If it is true that a poorly educated society fosters epidemics of delusion or that poor health coverage allows for epidemics of disease, then Jeremy Bentham would say that universal education and health care is a matter of utilitarian calculus, a way to ensure the greatest good for the greatest number.

Whichever of these approaches the reader finds most compelling (and I personally like many of them), we can be sure that the role of ethical thinking is vital for good public policy. We therefore need to have more discussions, more public awareness, and more debates about these underlying ethical choices, because the goals of sustainable development depend on the ethical positions we adopt.

II. United Nations Declarations, Covenants, and the MDGs

When the United Nations was formed at the end of World War II, one of the first great steps it took after its founding was the adoption by its member states in 1948 of the "Universal Declaration of Human Rights" (UDHR; UN General Assembly [UNGA] 1948). This is a remarkable and powerful document, adopted at a time of hope and in remembrance of the devastation of the war that had just ended. The prevalent idea of the UDHR was that by meeting the basic rights of all people in the world, one could not only ensure their dignity and improve their economic wellbeing but also help to prevent another global war. The UDHR is, in essence, the moral charter of the United Nations. (The United Nations has a legal charter that defines how the institution functions, which itself is an extremely important document of international law and practice.) The UDHR is the moral heart and soul of the United Nations and still offers inspiration and guidance to the world well over half a century after its adoption.

In 1948, the governments of the world agreed to the following in the UDHR:

> THE GENERAL ASSEMBLY proclaims THIS UNIVERSAL DECLARATION OF HUMAN RIGHTS as a common standard of achievement for all peoples and all nations, to the end that every individual and every organ of society, keeping this Declaration constantly in mind, shall strive by teaching and education to promote respect for these rights and freedoms and by progressive measures, national and international, to secure their universal and effective recognition and observance, both among the peoples of Member States themselves and among the peoples of territories under their jurisdiction. (UNGA 1948)

This preamble says that all of the member states should strive to teach, advocate, honor, and progressively achieve all of the rights of the declaration itself.

There are many rights in the document, and these demand serious study, but a few are worth highlighting. Article 22 calls for the right to social security; in other words, to a guaranteed income base that maintains human dignity and that allows individuals to meet the most basic human needs of water, shelter, clothing, and so on. Article 23 of the UDHR calls for the right to work and to a livelihood that enables individuals to support themselves and their families. Article 24 calls for the right to rest and leisure, so that one's employer cannot demand work around the clock or in burdensome and crushing conditions. Article 25 states that there is a universal right to a standard of living that is adequate for the health and wellbeing of the individual and of their family. The elements of the standard of living include food; clothing; housing; medical care; necessary social services;

and the right to security in the event of unemployment, sickness, disability, widowhood, old age, or any other lack of livelihood in uncontrollable circumstances. Additionally, mothers and children are entitled to special care and assistance. Article 26 holds that everyone has a right to education, which shall be free, at least in the elementary and fundamental stages. Elementary education shall be compulsory, a merit good, and should apply to everyone in the world.

Article 28 holds that "everyone is entitled to a social and international order in which the rights and freedoms set forth in this Declaration can be fully realized." In other words, the UDHR is not meant to be merely a statement of wishes but also a call for a political and social order in which the enumerated rights can be *progressively realized*. This is a key point. One could still be cynical and say the rights are just words on a page that governments do not have to implement. But it really does say more than that, and the effects have been powerful. What the document is saying is that not only do individuals have rights, but individuals collectively, in our nations and in a grouping of nations at the United Nations, also have the right to a system of government, of taxation, of spending and delivery of services in which the declared rights and freedoms can be fully realized.

Coming out of the UDHR were two more detailed international covenants that also helped to implement it. One is the International Covenant on Civil and Political Rights (ICCPR), and the other is the International Covenant on Economic, Social and Cultural Rights (ICESCR). These two covenants, adopted in 1966, cover the five main areas of human rights.

The ICCPR focuses on the rights of citizenship and protection from abuses of the state. Here are a few key points of the civil and political rights defined in the ICCPR (UNGA 1966a). Article 6, for example, holds that the law protects the right to life. Government cannot willfully and without due process take a life. Of course, in many countries of the world, capital punishment is illegal and was eliminated long ago. Yet many governments tragically continue to kill their own people with brutality and lawlessness.

Article 7 holds that no one can be subjected to torture. Yet governments have also violated this willfully, and torture still remains a terrible scourge. The United States itself engaged in waterboarding after the September 11 attacks. Article 8 holds that no person can be held in slavery. While slavery is illegal throughout the world, there is still human trafficking and people are illegally held in slavery. Governments have the responsibility to fight this and to free people who are being held against their will. Article 9 recognizes the right to liberty and security of individuals that governments must not violate. Article 16 recognizes the right of citizenship, defined as the recognition of all persons before the law. Article 18 declares the freedom of thought. Article 24 underscores the protection of children. Article 26 emphasizes the equal protection of the law from discriminatory application.

These rights are not realized to the extent that they ought to be, but they are still widely accepted in principle and widely championed by human rights advocates around the world. This is a start upon which the full realization of rights can be built.

Companion to the civil and political rights are economic, social, and cultural rights, adopted at the same time in a complimentary international covenant (UNGA 1966b). As in the UDHR, article 6 of the ICESCR recognizes for the right to work. Article 7 recognizes the right to just and favorable conditions of work, decent remuneration, and a safe working environment. Article 8 declares the right of individuals to form and join trade unions. This is obviously a right that is still not accorded to workers in many countries. Article 9 of the ICESCR calls for the right to social security. Article 11 calls for the right to an adequate standard of living, in which basic needs are met.

Article 12 is very interesting and relevant for the Millennium Development Goals (MDGs) and the Sustainable Development Goals. This article recognizes the right to "the highest attainable standard of physical and mental health." One cannot grant the right to health per se, because some people may suffer from conditions that are neither preventable nor treatable. Yet one can recognize the right to the highest attainable standard of health. Article 13 calls for the right to education. Article 15 calls for the right to take part in cultural life.

As with the original UDHR, the ICESCR realizes that the economic, social, and cultural rights will only be realized over time, in part as countries achieve economic growth and sustainable development. The standard is not the immediate realization of all rights but the progressive realization of the rights. Even though these rights are therefore not achieved at a moment of time, they still stand as a beacon, an inspiration, a standard, and a goal.

Most governments subscribe to both of the covenants. The United States did not subscribe to the ICESCR, and it is interesting to note why. While many Americans certainly agree with the ICESCR, the libertarian streak in U.S. public opinion does not accept the role of the state in meeting economic objectives. The libertarians argue that governments should let markets and individuals transact freely and that it is not the role of the state to implement an ethical framework beyond that of the protection of the individual and private property. Of course, most other ethical traditions reject this libertarian point of view, and libertarianism is a philosophical tradition mainly popular in the United States and a few other countries of Anglo-Saxon heritage (the United Kingdom and also some groups in Canada, Australia, and New Zealand).

In addition to the ICESCR, the United Nations has also adopted many more *specific* goals around meeting basic needs. The most important of these have been the Millennium Development Goals, or MDGs, which were adopted in September 2000 (UNGA 2000). The MDGs also took their inspiration from the UDHR and the international covenants, because the goal of the MDGs is to implement the human right to meet basic needs in all major dimensions (income, food, education, jobs, health, and safe living conditions). When the MDGs were launched in September 2000 as part of "United Nations Millennium Declaration," the UN member governments agreed that they would "spare no effort to free their fellow men, women and children from the abject and dehumanizing conditions of extreme poverty to which more than a billion are currently subjected." They committed to making the right to development a reality for everyone and to freeing the entire human race from want. Human rights were therefore at the core of the MDG agenda and remain at the moral heart of the United Nations and the new era of Sustainable Development Goals.

III. Divided Societies

Social inclusion aims for broad-based prosperity, for eliminating discrimination, for equal protection under the laws, for enabling everybody to meet basic needs, and for high social mobility (meaning that a child born into poverty has a reasonable chance to escape from poverty). Of course, no society achieves uniform equality and nobody would want that. People differ by luck, tastes, and work effort, so naturally these differences show up in variations in economic outcomes as well, such as in income, wealth, and job status. People expect these differences to a reasonable extent. But societies can have differences in economic outcomes far beyond anything that would naturally emerge from the normal range of differences between individuals, luck, and circumstances. And when people are not able to look after themselves by dint of nature or circumstance, we also expect (according to most ethical systems) that they will be helped by society to meet their needs with dignity.

Let us consider the evidence on the inequalities of income in different parts of the world. For that we turn to the Gini coefficient, a statistical measure of income inequality that varies between 0 and 1. A Gini of 0 signifies a society where every person has exactly the same income (that is, complete equality); a Gini of 1.0 signifies a society where one single individual holds all the income and the rest have none (that is, complete inequality). As we see in figure 2.5, there are some notable and systematic differences across the regions of the world in terms of the extent of inequality as measured by the Gini coefficient. On the whole, with the notable exception of Canada, the Americas are a part of the world with relatively *unequal* incomes, that is, with relatively high Gini coefficients. Western Europe, by contrast, is a region with relatively equal income distributions and therefore has relatively low Gini coefficients. Northern Europe, especially the Nordic countries (Norway, Sweden, Finland, Denmark, and Iceland), is home to the most equal societies in the world measured by the Gini coefficient.

There are long histories and legacies that have led to these various degrees of inequality. Consider the case of the Americas (with Canada the exception). Until Christopher Columbus united Europe and the Americas through trade and settlements after 1492, the Americas had been largely isolated from the Old World population. The original Amerindians had arrived around 15,000 years ago over the Bering land bridge that connected Asia and North America during the Ice Age (at a time when sea levels were much lower due to vast stores of water being held in the glaciers that covered the Northern Hemisphere). When the Ice Age ended and ocean levels rose, these Amerindian populations found themselves cut off from the Old World. The links would be restored only with Columbus's voyages, putting aside the brief sojourn of Leif Eriksson on the coast of Canada sometime around 1,000 C.E.

Columbus and the waves of Europeans who followed in the late fifteenth century and later had two huge advantages. The Europeans had far more powerful weapons, and they had pathogens they carried from the Old World that spread in epidemic waves throughout the unprotected Native American populations, leading to mass deaths. The Americas thereby developed as "conquest societies," in which Europeans dominated indigenous populations. Europeans also added one more crucial segment of society to the Americas starting in the sixteenth century: African slave labor, brought from West Africa mainly to the tropical and subtropical regions of the Americas, such as the Caribbean, the American South, and northeast Brazil. Slave labor was not brought in large numbers to Canada, which is one of the reasons why Canada is so different today in terms of inequality.

What emerged in the Americas was therefore a very complicated society of groups of widely varying wealth and power. A small and dominant European group ruled by force and imperial power over both the indigenous and slave populations of the New World. Over time, these three societal groups—Europeans, indigenous Amerindians, and African slaves—became far more complicated through the mixing of the three populations. So-called mixed populations of African slaves, European conquerors, and indigenous populations became numerous in many American societies.

From the sixteenth century onward, these groups faced completely divergent situations from the political, social, and economics standpoints. The Americas developed into a region of vast inequalities of wealth and power, with the small European populations tending to own the good land, while the native population was pushed into smaller land areas and deprived of traditional landholdings, often through brutal and insidious means. Very often the "law" of the Europeans was used to dispossess the indigenous populations, since European-based law did not recognize traditional communal rights to the land.

The violent and conflict-ridden history of the Americas thereby created huge and shocking inequalities. That legacy of inequality has, rather remarkably, continued to the present day. Slavery was thankfully eliminated throughout the

Americas in the nineteenth century, usually by law or decree, although in the United States only by civil war, and in Cuba and Brazil only as late as 1886 and 1888, respectively. Yet slavery's powerful legacy in the hemisphere continued into the twentieth century, as the descendants of slaves generally faced massive liabilities of poverty, poor health, lack of rights, and pervasive social discrimination and violence.

The dark red of the Americas in the Gini map (figure 2.5) reflects the long legacy of societies created by European conquest. This kind of historical perspective is necessary to understand the forces of inequality today in various parts of the world. In many places, the inequalities created in earlier centuries continue to cast a shadow. Intergenerational dynamics of income, in which poverty in one generation leads to poverty in the next, plays a huge role. Even where slavery ends, the legacy of slavery continues. Even when the massive brutality against Native Americans in the United States or the so-called First Nations in Canada ended by law, the legacy of poverty, poor health, discrimination in property holdings, and so forth, continues. Social discrimination, racial discrimination, and ethnic discrimination continue, alongside gender discrimination. Regional differences also persist, sometimes for geographical reasons such as climate, distance, and transport costs, yet sometimes as a social legacy as well.

I have already mentioned the shocking phenomenon across the world, that indigenous populations face wanton discrimination, usually building on brutal land confiscations and treatment in the past (UNPFII 2009). These indigenous populations are a window onto the human penchant for brutality against groups they do not understand, and groups that threaten their own wellbeing (or at least seem to do so). Indigenous populations around the world held land that was coveted by those who came later. And the newcomers typically lost no time in brutally expelling the indigenous groups from their native lands. These expulsions were backed by law, power, politics, and self-serving cultural stereotypes (typically that the natives were less than human).

There is generally a long-lasting legacy of this brutal state policy. The indigenous groups have often been forced to settle in marginal lands, such as forests, mountains, or desert regions. One common mistake is to assume that an indigenous group that inhabits a difficult environment (e.g., a desert region) does so as a matter of cultural tradition. More typically, the tribe has faced a displacement from more favorable lands, often a few generations back; a fact now forgotten by the general population because it is not discussed or treated in the history books. Maybe the group was once in the most fertile lands, but then when later waves of migration arrived, the indigenous groups were forced into the marginal lands.

The poverty rates of indigenous populations are very high around the world. This is not a small part of the overall global extreme poverty. Nobody has a precise count of indigenous populations, but an approximate estimate is around 370 million people, nearly 6 percent of the world's population (UNPFII 2009, 1). Since many of those 370 million live in extreme poverty and abject social exclusion, this is still a large portion of the global population to consider.

Figure 7.3 shows the poverty rates in some relatively poor countries in the Americas—Bolivia, Ecuador, Guatemala, Mexico, and Peru. In each case, the poverty rates among the indigenous populations (shown on the top) are higher than the poverty rates among the non-indigenous population (shown on the bottom). If one were to show the poverty rates of African Americans in the United States as well, those would be much higher than the white population. This is an example of how the indigenous populations in the Americas face continuing challenges from their long legacies of political, military, and cultural discrimination and violence.

The map in figure 7.4 of ethnolinguistic groups highlights another key point: around the world there are significant differences in the extent of ethnic diversity. Ethnic diversity is sometimes measured along linguistic lines. A map like this gives an indication of what is called ethnolinguistic fractionalization, a measure of similarity or difference in the spoken languages in a population. When fractionalization is high, inequality is often high as well, with some groups dominating others politically and economically.

In addition to inequalities across groups, there are also inequalities across individuals. There are inevitable variations across individuals in their work efforts, capacities, and luck, and these too create some degree of income inequality in any society. Yet the extent of inequality that results from such differences also depends on public policies. Does the state help every family to meet its basic needs? Does

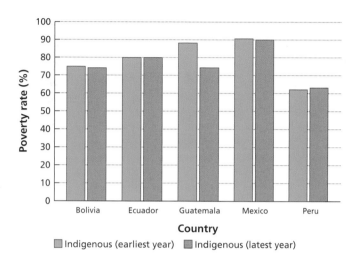

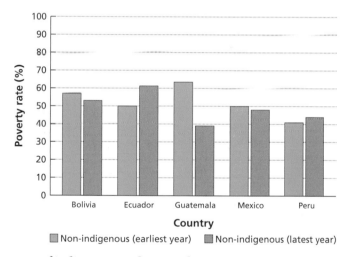

7.3 Poverty rates of indigenous and non-indigenous populations in Latin America (1980s to 2000s)

Source: Patrinos, Harry Anthony, and Emmanuel Skoufias. 2007. "Economic Opportunities for Indigenous Peoples in Latin America: Conference Edition." Washington, DC: World Bank. https://openknowledge.worldbank.org /handle/10986/8019. License: CC BY 3.0 unported.

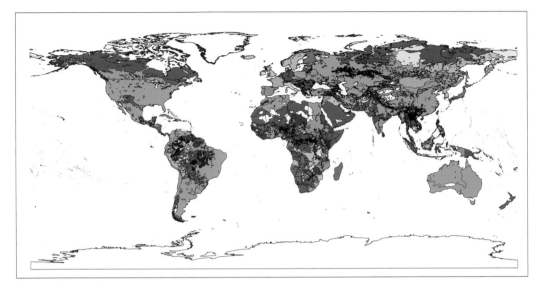

7.4 Global map of ethnolinguistic groups

Weidmann, Nils B., Jan Ketil Rød, and Lars-Erik Cederman. 2010. "Representing Groups in Space: A New Dataset."
Journal of Peace Research 47(4): 491–499.

the state ensure education for all children, even those from very poor families? Is the quality of education that the poor can access high enough to provide the basis for social mobility, or are the educational services for the poor so weak that children raised in poverty get trapped in an intergenerational cycle of poverty? The answers to these questions differ across societies.

This discussion highlights that we must address the challenges of social inequality and human rights across several dimensions. Race, ethnicity, power, conquest, and individual characteristics are all determinants of inequality in society. So too are the political responses, the extent to which power is used to reduce inequalities (e.g., through tax-and-transfer policies) or the extent to which power is used to exacerbate inequalities (e.g., through displacing indigenous populations from traditional lands). Inequality is therefore a legacy of power, history, economy, and individual differences, amplified or diminished through the powers of the state.

IV. Forces of Widening Inequalities

During the past twenty years, income inequality has risen markedly in the United States and many other countries. Income per capita has been rising, but most of the gains have accrued to those at the top of the income distribution. The U.S. Gini coefficient has risen from around 0.40 to 0.48 between 1970 and 2013, a very sizable increase.

There are at least three fundamental forces playing a role in the widening economic inequalities in the United States, several European countries, and many of the emerging economies around the world. One key factor is the rising gap in earnings between high-skilled and low-skilled workers. The returns to education have increased markedly, leaving those with less education behind. The rising earnings premium to education probably reflects the combined forces of globalization and technological changes, both of which have been to the disadvantage of less-educated workers. A second phenomenon has been the increased use of robotics, advanced data management systems, and other information technologies, which seem to be shifting income from labor to capital. Since capital ownership is highly concentrated among wealthy households, the shift from wages to capital income has also widened the inequalities of income across households. The third force has been the political system, which in the United States has amplified the widening inequalities caused by market forces. For example, wealthy campaign contributors have been able to use their political influence to get special privileges in the form of tax breaks, subsidies, or advantageous regulatory changes.

A useful starting point to understand these changes is the frequently viewed graph in figure 7.5, which shows the premium in income (the extra income) earned by college graduates in the United States relative to high school graduates. In 1973, a college graduate had a 30 percent premium relative to a holder of a high school diploma. That premium actually dipped during the 1970s to under 25 percent. However, from roughly 1979 onward, the college wage premium soared and has increased to around 45 percent. It is interesting that that increase started around 1979, because that is also a period when many powerful forces of globalization linked the high-income world with the emerging economies.

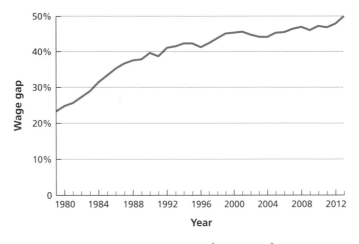

7.5 The college vs. high school wage premium (1979–2013)

Source: Economic Policy Institute Briefing Paper No. 378.

The globalization and economic integration that began in the late 1970s resulted in the globalization of production systems. Companies began to move work offshore to China, Mexico, and other locations with lower-wage workers. The jobs formerly held by American workers with relatively less education (e.g., a high school diploma rather than a college degree) were easier to shift abroad. The U.S. labor force in the manufacturing sector began to shrink notably as jobs moved toward Asia and the Caribbean basin. This hit lower-skilled workers hard as their jobs disappeared or could be saved only through wage cuts.

The United States had its peak of employment in the manufacturing sector in 1979 at around 19 million workers. From 1979 onward, there was a significant decline of manufacturing sector employment due to the shift of work to lower-wage economies. For many low-wage economies, and notably China, there was a huge benefit. A key part of their rapid economic growth was due to this gain in jobs from the United States and Europe. But for the less-educated workers in the United States, there was pain and retrenchment. As of today, there are around 12 million jobs still left in U.S. manufacturing.

Aside from globalization, another force at play is the information revolution and the progressive automation of many production processes. It is another force leading to higher productivity but also to fewer jobs in traditional manufacturing sectors. Technological change has dramatically remade the shop floor of major manufacturing sectors. Robotics is one of the most visible and remarkable aspects of the information technology and computerization revolution of recent years. Robotics can raise productivity dramatically and lower the costs of many goods and services; but just like the offshoring of jobs, robotics can take a toll on workers who used to carry out the kind of labor that can now be automated.

The third major factor is politics. When market inequality starts to rise, politics also enters the scene. In some political systems, government forces resist the widening inequality by providing extra help to lower-skilled workers, such as job training, tax cuts, or added family benefits. These countries may call on the higher-skilled workers to take on some extra societal responsibilities, such as increased tax payments to support the transfers to the lower-earning households. In this way, governments "lean against the wind" to narrow the inequalities caused by market forces.

Yet in some places, including the United States, political forces have tended to amplify rather than lean against the market forces. In the United States, this perverse political practice has occurred since the 1980s. Ronald Reagan entered office in January 1981 with the support of libertarian political forces. These political forces encouraged the cutting of federal government services, the reduction of federal taxes, and the substantial deregulation of the business sector, including Wall Street. The evidence shows that U.S. inequality started to rise significantly in this period, with the trend persisting till today.

With the support of deregulation, especially in the financial sector, and a weakening of trade unions (in part as the result of globalization), U.S. chief executive officers (CEOs) began to pay themselves mega-salaries. Typically a "compensation committee" of corporate board members signs off on the salary increase. Yet the CEO often handpicks the compensation committee, which in turn sets the CEO salary. The result is seen in figure 7.6, which shows the ratio of compensation of American CEOs to average workers. In the 1970s, the CEO compensation

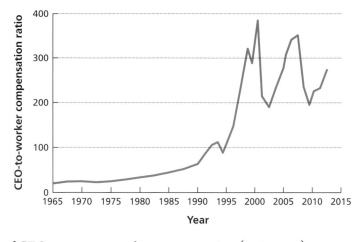

7.6 Ratio of CEO to average worker compensation (1965–2012)

Source: Economic Policy Institute Briefing Paper No. 367.

was roughly twenty times that of average workers, but then the CEO compensation packages exploded. With deregulation and other constraints on the CEOs weakened, the CEOs began to receive compensation packages that were literally hundreds of times the income of the average worker.

The result of this skewed compensation is shown in figure 7.7, which illustrates the income share taken home by the top 1 percent of America's richest households. It shows that for a century up to 1929, when the Great Depression started, the income share was between 15–20 percent. Then came the Great Depression and the New Deal. Government cracked down on many abuses in the financial sector, and the income share of the top 1 percent went down sharply. Tax rates for the top earners also went up sharply, discouraging CEOs from pursuing mega-compensation. From the 1940s to the 1970s, the income share of the top 1 percent was around 10 percent of total household income. But then the income share of the top 1 percent turns up steeply.

Market forces cannot be solely responsible for the rising share of the top 1 percent, because this gain at the top in the United States was not matched (at least in extent) in other high-income countries. It is politics in addition to market forces

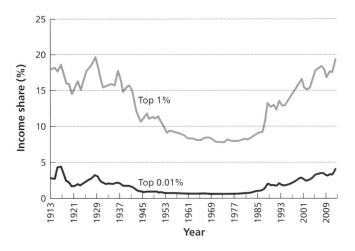

7.7 Income share of top 1 percent and top .01 percent, United States

Source: Alvaredo, Facundo, Anthony B. Atkinson, Thomas Piketty, and Emmanuel Saez. 20/05/2014. "The World Top Incomes Database." http://topincomes.g-mond.parisschoolofeconomics.eu/.

that account for this remarkable, and troubling, trajectory of inequality. What is remarkable is that the top 0.01 percent of U.S. households, just 12,000 households in total out of 120 million households, now takes home around 5 percent of the total income of American households, up from 1–2 percent in the 1970s.

With the rise of income inequality in the United States, social mobility has also declined. Poor children are growing up to be poor adults. The children of poor households simply can't break the poverty trap. Many drop out of school early, and few have the means to complete a four-year college education.

There are policies available to reestablish social mobility and social inclusion, but the political system would have to cooperate to make those policies effective. Such policies would include: tax increases on the richest households; greater social support for low-income households; increased training and educational opportunities for the poor; greater regulation of CEO compensation; and a crackdown on various tax breaks and offshore tax havens. So far the U.S. political system has remained impervious to such policies and instead has continued to cater mainly to the interests of the richest households and companies.

V. Gender Inequality

Gender inequality has been a long-standing feature of most societies around the world. Men have been in the paid labor force while women have traditionally carried out farm labor and home-based production while also raising the children. Laws and social customs bolstered this traditional division of labor, often making it impossible for women to own businesses or control their own incomes.

Fortunately, these long-standing gender inequalities are diminishing rapidly in many parts of the world. Indeed, I have seen such dramatic social changes from my own experience working in poor rural areas with traditional patriarchal (male-dominated) cultures. Even in such rural settings, girls are beginning to stay in school in larger numbers, and a few are graduating university and taking professional roles in their communities. In turn, these pioneering young women are changing mind-sets and serving as life examples for the younger girls following in their tracks.

We are therefore in a period of flux, where age-old practices of discrimination against females are changing, ideas are changing, economic demands are changing, and economic possibilities for girls and women are moving in the right direction. Sustainable development can help that process by promoting legal and administrative changes that empower girls and women. In turn, as girls stay in school longer and then enter the labor market with higher skills, many other benefits to development are soon realized. Fertility rates go down sharply (as girls marry later and also remain in the labor force). Households with fewer children invest more per child in education, health, and nutrition. Girls enter the labor force, providing a rising share of the working-age population in gainful employment.

The MDGs took on gender equality in a strong and direct way. MDG 3 calls on societies to promote gender equality and empower women. The specific target attached to MDG 3 is in the education sector. The goal calls on each country to eliminate the gender disparity in enrollments and graduation rates by the year 2015. This is occurring throughout the world.

Many social benefits will be achieved as countries fulfill MDG 3. First, of course, is the human right of women to equality and economic opportunity. In the UDHR, the ICESCR, and the ICCPR, the rights of girls and women are protected

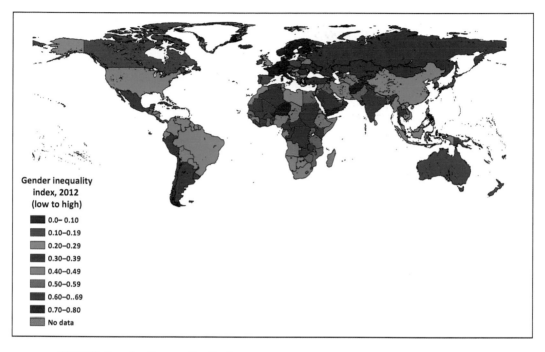

7.8 UNDP Gender Inequality Index

Source: United Nations Development Programme. 2013. Human Development Report 2013. *New York: United Nations Development Programme.*

as clearly as the rights of boys and men. Gender equality has been a key human right from the very start.

The economic gains are also enormous. The direct gains, of course, are the benefits of more educated women in the labor force. The indirect gains involve the children of those women, who are likely to grow up healthier, better nourished, and with a much greater chance of succeeding in school. Investing in a girl's education is also investing in breaking the intergenerational poverty trap.

In 2010, the United Nations Development Programme introduced a Gender Inequality Index (GII), shown in figure 7.8 (UNDP 2013a). Notice the especially high gender inequality (meaning adverse conditions for women) in tropical Africa and South Asia. In these two regions, women still lack political power and social standing.

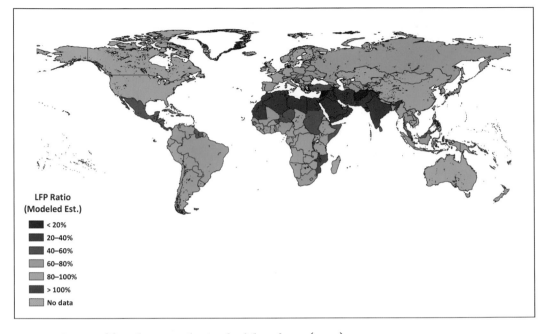

7.9 Ratio of females to males in the labor force (2012)

Source: World Bank. 2014. "World Development Indicators."

Like the Human Development Index, the Gender Inequality Index combines several indicators on a weighted basis to offer a quantified assessment of gender inequality in each country. This GII includes three categories. The first is *reproductive health,* including the maternal mortality rate (the rate at which mothers die in pregnancy or childbirth) and the adolescent fertility rate (the rate at which adolescent girls are bearing children). The latter rate is an indication of the extent to which girls are being forced out of their education and into early marriage and childbearing. The second category is *female empowerment,* measured by the share of total parliamentary seats held by women and by the enrollment rate of women in higher education. The third category is the *labor force participation of women,* depicted in figure 7.9 as the ratio of women to men in the labor force. Note how the countries of North Africa, the Middle East, and South Asia have especially

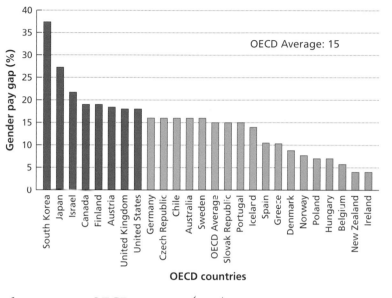

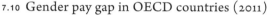

7.10 Gender pay gap in OECD countries (2011)

OECD (2014), OECD Family Database, OECD, Paris. http://www.oecd.org/social/family/database.

low ratios of women to men in the labor force. In many countries in these two regions, cultural practices discourage women from participating in employment outside the home.

Even in the high-income countries, where female labor force participation rates have risen markedly in the past thirty years, there are still significant gaps between the earnings of men and women, as shown in figure 7.10. Women may take time out from the labor market to raise young children, which can cause them to suffer a loss of earnings. Companies may discriminate and not invest in the skills or the promotion of women, because they believe the women will leave to raise families. Another reason is simply because of social norms. And, of course, a glass ceiling still exists in many cases; even with legal barriers eliminated, the old boys' club and the informal rules of the game still discriminate heavily against women.

However, there is important progress in gender equality, even in the poorest countries. The MDGs have helped narrow the gender gap, especially in terms of

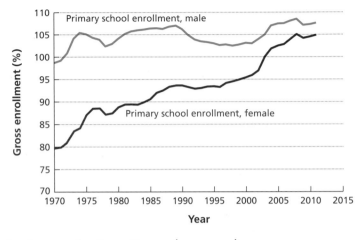

7.11 Global primary school enrollment (1970–2011)

Source: World Bank. 2014. "World Development Indicators."

education. At the primary education level, the enrollment gap between boys and girls has nearly closed, as we see in figure 7.11. In some parts of the world, girls' enrollments rates have actually overtaken those of boys at the secondary and tertiary levels. Latin America, which had a tradition and a reputation as a machismo, male-oriented society, is a place where girls have done very well in secondary and tertiary education.

What can and should be done to close the remaining gender gap? It is clear that, as always, a multidimensional approach should be deployed (Sustainable Development Solutions Network Thematic Group on Challenges of Social Inclusion 2013). Some of the remaining gender barriers are legal, others are cultural, and still others are a matter of tradition and inertia. Legal reform is often a remaining priority. Can women own and run businesses? Can they own and inherit property? Shockingly the answer is often "no," even today. What about the delivery of public services? Are girls and women receiving the public services they need? For example, are hygienic facilities for girls at secondary level available?

As for female representation in politics, some societies have added quotas that require a certain proportion of the votes on a party's proportional representation list to be occupied by women. When women enter the parliament and government there can be a very powerful effect on changing ideas, norms, and policies. Government financial support for maternity leave and child care can also play a large facilitating role to help women in the labor force. The Scandinavian countries have been especially successful and innovative in this regard. The rest of the world can learn from their example. In addition, early childhood development (ECD) programs, including high-quality child care, preschool, health care, and nutrition programs for young children, give an enormous boost to gender equality and to the wellbeing of young children and their mothers.

Finally, it bears emphasizing that around the world countless women face a terrible, unspoken burden of violence, whether it is rape, husbands beating wives, or other violent acts. UNICEF and other UN agencies have undertaken major efforts to bring public awareness to this kind of violence. Brave activists are pressing for better law enforcement, laws, and public leadership. In many societies, women have been viewed as if they were the mere property of their husbands. This inhumane approach is a fundamental denial of human rights and should come quickly to an end.

8

EDUCATION FOR ALL

I. Life Cycle Approach to Human Development

Economic development depends on investment. Countries achieve economic growth when they have the roads, ports, railways, fiber-optic systems, and power grids that give them the basis for developing industry and expanding the economy. Investments in all these types of infrastructure and industry are crucial. But the most important kind of investment that countries make is in their own people, especially investment in their children. Economists have come to use the language of investment when talking about education, health care, nutrition, and the other inputs to a healthy productive life. Economists speak of investments in "human capital," just like investments in the physical capital of roads and bridges. And just as a single business or an entire economy can accumulate physical capital, so too can an individual or an entire society accumulate human capital, meaning more education, job skills, improved health, and the like.

The concept of human development includes two related ideas. The first is the important fact that the abilities and health of an individual depend on a cumulative process, of good health and access to health care, living in a safe environment, education, building skills, and on-the-job experience (Sustainable Development Solutions Network [SDSN] Thematic Group on Early Childhood Development, Education and Transition to Work 2014).

The evidence is quite strong that as individuals accumulate more education, more on-the-job training, and more work experience, their productivity in the labor force (as indicated most directly by earnings) also rises. Similarly, investments in health accumulate. Investments in a child's health help to provide the basis for health as an adult. In this process of accumulating human capital, certain periods of life are most crucial, starting with neonatal survival within the first few weeks of life. Early childhood is extraordinarily important, not only because it is the time when we learn many of the necessary social and human skills, but because it is also the time of the formation of the brain itself.

The second important idea of human development is of the individual "life cycle." We should consider an individual over an entire life span. An individual's capacities, health, and productivity at any stage of the life cycle depend on the choices that are made at earlier stages of that life cycle (National Scientific Council on the Developing Child/National Forum on Early Childhood Policy and Programs [NSCDC/NFECD] 2010). Each stage of the life cycle sets up the conditions for the stages that follow. Even the health of parents before conception can be important. Bad health and poor nutrition can actually transmit across generations in addition to genetic influences, in a phenomenon called "epigenetics." The safety of the mother in pregnancy, a healthy childbirth, an infant's good health, and proper nutrition and deworming as necessary for young children are all extraordinarily important not only for an individual's survival through the health challenges of childhood but also for an individual's productive and prosperous life as an adult.

A child's cognitive development begins at an early age. Brain development in infancy depends on a loving and low-stress environment, bonding with caregivers, and hearing many words spoken by the caregivers. The formal education process, we now understand, should begin even before the start of primary school. More and more countries are introducing pre-kindergarten classes for young children as a way to bolster early learning and healthy brain development. From there, primary education for all children is now globally recognized to be a basic need and a basic right of all children, and is enshrined as Millennium Development Goal 2 (MDG 2). Yet in the twenty-first century world economy, universal primary education is surely not enough. All children need a secondary education

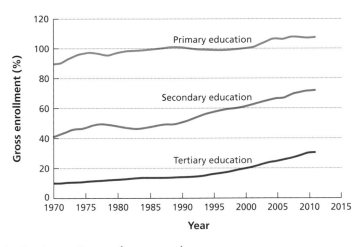

8.1 Global school enrollment (1970–2011)

Source: World Bank. 2014. "World Development Indicators."

followed by some form of vocational (skills) training or tertiary education. It is the very nature of our technological age that young people today will on average require more skills and training than did their parents.

Beyond secondary education there will be many tracks to job skills and further training. These might include vocational schools to learn a specific skill or higher education in the form of two-year programs (typically called an "associate's degree" in the United States) or four-year programs (typically called a "bachelor's degree"). With the advent of online education, more and more training may come from downloadable, freely available courses. And online education is likely to promote continuing education through adulthood as a lifelong strategy.

There has been recent significant progress in achieving MDG 2, which calls for universal access to primary education (World Bank 2014a). The data in figure 8.1 show that as of 2010, the *gross enrollment rate* at the primary level is above 100 percent. What does it mean to have a gross enrollment rate of more than 100 percent? The gross enrollment rate measures the number of children *of any age* attending primary education divided by the population size of primary-school-aged children. Since some older children (outside of the primary-school-age group) are

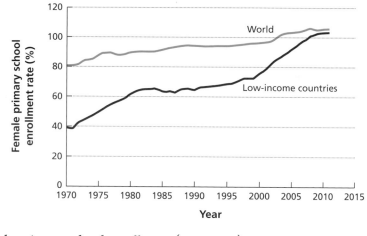

8.2 Female primary school enrollment (1970–2011)

Source: World Bank. 2014. "World Development Indicators."

sometimes enrolled in primary school (because they started late or were held back), there can actually be more children enrolled in primary school than there are children of primary school age. Back in 1970, the gross enrollment rate was around 85 percent. With the extra push of the MDGs, this has increased to more than 100 percent.

This very high primary school enrollment rate is also extremely important in terms of MDG 3, which calls for gender equality for education. Girls throughout the world at the primary school age are now by and large attending primary school. As we see in figure 8.2, as recently as 1990, the gross enrollment rate for girls was still only around 60 percent, compared with 90 percent for boys. As of 2010, the gender gap at the primary level has essentially closed.

While there has been great progress made at the primary school level, the progress in educational enrollment and attainment is much less at the secondary level and above. Figure 8.3 shows a world map of secondary education rates. Much of the world now has relatively high secondary school enrollment, but in tropical Africa and in parts of Asia where extreme poverty persists, secondary education levels remain inadequate. The sustainable development

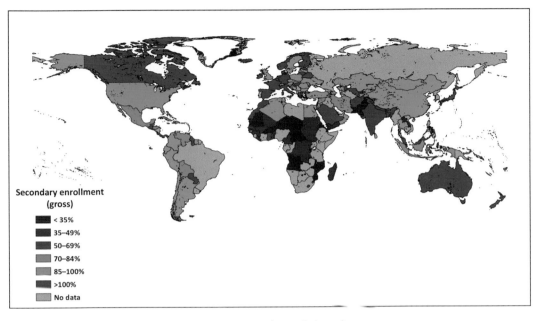

8.3 Global secondary school enrollment (gross) (2011)

Source: World Bank. 2014. "World Development Indicators."

goals for the period 2015–2030 should focus on ensuring universal secondary education as well as access to job-skills training beyond secondary education (SDSN Thematic Group on Early Childhood Development, Education and Transition to Work 2014a).

The situation with respect to higher levels of education is even more varied (World Bank 2014a). The poorest countries in the world still have very low tertiary education levels, often 10 percent or below, as shown in figure 8.4. The lack of adequate enrollment in higher education is becoming a major impediment to the economic progress of the low-income countries. At this stage of global economic development, every economy, rich or poor, needs a significant cohort of university graduates. The tertiary institutions are necessary to ensure that there are qualified teachers, sufficient numbers of technical workers, and a generation of skilled young people trained in public policy and sustainable development.

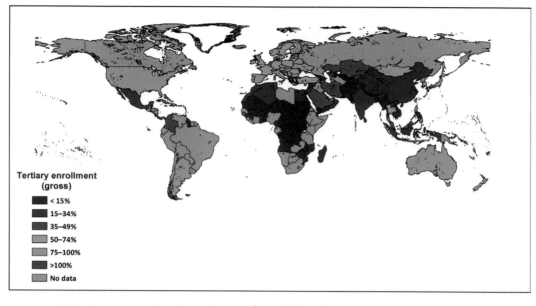

8.4 Global tertiary school enrollment (gross) (2011)

Source: World Bank. 2014. "World Development Indicators."

II. Early Childhood Development

One of the most significant advances in the understanding of human capital over the life cycle has been in the area of early childhood development (ECD). Twenty or thirty years ago, most of the focus was on the formal public education system, with little understanding of the crucial importance of the preschool environment, including the health, nutrition, physical safety, and preschool preparation of children ages 0–6. Research over the past twenty years has shown the startlingly important effects of early childhood, especially during the first three years, when the brain develops in many dynamic and important ways. If those first three years are a period of excessive environmental stress (e.g., a household marked by violence, noise, and lack of security), repeated illnesses or undernutrition, or the lack of adequate cognitive stimulus and educational preparation, a young child will likely incur liabilities that may be impossible to overcome during school years or later.

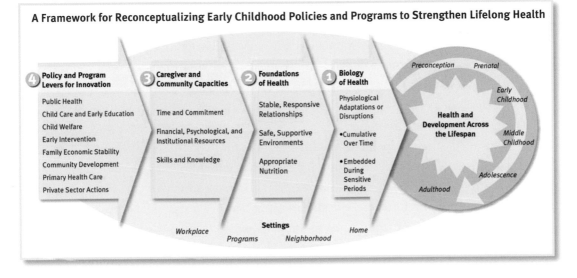

A Framework for Reconceptualizing Early Childhood Policies and Programs to Strengthen Lifelong Health

8.5 Health and development across the lifespan

Credit: Center on the Developing Child at Harvard University.

Investing in the early health, wellbeing, safe environment, and cognitive development of young children is therefore crucial for children's subsequent development. This is an area in which the concept of cumulative investment is essential. Scientists studying brain development of young children and overall physiological development have come to the conclusion that the cumulative amount of stress that a young child faces will shape the child's cognitive and physical development and conditions as an adult (NSCDC/NFECD 2010). If there is an overaccumulation of stress, the subsequent changes to the body's biophysical pathways can lead to a lifetime of physical and mental difficulties. This is shown in a graphic from the Harvard University Center on the Developing Child (figure 8.5), which emphasizes that the biology of health is cumulative over time and embedded during very sensitive periods, especially during brain formation.

Several studies have tried to take a look at the overall evidence on what great exposure to risk during this period really means in terms of raising a healthy child. The conclusion summarized in an important 2011 study in the *Lancet* emphasizes that exposure to biological and psychosocial risks, such as being in an unsafe (perhaps violent or noisy) environment, can affect brain development and compromise the subsequent development of the child both cognitively and physically (Walker et al. 2011). Inequalities of childhood development start at a very young age. By age six or seven, a child raised in an unsafe environment will already have huge disabilities and liabilities relative to those children fortunate to be raised in a safe and secure environment.

This means that reducing inequalities across children requires integrated and very early interventions in ensuring a safe environment, in cognitive development, in preschool learning, and in proper nutrition and health care for young children. The time for those investments is at a very young age, because these investments are very hard to make and much less effective if made as a correction for a child who is already five or six years old. If a child's growth is inhibited early on, the consequences can last for the entire life of that individual and for future generations to come; the consequences for society can be absolutely enormous.

Making those investments, which requires leadership of government, therefore has tremendous social returns in many ways, including raising national income because of the population's improved productivity. A leading economics scholar, Nobel laureate James Heckman, has studied this issue of investing in human capital throughout his career and has made a graph of the kind shown in figure 8.6 (Heckman 2008). On the horizontal axis is age, and on the vertical axis is the rate of return to investing in human capital. Those returns are absolutely the highest at the preschool age. As age increases, the returns that one can achieve by incremental investments in human capital are lower. Missing a year of investment in human capital when a child is two cannot be made up by that same investment when the child is six. The returns are very high when investments are made in the formative years of brain development, in early socialization and development of personality, in nurturing cognition and scholastic aptitude, and in ensuring physical wellbeing; this early intervention cannot simply be replaced by investments later on.

Wealthier families are likely to make the needed investments in the preschool years of their young children. They can bear the cost, and they themselves are more likely to be educated and therefore realize the large benefits of preschool. There are likely to be more books and even more words used in those households,

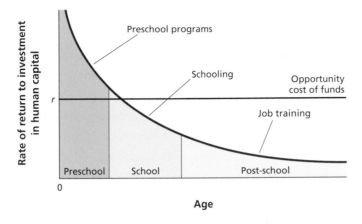

8.6 Rates of return to human capital investment

Source: Heckman, James J. 2006. "Skill Formation and the Economics of Investing in Disadvantaged Children." Science 312(5782): 1900–1902. Reprinted with permission from AAAS.

so children benefit not only from formal preschool away from the home but also from a highly supportive learning environment at home. It is the children of poor families who are likely to suffer the deficits of underinvestment in preschool. Their parents will not be able to afford the market costs of preschool (and will be unable to borrow if necessary to cover those costs). And the children will generally be less well prepared by their home life for learning outside the home.

This pattern suggests that poverty will repeat itself from one generation to the next. Low-income parents will tend to underinvest in their children, who will then grow up to be low-income parents themselves. High-income parents, by contrast, will invest heavily in their children (both at home and at preschool), thereby preparing their children for economic success as adults. The situation would seem to be grim for intergenerational income mobility.

Yet here is where government can play a crucial role. Government programs and financing can help children of impoverished families to get a decent start. Part of the issue is money: transfers to poor families (as are common in Scandinavia, for example) can ensure that poor households, like richer households, have the financial means to provide adequate health care, nutrition, environmental safety, and an enriched cognitive environment (e.g., with books and toys for learning) for young children. Public financing can ensure access to preschool programs that would be out of financial reach for poor families. And government programs can provide targeted support as well, for example, by training parents in parenting skills that can enhance the wellbeing, development, and cognitive enrichment of their children. When parents have little or no education, they are likely to need guidance and support to create an effective learning environment for their children. In careful studies, good parenting skills have been shown to play the important role we would expect in a child's cognitive development.

The upshot is that societies that make substantial investments in the preschool years, usually with ample public financing, end up with more upward social mobility for poor children, and therefore with more inclusive and productive societies. Societies that fail to invest in preschool are likely to have lower social mobility and a greater gap in lifetime attainment between children born to high-income and low-income households. If the government does not add its support to low-income families, poverty is likely to be passed from one generation to the next.

Figure 8.7 hints at this pattern. The figure shows the relationship between two key variables. On the horizontal axis is the proportion of young children in preschool programs. On the vertical axis is the gap in educational attainment of youth ages 15–19 who grow up in the richest 20 percent of households versus those youth who grew up in the poorest 20 percent of households. A point that is high on the vertical axis, say with a gap of four years, signifies a large difference in the school attainment of affluent youth versus poor youth. The graph depicts a *downward-sloping relationship*. Where there is more access to preschool, there is a smaller gap in educational attainment of rich versus poor youth. This is just as we would expect. Preschool offers upward mobility for poorer children, enabling them to come closer to the educational attainment of richer children. Investing in preschool lowers the extent of inequality and in the right way—not by pulling down any child from the top but by raising kids from the bottom who otherwise would not have the educational opportunity they need.

Table 8.1 offers an example of the estimated economic benefits of a specific preschool program in the United States: the Perry Preschool Program (Schweinhart et al. 2005). The Perry Preschool Program offered intensive help to poor children

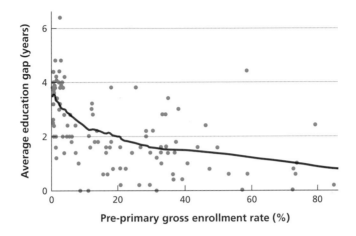

8.7 Preschool enrollment and the gap in educational attainment

Source: Engle, Patrice L., et al. 2011. "Strategies for Reducing Inequalities and Improving Developmental Outcomes for Young Children in Low-Income and Middle-Income Countries." The Lancet. 378(9799): 1339–1353.

Table 8.1 Economic benefits and costs of Perry Preschool Program

Child care	$986
Earnings	$40,537
K–12	$9184
College/adult	$−782
Crime	$94,065
Welfare	$355
Abuse/neglect	$0
Total benefits	$144,345
Total costs	$16,514
Net present value	$127,831
Benefits-to-cost ratio	8.74

Note: All values are discounted at 3 percent and are in 2004 dollars. "Earnings," "Welfare," and "Crime" refer to monetized value of adult outcomes (higher earnings, savings in welfare, and reduced costs of crime). K–12 refers to the savings in remedial schooling. College/adult refers to tuition costs.

From Heckman, James J. "Skill Formation and the Economics of Investing in Disadvantaged Children." *Science* 312(5782): 1900–1902. Reprinted with permission from AAAS.

in preschool. The costs and benefits of this program have been measured with great specificity and demonstrate that the returns to these preschool investments were very high, and in several ways. According to the data, the direct costs of the program per child amounted to $16,514, while the lifetime gains amounted to $127,831 (measured in present-value terms, by discounting future benefits according to the rate of interest). Thus the benefit-to-cost ratio is estimated to be 8.74, meaning that the program is enormously justified by the social benefits. And what are those benefits? One part is higher future earnings, estimated to be around $40,000 (not discounted). Yet another major part is less crime and less cost to the criminal justice system. This is estimated to be a whopping $94,065, even larger than the gain in earnings. The Perry Preschool Program demonstrated lasting benefits in so-called noncognitive personality skills, such as perseverance, motivation, study habits, and sociability.

In the United States, a misguided criminal justice system combined with chronic underinvestments in children has led to a tragically large population in prisons, roughly 2.4 million people. Many of those imprisoned are young men from poor families, who received an inadequate education and grew up in environments of great stress, little support, and little promotion of learning. Not only is poverty thereby passed from one generation to the next, it is passed on through an especially tragic and costly process: criminality and imprisonment. If the United States shifted resources from locking up young men to educating young children, it would experience a huge gain in fairness, productivity, and wellbeing of the society. The pure economic gains society-wide would be enormous, as the costs of excessive incarceration run to tens of billions of dollars per year.

Societies around the world are finally recognizing, based on rigorous evidence across pediatrics, psychology, physiology, and economics, that investments in ECD are the best investments they can make. These investments in young children not only lead to efficiency in the sense of high economic returns but to fairness and social inclusion as well. The children benefit not only with greater cognitive development but also with social (noncognitive) skills that are highly valuable for lifetime achievement. With a strong investment in ECD—combining health care, nutrition, environmental safety, and a preschool learning environment—all children rich or poor have a real chance to succeed and to become productive citizens.

III. The Rising Returns of Education and the Supply Response

It is crucial to make investments at all stages of the life cycle. There remains a significant gap between rich and poor countries, and rich and poor households, in access to secondary education. This also should be eliminated as part of the new Sustainable Development Goals covering 2015 to 2030. For children who are given an adequate start (preschool through secondary) there is also a huge return for completing a college degree. In figure 7.5, we saw the wage premium in the United States enjoyed by college graduates compared with high school–diploma holders. That wage premium has been soaring since 1979, from about 25 percent in 1979 to around 45 percent by 2010.

The rising relative earnings of college graduates is a market signal to children to stay in school and obtain a college degree—if they can afford it, and if they are adequately prepared by their experience up to college age. Alas, too many young people are in fact not college-ready when they finish high school, because they have lacked the benefits of an enriching household environment as well as the access to preschool and quality primary and secondary education. Many of these ill-prepared young people start college, take on financial burdens to pay for tuition, and then end up dropping out, with heavy debts and without the benefits of a college degree.

When the market sends a signal that an activity has a high rate of return, we would expect people to pursue that activity, and so we would expect more people in the United States to complete a bachelor's degree. While this is happening to a small extent, something odd is also happening. Figure 8.8 shows the percentage of adults in the United States who have finished a high school degree and who have finished a college degree. (There are two lines shown for each educational category: one for 25- to 29-year-olds and the other for all adults over 25 years of age.) As for high school completion, the rate has risen sharply over time to around 90 percent today. As for college completion, the curve has also sloped upward over time, but not so dramatically. From 1940 to around 1975, the proportion of 25- to 29-year-olds with a college degree increased significantly, from around 5 percent to around 20 percent. Yet starting around 1975 that upward sloping curve has a kink; there is almost a leveling off after 1975, even as the returns to college soared.

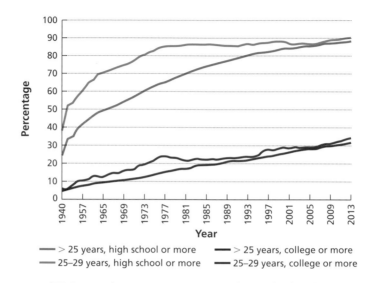

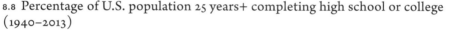

8.8 Percentage of U.S. population 25 years+ completing high school or college (1940–2013)

Source: U.S. Census Bureau Current Population Survey.

While a bit over 20 percent in 1975 have a four-year degree, by 2009–2010 the proportion is only around 30 percent. The numbers have really stopped increasing at anything like the rate we would expect, despite the strong signal of enormous benefits of a college degree. What is happening?

There are clearly bottlenecks on the supply side, and figure 8.9 gives some suggestion of one key bottleneck. Tuition costs are extremely high and continue to rise. Just when society ought to be helping young people to make an investment in higher education, very high tuition costs are holding back the supply response to a clear demand.

There are two troubling implications here. One implication is that a very large proportion of young people in the United States, a very rich country, are unable to enjoy the benefits of a completed higher education. Tuition costs are a huge barrier, and those tuition costs have not only soared compared with the past but also compared with tuition rates in other countries. The second troubling aspect is that college completion is disproportionately by kids from wealthier families.

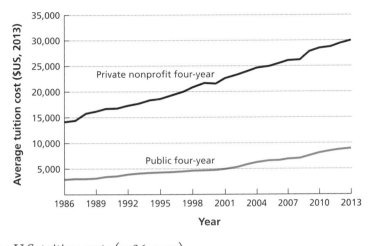

8.9 Rising U.S. tuition costs (1986–2013)

Source: Baum, Sandy, and Jennifer Ma. 2013. "Trends in College Pricing 2013." New York: The College Board.

College education is not equalizing incomes across the income distribution but exacerbating inequalities. Poor kids can't make it through college.

The gaps show up not only by income but also by race and ethnicity. In 2012 in the United States, the percentage of white, non-Hispanic people over age 25 with a four-year university degree was 35 percent; for African Americans, it was 21 percent; and for Hispanics, it was only 14.5 percent. It is poor kids, and especially poor non-white kids, who are not making it. There are no doubt several barriers standing in the way of the poorer kids completing a college degree: a less supportive home environment while growing up (with less-educated and lower-income parents and more single-parent households), less access to preschool, lower quality primary and secondary education, less college readiness upon high school graduation, and less ability to finance a costly higher education.

The United States has not deployed social policies (such as universal preschool or free higher education) to address these vast social inequalities, and the dangerous underinvestments in educational attainment. By virtue of its libertarian traditions, the United States tends to look to "self-help" market solutions. In the case of higher education, young people are told to borrow money to go to university, and

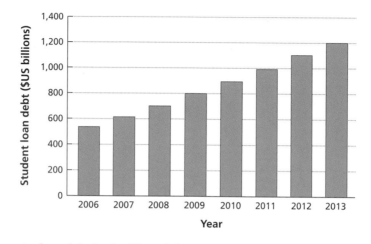

8.10 Rising student debt in the United States

Source: Federal Reserve of Economic Data.

so the government has helped to create a rapidly rising and increasingly problematic level of student debt, which now totals around $1 trillion. The double irony is, of course, it is the poor kids who take the loans, because the parents of the youth from wealthier families are able to pay tuition from the start.

Figure 8.10 shows how rapidly student debt has been growing in recent years, from an estimated $250 billion total in 2003 to a fourfold increase in 2013. As I've indicated, young people who will not complete a four-year degree owe much of this debt, and therefore will not earn the extra income necessary to repay what they've borrowed.

The United States therefore faces a triple challenge: highly unequal access to higher education; very little increase in the rate of college graduation since the 1970s; and a massive buildup of student debt. One solution in the future would be a decisive lowering of tuition costs, and one important innovation that can help that come about would be online education, as exemplified by the online course upon which this text is based. We can realistically envision a big breakthrough in access to higher education in the future, as well as new ways to learn and to combine online learning and "brick-and-mortar" campuses. Technological progress therefore offers us some realistic hope that higher education can soon reach a vastly higher proportion of young people in all parts of the world.

IV. Social Mobility

Education is a path to a more productive life as a citizen and an income earner, but we've noted that it can also be an amplifier of social inequality. If higher education is so expensive that only children from affluent families are able to pursue higher degrees, and if the returns to higher degrees are themselves high, then education becomes a bottleneck for the poor and a cause of widening inequalities. This seems to be the case in the United States today, a country that once prided itself as "the land of opportunity," but now is a society of high inequality and low social mobility.

Figure 8.11 illustrates the problem. The graph is based on a sample of eighth-grade children in 1988 (National Center for Education Statistics 2000). It divides the children into three categories according to their 1988 household income: the lowest quartile (the poorest 25 percent) of households; the middle two quartiles (from the 25th to 75th percentile); and the top quartile (from 75th percentile to the richest household). It then examines the educational attainment of these children as of the year 2000. There are four categories of educational attainment:

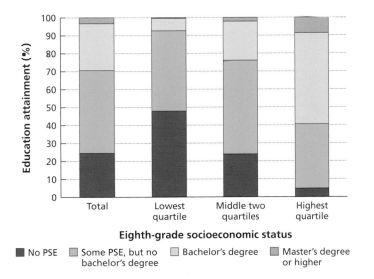

8.11 Educational attainment (2000) by eighth grade socioeconomic status (1988)

Source: National Center for Education Statistics.

no postsecondary education (PSE); some PSE but no four-year bachelor's degree; a bachelor's degree; a master's degree or higher.

The results are striking. Of the children in the poorest quartile, 48 percent had no PSE, and another 45 percent had some PSE but no bachelor's degree. Only 7 percent had a bachelor's degree, and none had a master's degree. Compare this with the children in the richest quartile. Now, only 4 percent had no PSE, 36 percent had some PSE but no bachelor's degree, while 51 percent had a bachelor's degree and another 9 percent had a master's degree or higher. The income inequality these youngsters faced in eighth grade will be replicated when they are adults by virtue of the fact that the children of high-income households have been able to avail themselves of the high returns to higher education, while the poorer children have not.

Figure 8.12 shows an extremely important and rather sobering relationship that is consistent with this finding (Corak 2013, 82). The horizontal axis shows the Gini coefficient (degree of inequality) for thirteen high-income countries; the higher the Gini coefficient, the higher the inequality. The countries with lowest inequality in this graph are, not surprisingly, the Scandinavian countries,

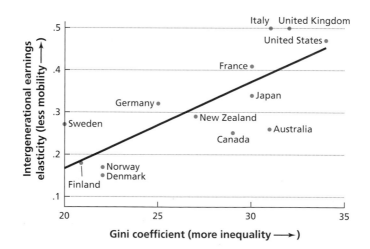

8.12 Great Gatsby Curve: inequality and intergenerational earnings mobility

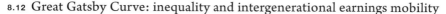

Source: Corak, Miles. 2013. "Income Inequality, Equality of Opportunity, and Intergenerational Mobility." Journal of Economic Perspectives 27(3): 79–102. Used with permission.

while the most unequal country is the United States, with the United Kingdom coming in second. The vertical axis shows an indicator of social mobility. For each country, the earnings of current workers are correlated with the earnings of their parents. If the correlation is high, it means that poor kids tend to grow up to be poor adults, and rich kids tend to grow up to be rich adults. In that case, social mobility is low. Alternatively, if there is little correlation in earnings between parents and children (so that a poor child has a reasonable chance to grow up to be a rich adult), then we say that social mobility is high.

We see a strong upward relationship in the graph. Countries (like the Scandinavian countries) that are relatively equal in income distribution have a high social mobility (a low correlation of earnings of parents and children), while countries that are highly unequal (like the United States) tend to have low social mobility. Parental income is a strong predictor of children's future income.

Two more charts help to make the same point (Corak 2009). Figure 8.13a shows a comparison of the United States and Canada. It considers poor households in which the father is among the lowest 10 percent of earners (the bottom decile), and asks where the sons end up in the income distribution. For sons born to low-income fathers, 22 percent also end up in the lowest decile and another 18 percent end up in the second-lowest decile. Thus, 40 percent of sons who are born to poor fathers end up in the bottom fifth of the income distribution (the bottom two deciles). In Canada, there is more upward mobility. Only 15 percent of the sons end up in the lowest decile and 13 percent in the second lowest, for a total of 28 percent in the bottom fifth. In Canada, being born poor is still predictive of staying poor, but there is much more upward mobility than in the United States.

Figure 8.13b shows the outcomes for sons born to rich fathers. In the United States, of sons born to fathers earning in the top 10 percent (the top decile), 26 percent also end up in the top decile and another 11 percent end up in the ninth decile (second from the top), for a total of 37 percent in the top fifth (the top two deciles). By contrast in Canada, only 18 percent end up in the top decile and 13 percent in the second decile, for a total of 31 percent. The difference is not as stark as with poor households, but the tendency is the same: Canada has higher social mobility than the United States.

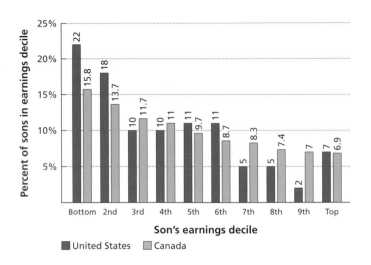

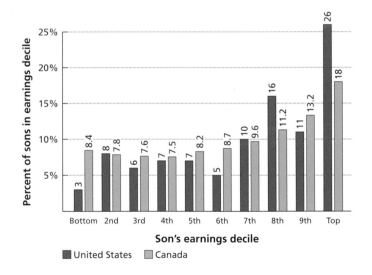

8.13A (*top*) Economic status of sons with low-earning fathers (United States vs. Canada) *Source: Corak, Miles. 2009. Chasing the Same Dream, Climbing Different Ladders: Economic Mobility in the United States and Canada. Washington, DC: Pew Charitable Trusts Economic Mobility Project.* © *2010 The Pew Charitable Trusts.*

8.13B (*bottom*) Economic status of sons with high-earning fathers (United States vs. Canada) *Source: Corak, Miles. 2009. Chasing the Same Dream, Climbing Different Ladders: Economic Mobility in the United States and Canada. Washington, DC: Pew Charitable Trusts Economic Mobility Project.* © *2010 The Pew Charitable Trusts.*

More equal societies, which generally also have a strong role of government in providing ECD and access to quality education at all levels, end up with greater intergenerational mobility. The social democracies of Scandinavia are thereby the outstanding achievers in social inclusion, promoting widespread prosperity and the highest observed extent of social mobility across generations.

V. The Role of Higher Education in Technological Advance

Higher education plays a key role in the two kinds of growth we discussed in chapters 3 and 4: endogenous growth and catching-up growth. Endogenous growth is economic growth based on new technological breakthroughs, such as the ongoing revolution in information and communications technology. These technological breakthroughs tend to be the result of intensive research and development (R&D) by highly skilled scientists and engineers with advanced degrees. Just as James Watt invented his steam engine in a workshop at the University of Glasgow, today's inventors are likely to be employed at universities, or in start-up companies linked to universities, or at high-tech companies in some kind of partnership with academic sciences and engineering. The greatest technological advances in the past fifty years, in computer science and applications, the Internet, fiber optics, genomics, advanced materials sciences, solid-state physics, aerospace, and much more have been the fruits of universities, national and international laboratories, and high-tech companies powered by highly trained individuals with advanced degrees in their respective areas.

Figure 8.14 shows the share of national income devoted to R&D in different parts of the world. We see that R&D is heavily concentrated in the high-income world. This has been true essentially for the two centuries since the start of the Industrial Revolution. A tremendously high share of the new innovations has come from a small subset of countries, including the United States and Canada, western Europe, Japan, and more recently, Korea, Singapore, and Israel. These countries account for the lion's share of scientific breakthroughs and patented intellectual property that underpins endogenous growth. Recently,

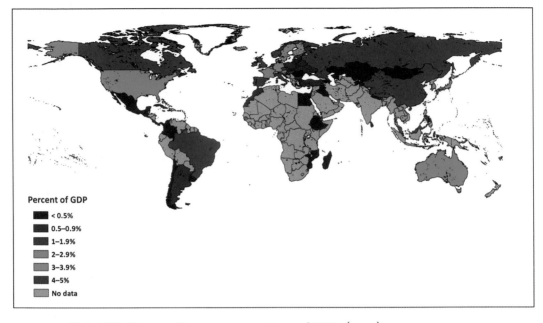

Percent of GDP
- < 0.5%
- 0.5–0.9%
- 1–1.9%
- 2–2.9%
- 3–3.9%
- 4–5%
- No data

8.14 Global R&D expenditures as percentage of GDP (2010)

Source: World Bank. 2014. "World Development Indicators."

China has increased its investments in R&D in a bid to join the group of high-innovation countries.

Research and development is underpinned not by a single institution (like a university) but by a full web of institutions, involving universities, national laboratories, and high-tech businesses. The complex interplay of these institutions in producing technological advances is called the country's "national innovation system." There are many tools to build a strong national innovation system, including public financing of scientific research, tax incentives for new innovations, prizes to honor and encourage scientists and engineers, national laboratories, and public and private philanthropic funding to support universities and other sites of innovative research activities. At the base of all of these institutions, however, is a strong system of higher education in the sciences, engineering, and public policy.

The second kind of growth is the adaptation of technologies from abroad. Sometimes these technologies do not require local skills in the importing country. Only a very few mobile phone users, after all, understand the advanced electronics of their phones. Only a few patients understand the biochemical mechanisms of their lifesaving medicines, and only a few farmers understand the genetic composition of their high-yield seeds. Nonetheless these technologies work wonders.

Yet some technologies cannot simply be used "off the shelf." The technologies have to be adapted to local use. This is almost always true of high-yield seed varieties, for example, which must be adapted to local pests and pathogens and local climate conditions. Machines imported from abroad don't require the advanced expertise of the original inventors, but they do require high skills to use the machinery. We can say that technology transfer from abroad typically requires at least some highly skilled specialists in the importing country. Universities are vital for preparing that skilled workforce. Universities are also vital for training the teachers who will train vast numbers of students to be ready to use the new technologies from abroad.

Universities are also critical for a third basic activity: helping society to identify and solve local problems of sustainable development. Every issue with which we are grappling—poverty, disease, climate change, new information technologies, and so on—requires locally tailored solutions, often based on sophisticated management systems. Suppose a country must shift its energy system from traditional coal-based power plants to low-carbon alternatives. What is the country's potential to develop wind or solar power? How safe is nuclear power? What about geothermal energy? Should the country deploy electric vehicles? These questions may well lie beyond the competency of government institutions. Universities may then be crucial partners in the national problem solving.

America has long promoted its universities for this kind of problem solving. One of the pioneering steps in the United States was the Morrill Act, a major piece of legislation passed by the U.S. Congress in 1862 and signed into law by President Abraham Lincoln. This wise legislation created "land-grant universities" in order to foster agricultural and mechanical advances based on science and technology. The U.S. federal government granted land and financing to every state to establish a new institution of higher education in agricultural and engineering studies. The Morrill Act said that this federal support would enable:

the endowment, support, and maintenance of at least one college where the leading object shall be, without excluding other scientific and classical studies, and
including military tactics, to teach such branches of learning as are related to agriculture and the mechanic arts, in such manner as the legislatures of the States may
respectively prescribe, in order to promote the liberal and practical education of
the industrial classes in the several pursuits and professions in life. (Morrill Act of
1862, Pub. L. No. 37-108 § 4)

What makes the Morrill Act so novel and important for America's economic
development is that these new institutions of higher education were set up not only
to train students but to work with the local communities in which they are located
to solve difficult problems and to develop technical capacities. These land-grant
universities established agricultural field stations and outreach programs run by
university-based scientists. The land-grant universities helped farmers across the
United States grapple with the problems of pests, crop productivity, soil nutrients,
climate, mechanical innovations, and the other keys to high-productivity agriculture. What a legacy these universities have created, and how wise of Abraham Lincoln and the U.S. Congress to realize, in the midst of a civil war, the importance of
building strong, practically oriented institutions of higher education.

It is with this history of the crucial role of universities in mind that I am especially honored that UN Secretary-General Ban Ki-moon asked me to assist him
in his global leadership challenge of promoting solutions to sustainable development, by helping to establish a new knowledge network based on the universities
around the world, so that these universities can be effective problem solvers in
their respective cities, nations, and regions. The new UN Sustainable Development Solutions Network (SDSN) is an outreach organization under the auspices
of the UN secretary-general that aims to link universities, businesses, and other
knowledge institutions around the world in the common challenge of finding
solutions to sustainable development. Countries and regions around the world
are now forming their own chapters of the SDSN and linking up with the global
SDSN. The hope is that the SDSN will effectively support universities around the
world to be available for problem solving in the same way the Morrill Act created
effective institutions for problem solving across the United States.

9

HEALTH FOR ALL

I. Universal Health Coverage

Good health stands at the center of sustainable development. Good health is at the center of wellbeing and is vital for everything else we hold dear. Good health aids children to not only survive disease but also to flourish, learn, and make their way through school and on to adulthood and the labor force. It also enables a workforce to be productive. Thus, good health enhances the ability of a community to develop human capital, undertake economic activities, and attract investment.

Health has also long been regarded as a basic human need and basic human right. Technically the goal is the "highest attainable standard of physical and mental health." Why do we refer to the "highest attainable" standard? Perfect health cannot be a human goal. Every human being will face disease and death. The highest attainable health refers to what is possible given current knowledge and technology. The world is far from this practical standard. Around 6 million children die each year before their fifth birthday, almost all of them in developing countries, and almost all as the result of disease conditions that are preventable or treatable; in other words, better health is attainable.

Since the beginning of the United Nations itself, the priority of health has been clear. In 1948, the Universal Declaration of Human Rights (UDHR) made it clear that health is a human right and basic need; and that even when these kinds of rights cannot be achieved immediately, they should be progressively realized. Yet time is progressing and we could still move faster. When the World Health Organization (WHO) was created, also in 1948, it declared in its central constitution that the highest attainable standard of health is a fundamental human right "without distinction of race, religion, political belief, economic or social condition." But we have yet to achieve this goal.

One notable global effort after 1948 to achieve universal health was launched in 1978 in Alma-Ata, now Almaty, Kazakhstan. World health officials gathered and adopted the important Alma-Ata Declaration, which called for universal health by the year 2000. Unfortunately when 2000 came around, there was not health for all. Instead, there were pandemics and poor health outcomes in many parts of the world. The HIV/AIDS pandemic was running rampant, with an estimated 36.1 million people infected with the HIV virus. Malaria, a tropical killer disease, rose tremendously in number of infections and in number of deaths, because the first-line medicine used in Africa to treat malaria lost its efficacy—the parasite had become resistant to the chloroquine drug. The year 2000 was a very bleak year for malaria, with a surging death toll—around 985,000 people. The year 2000 was also a bleak year for tuberculosis, another horrific scourge of humanity that claims millions of lives each year. The tuberculosis surge was partly riding on the HIV/AIDS pandemic, because immunocompromised individuals suffering from AIDS died in very large numbers from tuberculosis. Like malaria, there was a crisis of drug resistance. New and lethal strains were spreading multidrug-resistant tuberculosis (MDR TB) bacteria that were resistant to a wide spectrum of traditional medicines. This became extensively drug-resistant tuberculosis (XDR TB), against which even the most heavy-duty rescue medicines lost their effectiveness as the bacteria continued to evolve resistance.

In other words, the year 2000 did not meet the hopes and aspirations of the health ministers who had assembled in 1978, nor did it fulfill the promise of the UDHR, nor the constitution of the WHO. So it was notable that in 2000, the Millennium Development Goals (MDGs) centrally took up the challenge of fighting these spreading scourges of killer diseases. We have made a lot of progress since then. The time from 2000 until now is a period of tremendous progress.

It rekindles the hopes of 1978. Indeed, we see within reach the possibility of truly honoring the commitment to health as a basic human right.

It is notable that three of the eight MDGs, which seek to end extreme poverty, are centrally about health. MDG 4 is about reducing child mortality. MDG 5 is about reducing maternal mortality. MDG 6 is about controlling epidemic diseases, including AIDS and malaria. All of the other goals, such as ending poverty and hunger, having children in school, and gender equality, are also goals in which health plays an important role, both as a determinant of outcomes and also as one of the main objectives.

There has been significant progress in health since 2000, and especially since around 2005. The science of public health, breakthroughs in modern medicine, and breakthroughs in areas such as food production and urban infrastructure have led to important gains. However, there are still major challenges. I've already noted that millions of children, especially in the poorest countries, die each year of preventable or treatable causes that could have been ended with a well-directed, well-implemented effort. These lives could be saved. Children who are dying today or are left with life disabilities could, with proper organized responses, have happy, safe lives that could give them the many opportunities they desire.

Around the world, most people consider health to be a merit good—a good that should be accessible very broadly or indeed universally in the population. Health is generally considered to be such a good, not only because it is vital for each individual from a moral and ethical point of view but also from a practical point of view. Untreated disease is a serious societal threat that spills over to the rest of the population. Think of the devastation that comes when a highly communicable disease races through a country or continent, the way AIDS has. The time and money spent fighting such infections, and the death tolls that result, are a huge cost to communities both in lives and in finances. Governments have a necessary leadership role to ensure widespread coverage as well as the efficiency and orientation of the health care system.

The great progress of public health, like economic development itself, is one of great achievement in the modern era. At the time of the Industrial Revolution, worldwide *life expectancy at birth* (LEB) was perhaps 35 years (there are no records to precisely establish a worldwide number, but this is a reasonable

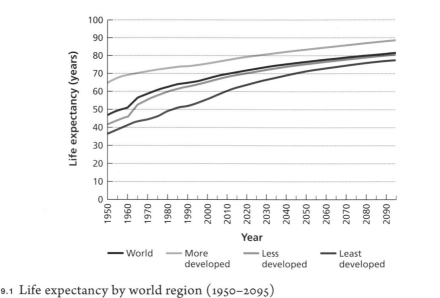

9.1 Life expectancy by world region (1950–2095)

Source: United Nations Department of Economic and Social Affairs Population Division (DESA Population Division). 2013. "World Population Prospects: The 2012 Revision." New York.

estimate). This does not mean that people were dying at the age of 35. It means that a newborn on average would reach an average age of 35, taking into account that a very large number of deaths were within the first five years of life. Therefore, if individuals reached the age of 20, there was a good chance they would reach the age of 50 or higher. Life expectancy at birth is therefore a measure made just at the time of birth. The life expectancy of an individual who has already survived childhood (say at age 20) can therefore be far higher than the LEB.

By 1950, just after World War II, there were some gains in the world's average LEB compared with the preindustrial era, but it is worthwhile to recall how much of the world was still mired in poverty in the middle of the last century. The United Nations estimated that during the five-year period from 1950 to 1955, the worldwide LEB was around 46 years. Figure 9.1 considers the trend in LEB since then for various groupings of countries.

As we see in the graph, the LEB for the entire world in 1950–1955 is estimated at around 47 years. In the developed regions, LEB was already around 65 years, while in the least-developed countries (LDCs) LEB was still only around 40 years, not much different from the estimated LEB for the preindustrial world. Note that in the landlocked developing countries, the LEB was even lower, at around 36 years. This very low LEB reflects the poverty and isolation typical of landlocked developing countries. It may also reflect the highland geography of many of those countries. As of 2010–2015, world life expectancy has increased to 70 years, roughly twice the life expectancy from the start of the Industrial Revolution. This is one of the great achievements of modern humanity, of modern science, and of economic development.

Nonetheless, there are still enormous health gaps between the developed and developing countries. In the developed countries, life expectancy is almost 80 years, whereas in the LDCs LEB is only 60 years, and less in many LDCs. In other words, there is roughly a two-decade gap in average life expectancy between the richest and poorest countries. These two decades of survival offer an indication of how much can be done to improve the health of the poorest countries.

In public health, it is now typical to speak of a full lifetime of 80 years, and to define excess mortality as a death under the age of 80. When an individual dies at age X (say 30 years), we can speak of the number of years of life that are lost, measured by $80 - X$ ($= 50$ years lost for a death at age 30). Most of the excess mortality in LDCs is caused by poverty in some way, directly or indirectly. From a medical and public health point of view, most "excess deaths" (that is, deaths under the age of 80) are the result of preventable or treatable conditions. There is therefore a strong chance that improved public health can spread throughout the whole world in the same way as economic development, thereby raising the life expectancy in the poor countries toward the LEB of the rich countries. We can speak of the convergence of public health in the same way we speak of the convergence of gross domestic product (GDP) per capita.

Figure 9.2 shows the worldwide relationship between the per capita income of a country and its LEB. The GDP per capita of each country is placed on the horizontal axis and the LEB is placed on the vertical axis. Notice the rising curve in this graph, showing the overall "best fit" relationship between these two variables. Richer countries have a higher life expectancy than poorer countries. Notice also that the curve flattens out at high income, however, meaning that once a country

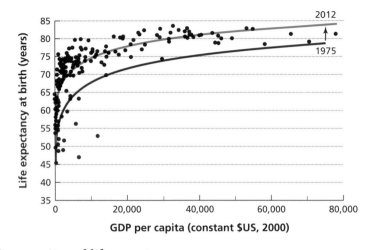

9.2 GDP per capita and life expectancy

Source: World Bank. 2014. "World Development Indicators."

has reached a certain level of development, there are not huge gains to LEB for further increases in GDP per capita. That top threshold seems to be around $20,000 (measured at international prices).

The solid line and the scatter plot are for the year 2012. A dotted line is also drawn showing the relationship that applied in 1975. Notice that the entire curve shifted upward, meaning that during the 30+-year interval, at any given level of income, an individual could expect to live longer in 2012 than in 1975. The main reasons are the technological and social improvements in public health, such as improved diagnostics, medicines, and surgical procedures; and also more healthy lifestyles (such as reduced cigarette smoking) in some populations. Of course, many of the advances are outside the health sector itself. As one simple example, people can now use their mobile phones to call for emergency help if necessary. The spread of literacy and public awareness about health and disease also enable people to promote their own health and survival.

There is another extremely important, yet subtle, lesson in figure 9.2. Note that the very steepest part of the curve is at very low incomes (that is, on the left-hand side of the graph). When countries are very poor, even small incremental changes in income (say, from $1,000 per year to $2,000 per year) lead to very steep gains in life expectancy, while the gains in health as the result of higher incomes tend to level off at higher incomes. For instance, according to the relationship depicted by the solid line, a rise in income from $1,000 to around $3,000 is associated with roughly 10 years of added life expectancy, from 45 years to 55 years. Yet going from $31,000 to $33,000, the same *absolute* gain in income, is associated with less than one added year of LEB.

Now, some people might claim that figure 9.2 shows that the best, or perhaps the only, way to higher life expectancy is through overall economic development. Get rich and get healthy as well. But that would be a mistaken interpretation. The fact that very small changes in the income of the poor can lead to huge changes in health outcomes suggests an alternative interpretation: that modest but targeted investments in public health for poor people can make a profound difference for their health outcomes.

Consider the following example, which is quite realistic. Suppose that health care depends on income in the following way. Typically a poor country will raise around 20 percent of its GDP in domestic revenues and will spend roughly 15 percent of those revenues on public health. The other 85 percent of the budget will be spent on other needs like education, roads, power, water, public administration, interest on the public debt, and so forth.

Now consider what happens when income rises from $1,000 to $3,000. Total taxes rise from $200 per capita to $600 per capita (20 percent of GDP). Spending on public health therefore rises from $30 per capita to $90 per capita (15 percent of the total revenues). The extra $60 per person of public spending on health can make a huge difference, by ensuring proper coverage of all core health interventions, such as vaccinations of children, safe childbirth, malaria control, and AIDS treatment for those individuals infected with the HIV virus.

Now here is the exciting point. Figure 9.2 implicitly suggests that even a small incremental sum, as small as $60 per person per year, could enormously boost public health. This is correct. Yet the poor country may simply lack the budgetary means to provide that increment, since it would exceed its budgetary revenues. What if a foreign funding source, such as the Global Fund to Fight AIDS, Tuberculosis and Malaria (GFATM), picked up the modest extra costs? The results

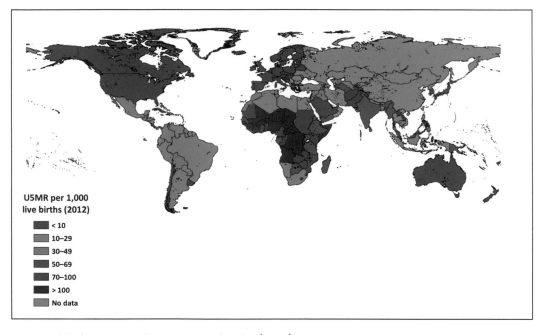

9.3 Under-5 mortality per 1,000 births (2012)

Source: World Bank. 2014. "World Development Indicators."

could be enormous, even historic. Lives could be saved by the millions; and with better health, the economy could actually enjoy a surge of economic growth. Soon enough, it would be able to cover the health bills on its own, out of its own revenues, once it reached a somewhat higher level of GDP per capita! This is the basic idea behind the successful concept of development assistance for health, which I discuss in greater detail later in the chapter.

Figure 9.3 maps out the under-5 mortality rate (U5MR), which signifies how many children under the age of 5 die for every 1,000 live births (World Bank 2014c). The world average for the 5-year period 2010–2015 is 52 per 1,000, according to the World Bank estimates. For the developed countries it is 7/1,000, and for the developing countries it is 57/1,000. For the LDCs, it is 99/1,000. Among the world's regions, the U5MR is the highest in sub-Saharan Africa (110/1,000), followed by South Asia (55/1,000). These two locations are the epicenters of the challenges of extreme poverty and health.

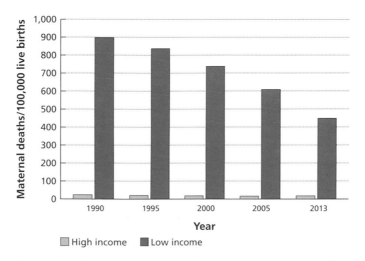

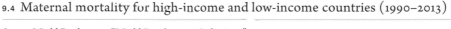

9.4 Maternal mortality for high-income and low-income countries (1990–2013)

Source: World Bank. 2014. "World Development Indicators."

Figure 9.4 shows another very crucial aspect of public health: maternal mortality (World Bank 2014b). This is measured as the number of pregnancy-related deaths (typically at childbirth but also earlier during pregnancy) for every 100,000 live births. Pregnancy-related deaths have huge variation between countries, since very few women die of pregnancy-related causes in rich countries (16/100,000) compared with poor regions like sub-Saharan Africa, where around 500 women die for every 100,000 births. However, the maternal mortality rate (MMR) in the low-income countries is falling sharply. It was around 900 deaths per 100,000 live births back in 1990, but as of 2013 is at 450 deaths per 100,000 live births, an enormous achievement in saving women's lives.

In general, the reasons for death in rich and poor countries differ. The poor die from many of the same causes that the rich do: cancer, cardiovascular diseases, and metabolic disorders such as diabetes. But the poor also die of conditions that rich people no longer die of, especially communicable diseases such as measles, malaria, or other kinds of infections. One basic principle of health is that undernutrition leads to the weakness of the immune system to resist infections (known as immunosuppression). For that reason, children in very poor countries, who are

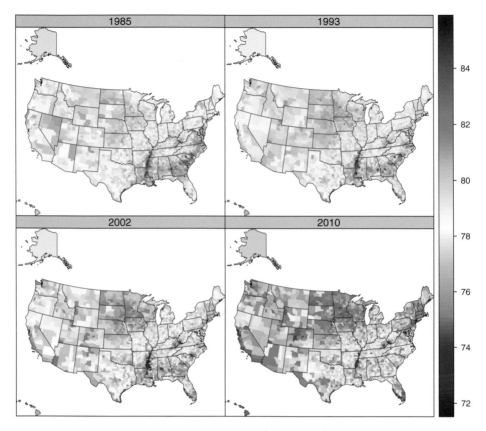

9.5 Life expectancy in United States, female (1985–2010)

© 2013 Wang et al.; licensee BioMed Central Ltd.

more likely to be undernourished, die of diarrheal diseases or respiratory infections that would not kill a better-nourished child in a richer country.

The disparities between rich and poor, and across ethnic groups, also apply within countries as well. The United States, already noted as having the highest income inequality among the high-income countries, also has significant disparities in life expectancy. Figure 9.5 shows the measured life expectancy for women across counties of the United States. The northeastern seaboard of the United States, including Boston and New York City, has high life expectancies. However, counties of the Deep South of the United States, in states such as Alabama or Georgia, have several years fewer in life expectancy. African Americans have significantly fewer years of life expectancy compared with white, non-Hispanic Americans. The gaps cross both class and racial lines perhaps also reflect direct effects of geography as well.

Various statistics are used to measure health outcomes. Life expectancy at birth, the U5MR, and the MMR are three important statistics we have already noted. Another important and very widely used concept is disability-adjusted life-years (DALYs; Sustainable Development Solutions Network Thematic Group on Health for All 2014, 11). As I noted earlier in referring to life-years lost, a normal life span is 80 years. If an adult dies at age 60, then 20 years of life are lost. DALYs count not only premature mortality but also disabilities. A year of life with a disability is counted as a part of a year of life lost, with the portion dependent on the severity of the disability. Thus, living a year with a very serious disability such as paraplegia, schizophrenia, or blindness, is counted as a substantial fraction of a life year lost, perhaps up to 0.9 years lost for each year lived with the condition (depending on the specifics).

The DALYs of a population is the sum of life-years lost plus the years lost to disability. It is standard to divide the total DALYs by the population to calculate the DALYs per capita, just as we divide GDP by the population to calculate GDP per capita. DALYs per capita are used to understand the varying disease burdens across the world, as demonstrated in figure 9.6. Every colored segment on the bar refers to the DALYs per capita of a *specific category of disease*. The bar at the top of the chart is the DALYs per capita of the high-income countries by disease category. The other bars refer to various regions of developing countries. A large bar means that that the total DALYs per capita is high; that is, that the disease burden is very high, with considerable premature death and disability. Clearly, sub-Saharan Africa has by far the highest disease burden, while not surprisingly the high-income countries have the lowest disease burden of any grouping.

The bars are divided by colors in an interesting way; they refer to different disease categories. Sub-Saharan Africa, for example, has the highest disease burden of HIV/AIDS (the dark portion at the left side of the bar). This is not surprising, as sub-Saharan Africa is the epicenter of the global AIDS pandemic.

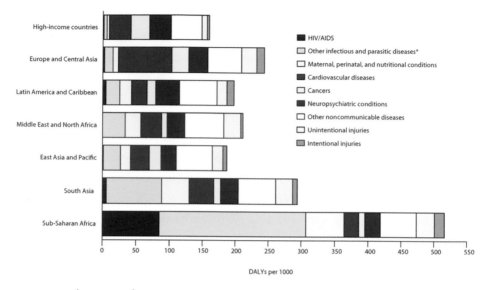

9.6 DALYs (per 1,000) by region

Source: Alan D. Lopez, et al. 2006. "Global and Regional Burden of Disease and Risk Factors, 2001: Systematic Analysis of Population Health Data." The Lancet. 367(9524): 1747–1757. doi:10.1016/S0140-6736(06)68770-9. With permission from Elsevier.

The light blue portion of the bar to the right of the HIV category shows the DALYs caused by other (non-HIV) infectious and parasitic diseases, such as malaria or typhoid. Here too the DALYs per capita in sub-Saharan Africa is "off the charts" compared with the other regions.

Compare the bar for sub-Saharan Africa with the top bar for high-income countries. In the top bar we see that HIV/AIDS is a tiny sliver. This does not mean that there are no deaths and disease from HIV/AIDS in the high-income world, only that the burden of HIV per person is low compared with the massive burden in sub-Saharan Africa. In the high-income countries, the category of communicable and parasitic diseases is virtually imperceptible, because it is so low on a per capita basis. Many of the communicable diseases are either completely absent (e.g., malaria) or are prevented by standard public health measures such as immunizations (e.g., measles). Other communicable diseases, such as diarrheal and respiratory infections, are generally treated by antibiotics and are rarely fatal in the high-income countries. The good news is that the field of public health already offers the tools to shrink the dark blue and light blue bars in sub-Saharan Africa.

The very light blue bar to the right of the communicable disease category is "maternal, perinatal, and nutritional conditions." This category combines deaths due to pregnancy and childbirth and deaths due to severe undernutrition. Once again, we see that Africa has an extraordinarily high disease burden in this category, far higher than in almost all of the rest of the regions. Only South Asia, like Africa, has a very high disease burden for pregnancy- and nutrition-related diseases.

Notice by comparing the DALYs per capita across the disease categories and regions that sub-Saharan Africa's extraordinary disease burden (in comparison with all the rest of the regions) is heavily concentrated in the first three conditions: HIV/AIDS, other communicable diseases, and pregnancy- and nutrition-related conditions. For all of the rest of the conditions, such as cardiovascular diseases or cancers, the DALYs per capita in Africa and the rest are not very different. The key point is that Africa's high disease burden is not a general phenomenon but a specific one: it refers to infectious diseases (including HIV), pregnancy-childbirth, and nutrition. By focusing on these main conditions, it would be possible to have Africa's public health conditions converge substantially with the disease conditions in the rest of the world.

There is only one other area that stands out: cardiovascular disease in eastern Europe and central Asia, shown as the red category in the graph. We know that middle-aged men in Russia and some other parts of the former Soviet Union are suffering massive disease burdens from heart attacks and other disorders. There is no single accepted explanation for this, but it probably relates heavily to lifestyles: heavy tobacco and alcohol use, binge drinking, little exercise, and a high-fat diet.

II. Poverty and Disease

Health is a challenge for all societies. In the rich world, where health conditions are far better than in poor countries, the problems of high and rising health care costs are front and center of the political and economic agendas. In the poorest countries, the challenges are also about finances, but much more importantly are about the still very high burdens of disease and the tragic loss of lives.

Poverty and ill health go together (Sachs 2005). This linkage must be understood to address the challenges of heavy disease burdens in poor countries, and the explanations need to be used to identify tools to break the poverty trap itself. There is a two-way causation whereby poverty contributes to disease, while disease also contributes to poverty. An individual suffering from disease cannot work at full capacity and so loses income. A community or country with a heavy disease burden similarly loses income from the resulting lack of productivity and the other high costs of health care.

Disease contributes through many channels. First, diseases of young children adversely impact the entire life-cycle development of a young child. Repeated illnesses at a young age are likely to set a child back in terms of readiness for school, capacity to learn, and even vulnerability to diseases as an adult. For example, adult cardiovascular diseases can be exacerbated by poor nutrition at a young age. Human development across the entire life cycle is strongly affected by health and disease early on.

Over the long term, a high death rate for children can contribute to adverse demographic problems. Consider what happens when 200 of every 1,000 children born do not reach adulthood. Parents know the high risks of children's deaths and respond by having very large numbers of children. For example, if low-income parents want to be sure that they raise a surviving son, either for cultural reasons (such as to perform funeral rites) or pragmatic economic reasons (such as to support the parents in old age), they may choose to have very many sons to ensure that at least one survives. (Note that the preference for boys still applies in many cultures and notably in many low-income settings in Africa and Asia.) Yet aiming to have three or more sons might mean six or more children in total, sons and daughters, leading to extremely rapid population growth and very low levels of *investment per child* in health, nutrition, and education.

A high disease burden can have a strong direct adverse effect on a local or national economy as well. It is not uncommon for a whole farm community to be knocked out by malaria and thus physically unable to harvest the crops necessary for survival. This kind of epidemic can have very serious harmful consequences for poverty, destitution, and hunger. Another obvious negative economic consequence of ill health is the direct cost of health care itself. Treating diseases can seriously deplete an already squeezed household budget. What would seem to be modest outlays of a few dollars per disease, an amount that would hardly be noticed in the rich world, can actually eliminate the margin of survival for a very poor family.

Finally, it is obvious that investors are not very keen on investing in areas with high malaria or AIDS prevalence. If a potential investor is aware that the labor force is likely to be sick and that there will be a need to replace absentee or dying workers and to cover their health care costs, the potential returns on an investment are lower, perhaps decisively so. Nor is it helpful for tropical tourism if the risks of infection with malaria, dengue, or other tropical disease are high.

The arrow of causation also goes in the reverse direction, from poverty to disease. That is the direction of causation that people most naturally consider: poverty leading to poor health. There are many ways that poverty causes disease and premature mortality. Most obviously, the poor often cannot afford a doctor, even when an illness is serious. They may be unable to afford the needed medicines or the transport to get to a hospital. Yet there are many other, less obvious ways in which poverty contributes to a heavy disease burden.

One that I've already mentioned is the immunosuppression caused by undernutrition. Children who are not fed well, who lack basic micronutrients, and thereby have degraded immune systems are consequently unable to resist disease and often succumb to infections that would be mild irritants for well-fed and affluent children.

Poverty also tends to create a more dangerous home environment. Poor families in rural areas typically do not have reliable sources of safe drinking water and sanitation. This means that many diseases are spread by open defecation, which leads to unsafe drinking water. Diseases like cholera are notoriously spread in this manner. Poor communities have poor infrastructure (piped water, sewerage, electricity to refrigerate or heat foods safely). Even the physical structure of housing can make a very big difference. Many poor people live in adobe huts with thatched roofs, where there is often a gap between the thatch and the walls through which various disease-carrying insects and other animals can easily enter the household.

Poverty is also related to the ecological conditions of disease. Worm infections caused by hookworm, *Trichuris*, and *Ascaris* are found in warm, moist tropical climates, so poor people living in those areas are especially vulnerable to debilitating worm infections. Malaria is also a disease of the tropics, since malaria only transmits in warm temperatures (generally above 18° Celsius or 64° Fahrenheit). As a result, the poor people living in tropical regions have that extra burden of disease. One can say that malaria is a condition of poverty, but it is really a condition of the geography where many poor people live. I should add that Africa's malaria burden is especially deadly, and it is estimated that sub-Saharan Africa accounts for up to 90 percent of the global malaria deaths. This seems to be caused by three main factors: very warm temperatures, specific species of mosquitoes that are indigenous to Africa, and enough rainfall to support breeding sites of the mosquitoes.

Poverty is also associated with illiteracy, innumeracy, and therefore poor health knowledge and weak health-seeking behaviors. Illiterate people may have a great difficulty understanding how to fill a prescription or adhere to a drug regimen. Health-seeking behavior is extremely important for all of us. Poor people with less formal education have a harder time and lesser ability to ferret out the right kinds of advice and help. This is why trained local health workers, such as Community Health Workers (CHWs), can be vital in helping poor people to understand how to keep themselves safe (e.g., by using an antimalaria bed net) and how to make the connections with the health system itself in the event of an emergency or special need (e.g., safe childbirth or immunizations for children).

People living in poverty are also very vulnerable to being pushed into dangerous work, such as young women who because of their poverty are cajoled or forced into sex work, and thereby become highly vulnerable to AIDS and other sexually transmitted diseases, as well as to job-related violence and other life-threatening conditions.

There are therefore two directions of causation: poor health leads to poverty and poverty leads to poor health. Whenever the arrows work in both directions, there is the possibility of a *vicious spiral*. A household can get trapped in poverty and thereby succumb to disease, which in turn lowers the earning power and further traps the household in debt and poverty, and so forth.

Yet whenever there is a vicious spiral, there is also generally the possibility of a *virtuous spiral* as well. Disease control can raise household and community income, and this higher income can lead to further improvements in health. The result is an upward spiral of health and wealth, with each leading to the other. It is part of the job of public health policy to help break the vicious circle that traps people in a poverty and disease trap and to create a positive spiral of health and development.

Breaking the Vicious Spiral of Extreme Poverty and Disease

How best can we intervene to stop the vicious spiral of disease and poverty and turn it into the virtuous spiral of health and wealth? The first thing to do is to implement what the health ministers called for back in 1978 in Alma-Ata: a primary health system for all. The second step is to help poor communities to achieve better nutrition. Part of that may occur through dietary fortification and supplementation or through better community knowledge about healthful diets. Part will also come through improvements in the agricultural sector. More productive farmers growing more food will support healthier diets, and the community may generate a surplus that can be used, for example, for school food programs. Agricultural interventions can thereby play a role not only in improving agriculture and raising household income but also in reducing the overall disease burden, once again illustrating the integrated nature of sustainable development.

Local infrastructure is another kind of investment in public health, in areas such as safe drinking water, sanitation, power, roads, and communications. Investments in electricity are needed to run the refrigerators in clinics that keep the vaccines cooled or to power the instrumentation needed for emergency obstetrical care. (There are of course dozens or hundreds of high-priority uses of reliable electricity.) Investments in water infrastructure help pump the irrigation water that improves agriculture. Roads are vital to be able to reach health facilities. Phones and the Internet have multiple purposes, ranging from simple text messaging to telemedicine. These are not health interventions per se, but communities that have access to phones and the Internet are also very likely to have access to better health.

There is some important progress in expanding access to a rudimentary health system in low-income countries. Table 9.1 shows the data on the mortality rates of children under five years of age. Back in 1990, 12 million children under the age of 5 died. By 2010, the number was 7.6 million. As of 2012, it was down to 6.6 million. This decline is a huge victory for public health, but it is still 6.6 million too many, since almost all of these deaths are caused by preventable or treatable diseases.

In table 9.2, those same mortality rates are put in terms of numbers per thousand live births (the U5MR) rather than aggregate numbers. In 1990, on average, 88 per 1,000 children in the world did not live to their fifth birthday. By the year 2010, that rate had declined to around 57 per 1,000 children. It is still a high number that can and should be cut significantly, but the U5MR is indeed falling sharply.

More than 6 million of the 7.6 million total deaths in 2010 were in just two regions, sub-Saharan Africa and South Asia. These are the continuing epicenters not only of overall poverty but also of the disease burden and of preventable and treatable diseases. Where are poor children succumbing and to what kinds of conditions?

This question is addressed by figure 9.7, which estimates the worldwide causes of deaths of children under 5 years of age. The overwhelming message of this chart is that there is still a very high burden of communicable diseases, including diarrheal disease, pneumonia, measles, malaria, and other infectious diseases. Another major category, death at childbirth, includes birth asphyxia, trauma, prematurity, and severe neonatal infections. Many of these causes can be attributed to a lack of sanitary birth conditions or simple equipment to help newborns breathe, and thus could be reduced substantially at low cost.

This kind of epidemiology is a good start in terms of understanding what to do. There is a poverty-disease trap, wherein poverty feeds disease and disease feeds poverty. This disease burden is heavily concentrated in sub-Saharan Africa and South Asia and is significantly composed of diseases that are either communicable, birth related, or nutrition related. Now that we better understand the specific problems, an effective starting point for addressing the continuing health crisis in low-income settings is to build a primary health system that is responsive to these specific challenges.

Table 9.1 Deaths of children under five (1990–2010) (thousands)								
Region	1990	1995	2000	2005	2009	2010	Decline (percent) 1990–2010	Share of global under-5 deaths (percent) 2010
Developed regions	227	151	129	112	102	99	56	1.3
Developing regions	11,782	10,550	9,446	8,355	7,654	7,515	36	98.7
Northern Africa	304	210	153	121	100	95	69	1.2
Sub-Saharan Africa	3,734	3,977	4,006	3,956	3,752	3,709	1	48.7
Latin America and the Caribbean	623	511	397	305	237	249	60	3.3
Caucasus and Central Asia	155	119	86	80	79	78	50	1.0
Eastern Asia	1,308	845	704	423	349	331	75	4.3
Excluding China	29	46	30	16	17	17	41	0.2
Southern Asia	4,521	3,930	3,354	2,829	2,588	2,526	44	33.2
Excluding India	1,443	1,233	1,060	875	837	830	42	10.9
Southeastern Asia	853	696	530	453	368	349	59	4.6
Western Asia	270	247	201	173	167	165	39	2.2
Oceania	14	15	15	14	14	14	0	0.2
World	12,010	10,702	9,575	8,467	7,756	7,614	37	100.0

Source: United Nations Children's Fund. 2013. *Levels and Trends in Child Mortality Report 2013.* New York: United Nations Children's Fund.

Table 9.2 Levels and trends in under-5 mortality rate (1990–2010) (deaths per 1,000 live births)

Region	1990	1995	2000	2005	2009	2010	MDG target 2015	Decline (percent) 1990–2010	Average annual rate of reduction (percent) 1990–2010	Progress toward Millennium Development Goal 4 target 2010
Developed regions	15	11	10	8	7	7	5	53	3.8	On track
Developing regions	97	90	80	71	64	63	32	35	2.2	Insufficient progress
Northern Africa	82	62	47	35	28	27	27	67	5.6	On track
Sub-Saharan Africa	174	168	154	138	124	121	58	30	1.8	Insufficient progress
Latin America and the Caribbean	54	44	35	27	22	23	18	57	4.3	On track
Caucasus and Central Asia	77	71	62	53	47	45	26	42	2.7	Insufficient progress
Eastern Asia	48	42	33	25	19	18	16	63	4.9	On track
Excluding China	28	36	30	19	18	17	9	39	2.5	On track
Southern Asia	117	102	87	75	67	66	39	44	2.9	Insufficient progress
Excluding India	123	107	91	80	73	72	41	41	2.7	Insufficient progress
Southeastern Asia	71	58	48	39	34	32	24	55	4.0	On track
Western Asia	67	57	45	38	33	32	22	52	3.7	On track
Oceania	75	68	63	57	53	52	25	31	1.8	Insufficient progress
World	88	82	73	65	58	57	29	35	2.2	Insufficient progress

Note: "On track" indicates that U5MR is less than 40 deaths per 1,000 live births in 2010 or that the average annual rate of reduction is at least 4 percent over 1990–2010; "insufficient progress" indicates that U5MR is at least 40 deaths per 1,000 live births in 2010 and that the average annual rate of reduction is at least 1 percent but less than 4 percent over 1990–2010. These standards may differ from those in other publications by Interagency Group for Child Mortality Estimation members.

Source: United Nations Children's Fund. 2013. *Levels and Trends in Child Mortality Report 2013.* New York: United Nations Children's Fund.

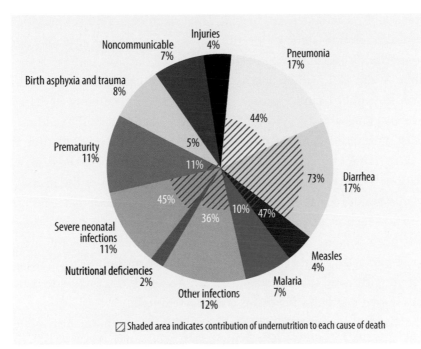

9.7 Major causes in death for children under five (2004)

Reproduced, with the permission of the publisher, from "Global Health Risks: Mortality and Burden of Disease Attributable to Selected Major Risks." Geneva, World Health Organization, 2009 (Fig. 8, Page 14 http://www .who.int/healthinfo/global_burden_disease/GlobalHealthRisks_report_part2.pdf, accessed 01 July 2014).

III. Designing and Financing Primary Health Systems in Low-Income Countries

The starting point for designing an appropriate health system is in the science of public health. What is the field of public health? When we think of medicine, we think of doctors and nurses who treat individual patients. Public health can be called population health. It tackles the treatment and health of a population rather than an individual, partly through the work of doctors and nurses but through other tools as well. Attention to safe drinking water, community access to antimalaria bed nets, and widespread coverage of effective vaccines are examples of critical tools in public health beyond what doctors and nurses provide. The public health solutions differ by location and income level, because they depend on the specific disease burden in any location.

Public health is a highly effective specialty and can be very systematic in practice. The public health approach starts with the epidemiology of disease, as we have just done with deaths of children under 5 years of age. The public health specialist must understand the disease burden in a particular population. What is the DALY count for each disease? What is the prevalence of the disease in a population? What is the incidence (number of new cases) of the disease in a given time period? Epidemiology measures the disease burden in a systematic way and also focuses on the transmission mechanisms. Is the disease transmitted person to person? Does an intermediate insect vector pass the infection from one person to another (such as in the case of mosquitoes and malaria)? Epidemiologists need to understand the various categories of disease; who the diseases burden; and how the diseases are transmitted.

The second stage of good public health thinking is to examine the feasible and desirable interventions, both within the health sector and in the closely related sectors of nutrition, agriculture, and infrastructure. What exactly can and should be done? We can think of these as the "apps" of public health; in the field of public health, they are generally called *interventions*. The interventions are systematic packages of diagnosis, treatment, and follow-up that address particular problems such as malaria or neonatal resuscitation.

The third stage of good public health is "systems design." Suppose you have created a list of needed interventions: for example, every child should get immunizations, every household should have bed nets, and every mother should have a specific number of antenatal visits. Then, systems design creates a way for the public health system to deliver these interventions properly and effectively. Who should do the work for each intervention? Who should give immunizations: doctors, nurses, clinical officers, or the CHWs? How should antimalaria bed nets get distributed to best ensure that households know how to use them? Which organizations or agencies should guarantee the supply chain of medicines? Who should perform and interpret a diagnostic test? Who manages the health workforce for its competence, honesty, and hard work?

These are the kinds of challenge that a business faces in running an organization. Running a public health system is also a quite complicated challenge. Building that system from scratch in a very poor setting is extremely difficult and important work that brings up issues of training, recruitment, job designations, and so forth. This system has to be translated into actual management and implementation. There are many things to do. If you are in charge, you may have to invest in a new operating theater for emergency obstetrical care, build an examination room, design a new clinic, and drill a new borewell for safe drinking water. You may have to train a local public health labor force, usually from a starting point of woefully few health professionals.

The public health system centrally involves the individuals in the community who seek health care and need to be connected with the health facilities. There should be trust and confidence in the CHWs. Households should understand health risks and health care–seeking behavior when disease strikes. Oversight of key personnel is crucial, as it is in any organization. How do you as the public health official make sure the work is actually getting done and that budget funds are used properly? How do you make sure the disease burden is as expected and thus your interventions are properly targeted?

Additionally, a good system not only monitors outcomes for management purposes, but also evaluates and assesses the performance of the health system. The health system should use timely information to interrogate its own performance. For example, if the data show that a mother died in childbirth, the key health personnel should investigate the death, determine whether it was in fact preventable, and if so, understand what kinds of adjustments will be needed to prevent a recurrence in the future. This kind of feedback system is absolutely crucial for practical success.

Next comes financing. It is not surprising that money looms large in this issue of health care in low-income settings. These are places where the absence of money plays a pervasive role in the continued high prevalence of preventable and treatable diseases. Even if a health system is designed with high efficiency, smart epidemiology, and cost-effective interventions, a poor community will need help to afford the health system to address the disease burden. A finance strategy will be vital.

Some feasible interventions that are based on the local epidemiology include interventions for malaria. A mosquito transmits the single-celled *Plasmodium* pathogen from individual to individual. How does the health system best protect the population? This could be done through medicines that kill the pathogen in an infected individual. It could be done through prevention such as the use of larvicides that kill the larvae of the *Anopheles* mosquitoes before they mature to become malaria-transmitting mosquitoes. It could also be done through the mass distribution of antimalaria bed nets that block mosquitoes from biting or through spraying the inside walls of houses to kill mosquitoes that come inside for blood meals. In fact, an effective control system will deploy several of these strategies together, not just one alone.

There are many examples of how good epidemiology can be connected with a strong set of highly effective and low-cost interventions. The results can be very striking and remarkably positive. In 2013 dollars, the total cost of a basic primary health system that addresses the previously described diseases might be no more than $60–90 per person in the community per year. This is a remarkably low cost to address all of those disease conditions, but the irony is that even $60 may be too much for a government in a very poor country. This is the part that is hard for those in the rich world to appreciate, because most people in high-income countries assume intuitively that if diseases with low costs of control nonetheless continue in very poor communities, it is perhaps because the host government is not trying very hard to deal with the issue. That inference is simply not correct. (I am not denying that some governments in low-income countries don't try very hard; *I am saying that even exemplary governments in such countries also need external financial assistance.*)

It is often hard to remember the financial plight of the very poor. A country such as Malawi is at around $400 per capita at current market exchange rates. As I noted earlier, a country like Malawi will typically collect around 20 percent of its national income in taxes. Twenty percent of $400 leaves the government's domestic revenue collection at $80 per person per year—that will be used to finance the government, the national defense, the police, the roads, the power sector, the roads and ports, the water and sanitation, the schools, environmental conservation, and yes, the public health system. Budget experts have recommended that

Table 9.3 Poverty, budgets and the health sector			
Country	GNP per capita (current US$)	Government health spending (current US$)	Health spending as share (%) of GDP	U5MR
Ethiopia	380	18	3.8	68
Niger	390	25	7.2	114
Mozambique	510	37	6.4	90
Sweden	56,120	5,319	9.6	3
United States	52,340	8,895	17.9	7
Brazil	11,630	1,056	9.3	14
Mexico	9,640	618	6.1	16
Thailand	5,210	215	3.9	13
France	41,850	4,690	11.7	4

Source: UN data.

governments in low-income settings devote around 15 percent of the total budget to the health sector, but at $80 per capita, 15 percent comes to just $12 per capita per year devoted to public health! In Europe and the United States, the public health budget is $3,000 per person per year and $4,000 per person per year, respectively. Table 9.3 illustrates the various levels of gross national product (GNP) per capita, health spending, and health outcomes (through the U5MR) for low-income, middle-income, and high-income countries.

The point is that even trying hard with its own revenues, a poor country cannot reach the necessary $60–90 per person per year on its own. As I emphasized earlier, this is why official development assistance (ODA) is so important for public health. Such ODA should be carefully, scientifically, and professionally targeted toward improving the health and health systems of poor people. I advocated for this kind of development aid when I chaired the Commission on Macroeconomics and Health for the WHO in 2000–2001, and I have championed such aid for more than a dozen years as special advisor to former UN Secretary-General Kofi Annan (2001–2006) and UN Secretary-General Ban Ki-moon (2007 to the present).

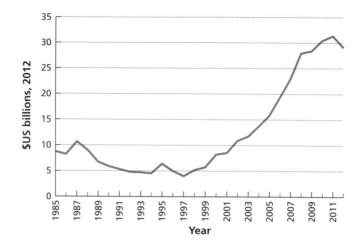

9.8 Global official development assistance for health and population (1985–2012)

Source: World Bank. 2014. "World Development Indicators."

A case in point of effective aid implementation is the Global Fund to Fight AIDS, Tuberculosis and Malaria, or GFATM, which I was very honored to help design and launch in 2002. Based on the extra ODA supplied by the GFATM, poor countries have achieved huge improvements in the control of all three diseases since the GFATM's establishment. The Global Alliance for Vaccines and Immunization (GAVI) is another great program that provides international financing for immunization coverage. The U.S. President's Emergency Plan for AIDS Relief (PEPFAR)and the President's Malaria Initiative (PMI) have also been quite successful in improving health outcomes in poor, disease-stricken countries.

These kinds of targeted, focused programs, backed by ODA, have been hugely effective and very successful, because the funds have been carefully monitored, assessed, and evaluated at each step of the way. Figure 9.8 is heartening. It charts the takeoff of ODA for health, especially bilateral assistance, from high-income governments to low-income country governments after the year 2000. With this added funding, malaria is coming down, the AIDS burden is coming down, treatment for tuberculosis is expanding, and U5MRs and MMRs have come down considerably. These successes demonstrate that ODA for public health works very well. Yet the poorest countries still do not have the full financing they need. The national budget gaps remain significant, perhaps $20–30 per capita (depending on the country), even though the financing gaps have been partly covered in recent years by the step-up in ODA.

I have recommended over the years that the rich countries should be giving around $40 per person in the recipient countries for these purposes. We are perhaps at the halfway mark or even a little above halfway, and this has made a big difference, but there still is an important financing gap that needs to be closed to achieve our long-standing and crucial objective of health for all.

IV. Ten Recommended Steps to Health for All in the Poorest Countries

The period of the MDGs since 2000 has been an exciting one for public health. Malaria is a prime example. Malaria is a not only a lethal disease, but it spreads very widely and in many parts of Africa. In many regions it is holoendemic, which means the entire community is typically infected much of the year unless active control measures are put in place. Controlling malaria is therefore a great feat; and malaria is being controlled right now in sub-Saharan Africa thanks to the focused efforts of organizations like the GFATM and the U.S. PMI. It is also thanks to great advances in public health and in new technologies, such as insecticide-treated bed nets; CHWs who go directly into the communities, obviating the need to bring sick or dying individuals many kilometers to a clinic; and new diagnostic tests, where instead of a microscope that reads the blood to detect the *Plasmodium* pathogen, a prick of the finger allows a CHW on the spot to know within a few minutes whether a child is infected with malaria. New medicines such as artemisinin are replacing medicines like chloroquine that have lost their efficacy to the parasite's drug resistance. All in all, malaria deaths of children under 5 years of age have declined by around 50 percent since the year 2000.

By now the poor countries have gotten more than halfway to building their primary health systems, and we should take inspiration from the recent health successes and move forward to fulfill the commitment to health for all that was made back in 1948.

I have ten basic recommendations for how we could achieve the full breakthrough in the next decade.

The first is financial. The total need for ODA for health is around $40 billion a year, which is roughly $40 per person for the 1 billion people who need the help. Since there are also about 1 billion people in the rich world, we need around $40 from each person per year in the rich world in order to save millions and millions of people each year in the poor countries. Forty dollars per person in the rich world is the bargain of the planet. It comes out to just 10 cents for every $100 in the rich world. Current aid flows for health are roughly about half of what is needed.

Recommendation 1. Rich countries should devote 0.1 percent of GNP ($40 billion per year as of 2009) to health assistance for poor countries in order to close the financing gap of the primary health system.

The second recommendation is to put that money into highly effective organizations. My own recommendation would be to transform the GFATM, which has done such an outstanding job, into a more general global health fund and to channel about $20 billion per year of the $40 billion per year (that is, half) through the global health fund. This expanded fund would help poor countries to establish their basic health care systems with all of the components discussed earlier.

Recommendation 2. Half of that money should be channeled through the GFATM, which would become the global health fund.

The third recommendation is that the low-income countries contribute around 15 percent of their total national budgets for funding health. (The 15 percent target is known as the "Abuja target.")

Recommendation 3. Low-income countries should fulfill the Abuja target of allocating at least 15 percent of domestic revenues to the health sector. Total spending (domestic and external funding) should be at least $60 per person per year in order to ensure basic health services.

The fourth recommendation is to finish up the job of comprehensive malaria control. Malaria is a pernicious disease, a killer, and a burden on development;

but we are so close to having it under control that attention must be paid to it. The world should adopt a plan for comprehensive malaria control. That would cost roughly $3–4 billion a year, which would enable the poor countries to finish up the building of supply chains and the funding of CHWs, rapid diagnostic tests, medicines, and so forth, to really get the job done.

Recommendation 4. The world should adopt a plan for comprehensive malaria control, with an (near) end of malaria mortality by 2015 (estimated cost $3–4 billion per year).

The fifth recommendation is that the leading donor countries should fulfill their long-standing commitment to providing universal access to antiretroviral medicines (ARVs) for individuals infected with HIV. These medicines work, and the treatment of poor people will save their lives. More treatment would also mean much lower transmission of the disease, because when an HIV-infected individual is treated with antiretroviral medicines, the viral load (the concentration of the virus in the body) diminishes sharply. This makes it much less likely for the virus to be transmitted from the infected individual to another person.

Recommendation 5. The G8 should fulfill the commitment to universal access to ARVs.

The sixth recommendation is that the leading donor countries should also fulfill their commitment and partnership with the poor countries in fully funding the Global Plan to Stop Tuberculosis (TB). This too has a financing gap on the order of $3–4 billion a year, approximately $3–4 per person per year in the high-income world. Roughly a cup of coffee once per year at a favorite coffee shop in New York City or London would be what is needed incrementally to build the requisite funding!

Recommendation 6. The world should fulfill the Global Plan to Stop TB, including closing the financing gap of $3–4 billion per year.

The seventh recommendation is that the world, especially the donor countries in their financing and the poor countries in their implementation, should guarantee access to sexual and reproductive health services. This would include emergency obstetrical care for safe childbirth, antenatal care for safe pregnancy, and modern contraceptives, as households would like. Many women around the world do want to use modern contraceptives and have fewer children, but they lack the access or the funds to be able to afford this on a market basis. These women need full funding of family-planning services, contraception, emergency obstetrical care, and pregnancy care, all of which need to be accomplished on a very low budget so these services can be made universal.

Recommendation 7. The world should fulfill the funding for access to sexual and reproductive health services, including emergency obstetrical care and contraception, by the year 2015, partly through the global health fund and partly through the UN Population Fund.

The eighth recommendation is for the global health fund to take up the "neglected tropical diseases." These neglected tropical diseases do not appear in the headlines as frequently as malaria or AIDS. They include hookworm; the worm infections ascariasis, trichuriasis, onchocerciasis, and schistosomiasis; and lymphatic filariasis, which is another vector-borne tropical disease with terrible consequences, but that is controllable through bed nets and with ample effort and organization. Another neglected disease is blindness from cataracts, which can now be treated with surgery at remarkably low cost.

Recommendation 8. The global health fund should establish a financing window for seven controllable neglected tropical diseases: hookworm, ascariasis, trichuriasis, onchocerciasis, schistosomiasis, lymphatic filariasis, and trachoma.

Recommendation nine calls on the global fund to establish special financing to complete the health systems, for example, the training and deployment of CHWs. This would be a crucial part of the transformation from a disease-targeted fund to a general global health fund that provides a broad base of services. My colleagues and I have also called for the United Nations to help African nations to deploy 1 million CHWs to Africa by 2015 as a major boost for achieving the MDGs.

Recommendation 9. The global health fund should establish a window for health systems, including the mass training and deployment of 1 million CHWs in Africa by 2015.

Finally, the tenth recommendation is to address the number of noncommunicable diseases that have been typically overlooked in many of these urgent MDG-related efforts. Dental care, for example, is very basic and important but often not present in poor countries. Eye care, mental health counseling, and mental health interventions are also crucial for the massive burden of depression, which is pervasive around the world. There are many cardiovascular diseases for which undiagnosed high blood pressure and hypertension can cause loss of life for adults if untreated, but where the adverse consequences can be controlled if a diagnosis is made in time. A number of cancers can be addressed at a very low cost. Campaigns against tobacco use are a crucial part for any good public health system, because tobacco remains a massive killer. This is a behavioral challenge, but one that needs to be met, because it is one of the most effective ways to save lives.

Recommendation 10. The world should introduce primary health care (mass prevention and treatment) for noncommunicable diseases in the areas of oral health, eye care, mental health, cardiovascular disease, and metabolic disorders, This would include taking measures to address lifestyle (smoking, trans fats, urban design for a healthy environment), surveillance, and clinical care.

The moral of the story is that the world is close to achieving primary health for all. The MDGs have hugely spurred the effort. We can now see a pathway to health for all—the next phase of the global development objectives and the Sustainable Development Goals. Universal health coverage will feature prominently in the next phase of the goals, and we will have the opportunity to complete what we have started—to finally achieve health as a basic human right.

V. The Continuing Challenges of Health Coverage in the High-Income Countries

We have analyzed the challenge of health in the very poorest parts of the world. One might assume there is no longer a challenge at the opposite end of the income spectrum in the upper-income countries. Life expectancy is high and the health care systems are technically very sophisticated. In high-income countries, annual public spending for health care is typically around $3,000–4,000 per person per year in the public budget. In the United States, the sum of public and private spending for health is $8,000 per person per year.

What then is the problem? The problem, especially in the United States, is that $8,000 per person per year is incredibly expensive. It is so expensive that it is a major burden on the economy (around 18 percent of the total GDP, the largest single sector of the economy!), a major burden on the budget, and a major burden on poor people who are often priced out of the health care market entirely, as well as on those in the middle class who may be burdened with crushing health care costs. In the United States there is high inequality of income alongside expensive health care and a rather limited social safety net provided by government programs.

The puzzle I want to explore is one that is specific to the United States, in one sense. Why is the U.S. health care system so expensive compared with others? Even though the problem is mainly in the United States, the lessons of the U.S. case are globally applicable and therefore important to understand.

One of the main reasons why the U.S. system is so expensive is that it is a privately oriented health-delivery system. Yet there is the common insistence in the U.S. debate that the private sector is more efficient, and the public sector is inefficiently bureaucratized and very costly. Nonetheless, in the United States, where the private sector role in health care is the largest of any high-income country, the costs per person are also the highest. It is important to understand why this is the case, because it gives further understanding of the proper boundaries between public and private sectors, especially in health systems. In general, it helps overcome a presumption that the free market is always the best solution. The results of this free-market orientation in U.S. health care are peculiar, to say the least.

From the start, it should be understood that health is not a typical market commodity. For one reason, it is a merit good: most of us agree that access to health care should be universal. Health is a basic human right. Health therefore starts out in a very specific situation, similar to education. Public provision of these services is important if for no other reason than to help ensure that the poor are able to gain access to these merit goods just like the rich.

But the problem goes even deeper than that. The economist Kenneth Arrow, the great Nobel laureate, in a seminal 1963 article observed that the health sector cannot really operate like a competitive market sector because there is a fundamental problem (Arrow 1963). Patients generally do not know what is best for them medically. They must rely on their doctors, who presumably do know (or should know!). Arrow noted that such an asymmetry of information violates one of the basic assumptions of the proper functioning of a free-market economy, which is that consumers and suppliers in a transaction have similar information.

What happens when only the suppliers have the information? Your doctor says you need a test and generally you listen to what the doctor says. When there is asymmetric information, and the supplier is the one who has the knowledge, then it is possible that the consumer can end up being overcharged and overbilled if the incentives facing the suppliers are not properly designed. Unfortunately the U.S. system has this problem. Too many health care providers have the incentive to overcharge patients, either by fraud, overexcessive use of procedures, or use of monopoly power (such as when there is only one hospital in the community).

There are other problems with treating health care as if it were a free-market commodity. Health care requires insurance because of the possibility of expensive health care costs in the event of a serious disease. Yet if individuals know their conditions while the health insurer does not, the insurer will find the health premium is too low. The health insurer will raise the premium, so healthy people will not buy insurance and only very sick people (who expect to use the insurance coverage) will do so. The pool of insured individuals will increasingly include only those who are already sick or who have good reason to believe they are likely to become sick. The insurance market will shrink, and even collapse, in what is commonly called an "insurance death spiral."

The U.S. system is characterized by substantial fraud, excessive care, waste, and monopoly power of local health providers. Insurance markets do not work well and leave many people outside the private system. The Affordable Care Act, also known as Obamacare, addresses some of the problems, but does not really get to the crux of the excessive health care costs in the United States (42 U.S.C. § 18001 [2010]). The U.S. system therefore remains far more expensive than other systems

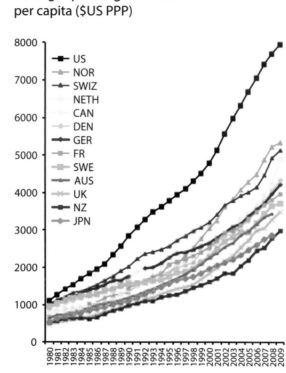

Average spending on health per capita ($US PPP)

Legend: US, NOR, SWIZ, NETH, CAN, DEN, GER, FR, SWE, AUS, UK, NZ, JPN

9.9 Average spending on health per capita (1980–2009)

Source: Squires, David A. 2012. "Explaining High Health Care Spending In the United States: An International Comparison of Supply, Utilization, Prices, and Quality." The Commonwealth Fund, Pub. 1595, Vol. 10.

that rely much more on regulated prices and much less on private, unregulated health providers.

In figure 9.9, the dotted black line at the top is the cost of spending in the United States per person. It is soaring. Back in 1980, the average spending on health in the United States was about $1,000 per person. By the year 2009, it was $8,000 per person. Norway, another rich country, is next, but at a much lower level of around $5,500 per capita. In general, the rest of the rich countries are clustered around this lower level. Typical spending is around $4,000 per person per year

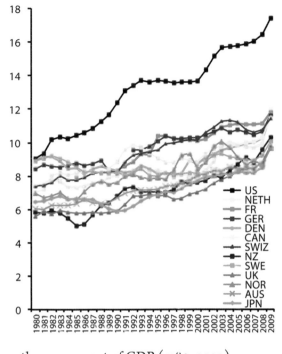

9.10 Total health outlays as percent of GDP (1980–2009)

Source: Squires, David A. 2012. "Explaining High Health Care Spending In the United States: An International Comparison of Supply, Utilization, Prices, and Quality." The Commonwealth Fund, Pub. 1595, Vol. 10.

outside of the United States—half of the U.S. level (Organization for Economic Co-operation and Development [OECD] 2011).

Figure 9.10 shows spending divided by GDP to get the share of health outlays as a percentage of income. Back in 1980, the United States was spending about 9 percent of its national income on health. By 2009, that had doubled to 18 percent of national income. In 1980, all of the countries including the United States were rather tightly clustered between 6 percent and 9 percent of national income. As of 2009, for most countries the spending is about 10 percent of GDP, but the United States has separated from the pack and become the most expensive health care system in the world. In general, health costs and health outlays as

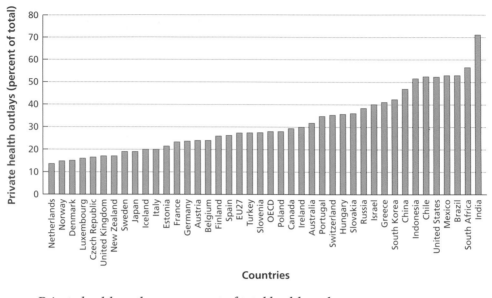

9.11 Private health outlays as percent of total health outlays

Source: Organization for Economic Co-operation and Development. 13/05/2014. "Aid (ODA) by sector and donor." OECD Publishing. http://www.oecd.org/dac/stats/data.htm.

a share of national income have been increasing, yet not at the same pace as the United States (OECD 2011).

One of the leading scientific organizations that studies the U.S. health system, the Institute of Medicine of the U.S. National Academies of Science, found something extraordinary in a recent work (Institute of Medicine 2013). The waste, fraud, and abuse in the health care system amounted to as much as 5 percent of U.S. national income, about $750 billion a year. This is from overbilling, waste of resources, repeated tests, outright fraud, and high management costs. To put it into context, the United States is spending 18 percent of its GNP in health while only getting 13 percent of national income in real value.

The distinctiveness of the U.S. system is the size of the private sector, as illustrated in figure 9.11. The figure shows the share of total health spending that comes from private sources (e.g., private payments to an insurance company) as opposed to public (budget) sources. We see that the United States is the only

		Price for 30 most commonly prescribed drugs, 2006–2007 (U.S. set at 1.00)			Primary-care physician fee for office visits, 2008		Orthopedic physician fee for hip replacements, 2008	
Table 9.4 Drug and physician prices								
	Brand name	Generic	Overall	Public payer	Private payer	Public payer	Private payer
Australia	0.4	2.57	0.49	$34	$45	$1,046	$1,943
Canada	0.64	1.78	0.77	$59		$652	
France	0.32	2.85	0.44	$32	$34	$674	$1,340
Germany	0.43	3.99	0.76	$46	$104	$1,251	
Netherlands	0.39	1.96	0.45	—	—	—	—
New Zealand	0.33	0.9	0.34	—	—	—	—
Switzerland	0.51	3.11	0.63	—	—	—	—
United Kingdom	0.46	1.75	0.51	$66	$129	$1,181	$2,160
United States	1	1	1	$60	$133	$1,634	$3,996
Median	0.43	1.96	0.51	$53	$104	$1,114	$2,052

Source: Squires, David A. 2012. "Explaining High Health Care Spending In the United States: An International Comparison of Supply, Utilization, Prices, and Quality." *Issues in International Health Policy* 10: 1–14.

high-income country where private payments are more than half the total outlays (OECD 2011). The United States is both the most expensive and the most private system. This is not a coincidence. It is the mistaken reliance on the private market system, one that does not satisfy the assumptions of an efficient, competitive marketplace, that explains America's overpriced system.

One can see this in table 9.4, a systematic comparison of U.S. costs with the costs in other high-income countries for some specific commodities and services (Squires 2012, 6). For example, the U.S. cost for the 30 most commonly prescribed medicines is set at an index of one. Then the cost in New Zealand is calculated at 0.33, which means one-third of the U.S. cost. The cost in Australia is 0.4, roughly one-half of the U.S. costs of those medicines. The cost in the United States of a visit to a physician paid by a public sector program ($60) is comparable to what is paid in other countries. However, the cost that is paid by a private payer or

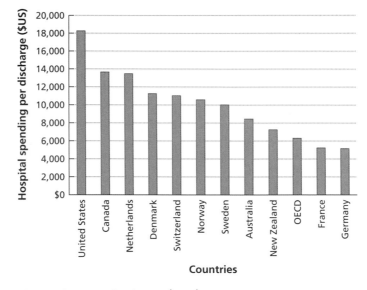

9.12 Hospital spending per discharge (2011)

Source: Squires, David A. 2012. "Explaining High Health Care Spending In the United States: An International Comparison of Supply, Utilization, Prices, and Quality." The Commonwealth Fund, Pub. 1595, Vol. 10.

private insurance company is over twice the amount ($133) that is paid by the public payer. The U.S. system costs are still simply out of sight.

Figure 9.12 shows the spending for each patient discharged from a hospital (Squires 2012, 6). How much was spent on a patient in the United States? Remarkably, the cost per hospital discharge was $18,000! In France and Germany, it was less than one-third of this amount. The average of the OECD, a high-income country group, was one-third of the average discharge cost in the United States.

Another disparity is in salaries. Doctors in the United States make far more than doctors in other countries (Squires 2012, 7). In 2008, orthopedic physicians in the United States made $440,000, while those in Germany made less than half that, $202,000. America's high salaries are shown in figure 9.13, which charts physician incomes across selected countries.

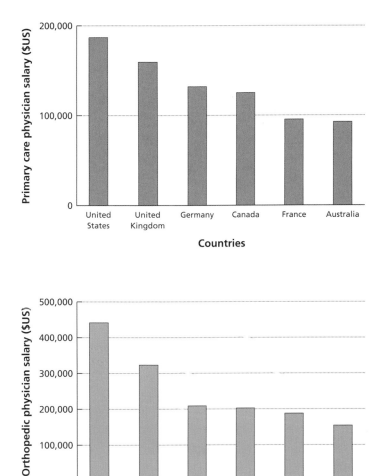

9.13 Comparison of physician incomes (2008)

Source: Squires, David A. 2012. "Explaining High Health Care Spending In the United States: An International Comparison of Supply, Utilization, Prices, and Quality." The Commonwealth Fund, Pub. 1595, Vol. 10.

Table 9.5 Lobbying outlays (1998–2014)	
Sector	Total (1998–2014)
Miscellaneous business	$ 6,078,960,580
Health	$ 6,008,970,746
Finance/insurance/real estate	$ 5,987,751,654
Communications/electronics	$ 4,928,429,581
Energy/natural resources	$ 4,462,152,496
Other	$ 3,184,831,387
Transportation	$ 3,059,318,830
Ideology/single-issue	$ 1,965,929,158
Agribusiness	$ 1,809,056,698
Defense	$ 1,730,989,688
Construction	$ 656,221,136
Labor	$ 594,160,648
Lawyers and lobbyists	$ 415,404,522

Source: Center for Responsive Politics (opensecrets.org).

The U.S. system is remarkably expensive, not because it is delivering a huge range of services that other countries are not, and not because the outcomes are better, but because the unit costs of the interventions and medicines are much higher than in the other countries.

There is one final dimension to this issue: political economy. The health sector in the United States is politically powerful, so it is able to resist effective regulation or replacement by a public system. Indeed, the private health care industry should be considered one of the four most powerful lobbies in the United States, alongside Wall Street, the military-industrial complex, and the oil industry. Indeed, during 1998–2014, the lobbying outlays of the health industry exceeded those of every other industry except "miscellaneous" (which means a grab bag of several industries), as we see in table 9.5. This powerful health care lobby has so far prevented any remedial action on the excessive costs.

Rank	Sector	Amount	% to Democrats	% to Republicans
1	Finance/Insurance/Real Estate	$ 172,827,317	33	56
2	Other	$ 113,552,639	45	39
3	Ideology/Single-Issue	$ 104,239,447	36	42
4	Miscellaneous Business	$ 88,473,842	36	56
5	Lawyers & Lobbyists	$ 71,300,161	60	33
6	Health	$ 65,375,066	41	54
7	Labor	$ 48,685,905	50	7
8	Communications/Electronics	$ 44,451,767	55	39
9	Energy/Natural Resources	$ 41,260,278	21	73
10	Agribusiness	$ 29,485,708	26	70
11	Construction	$ 29,149,051	26	63
12	Transportation	$ 26,760,118	28	69
13	Defense	$ 14,054,121	41	59

Table 9.6 Campaign financing by industry (2011–2012)

Source: Center for Responsive Politics (opensecrets.org).

The health sector is also a major contributor to campaign financing. The health sector was number five on the list of industries in terms of the size of campaign contributions during the most recent campaign cycle (table 9.6). In the complete campaign cycle of 2011 to 2012, management and employees of the health sector contributed approximately $260 million to campaigns. This means politicians are attentive to the interests of this concentrated group, and not necessarily to the interests of the taxpayers or the citizens more generally.

What are some of the reform options?

The first would be to move to a single-payer system as in Canada. Canada's health system is high quality and far lower in cost compared with the U.S. system. The governments of Canada's provinces cover most costs of health care, so private outlays are below 30 percent of the total.

A second possibility is what is called an all-payer system. Health payments would still come from private employers, but regulations would prevent the exercise of monopoly power in pricing. In particular, there would be a single, publicly known price for every health service. Hospitals and doctors would not be allowed to "price discriminate," that is, to charge a monopoly price when they can get away with it (as they can do now).

A third possibility is called capitation. Rather than paying service by service, the hospital or doctor would receive a fixed amount of money per patient per year, irrespective of the specific services that need to be provided during the year. The fixed amount would be according to the "expected costs" of providing quality health care. If the actual costs are higher, then the hospital would bear the excess. There would be no way to overcharge per patient and no incentive to call for excessive services. Of course, there might be trouble in the other direction, with the doctor or hospital refusing to carry out a procedure or surgery because there is no added financial incentive to do so. To make such a system work, the health care provider would be responsible for quality care and would be subject to performance review.

Finally, technology can be the friend of lower costs, with information technology, smarter systems, and even patients monitoring their key vital signs at home; this could all be used to bring costs down sharply. In some places, health care providers are already using extensive distance monitoring of patients, with clinical information automatically transmitted and read at a distance. With these new technologies it is also possible that low-cost CHWs can provide some of the services currently provided by high-cost doctors and nurses.

There are plenty of paths to reform. Incentives must be partly changed, and new technologies must be deployed. Of course, that kind of reform also depends on politics. If the lobbies get their way, there will continue to be greatly inflated costs. If the U.S. system is run for the public's benefit, then tremendous good can be done to reach more people and improve health outcomes at considerably lower cost.

10

FOOD SECURITY

I. Sustainable Food Supply and the End of Hunger

One of the most complicated unsolved problems of sustainable development is how the world will feed itself. This problem is an ancient one. Yet many people thought it had been solved with great breakthroughs in food productivity based on scientific advances. Especially after the Green Revolution of high-yield crop varieties that took off in the 1960s, it seemed very likely that food production would inevitably stay ahead of the growing world population. Now we have some serious doubts. Not only are we coming to realize that a large portion of humanity is poorly fed, but we are also realizing the seriousness of the threats to global food security that lie ahead.

We cannot say we have not been warned. The warnings have been with us for more than two centuries, starting in 1798 with Thomas Robert Malthus who, in *An Essay on the Principle of Population,* posed the basic challenge of food security for a growing population. Malthus's basic point was that any temporary boost in food production capable of relieving food insecurity would cause a rise in the population to the point that humanity was once again reduced to a condition of food insecurity. Malthus would look at our current overall global food surplus and warn, "Yes, that's all fine and good, but what will happen when the population surges from 7.2 billion people today to more than 10 billion people by the end of the century?" He would also note that many people even today live in chronic hunger.

When Malthus posed the challenge of feeding the world population, there were around 900 million people on the planet. Since then, the population has increased by a factor of eight. With 7.2 billion people on the planet and with the global population continuing to grow by around 75 million people per year, the challenge of feeding the planet is with us again. The problem is even more complicated than Malthus could have imagined, for four main reasons:

1. A significant share of the world population today is malnourished.
2. The global population continues to grow.
3. Climate change and other environmental changes threaten future food production.
4. The food system itself is a major contributor to climate change and other environmental harms.

Let us look first at the issue of malnutrition (SDSN 2013c). Malnutrition is a pervasive problem: around 40 percent of the world's population is malnourished in one way or another. One major component of malnutrition is *chronic hunger*, or undernourishment. The Food and Agriculture Organization (FAO) defines chronic hunger as the insufficient intake of energy (calories) and proteins. Hundreds of millions of people are afflicted by chronic hunger and have only the energy for mere survival. The FAO estimated the number at 870 million people for the years 2010–2012.

There is another kind of less visible malnourishment that is sometimes called *hidden hunger*, or micronutrient insufficiency. The calories and proteins may be sufficient, but micronutrients like vitamins or particular fatty acids are not adequately present in the diet. Such micronutrient deficiencies result in various kinds of ill health and vulnerability to infection and other diseases. Key micronutrient deficiencies prevalent in many low-income countries include vitamin A, vitamin B_{12}, zinc, iron, folate, omega-3 fatty acids, and iodine.

The third kind of malnutrition, which is now at epidemic proportions in many parts of the world, especially the richest countries, is the excessive intake of calories leading to *obesity*, meaning weight is far too high for height. Technically, obesity is often defined as a body mass index (BMI) greater than 30, where BMI equals the weight in kilograms divided by the height measured in meters. Overweight is defined as a BMI greater than 25. It is estimated that roughly one-third of all adults in the world are overweight, and around 10–15 percent are obese.

Adding it all up, the numbers are staggering. Around 900 million people are chronically hungry. Perhaps another 1 billion more have enough macronutrients (calories and proteins) but suffer from one or more micronutrient deficiencies. Roughly 1 billion more are obese. In total, around 3 billion people are malnourished out of a world population of 7.2 billion people, meaning that a staggering 40 percent of the world is malnourished.

This is not exactly where we would want the world to be more than 200 years after Malthus warned us about the chronic crisis of food insecurity. We do indeed have a crisis of food insecurity. Sometimes it is hunger; sometimes micronutrient deficiency; and sometimes excessive caloric intake with unhealthy diets, such as diets too heavy in sugars and carbohydrates. Any serious focus on a sustainable and secure food supply for the world must view the nutrition crisis in all its dimensions, from those who lack the basic caloric intake to those who suffer from obesity and the ill effects that come from that condition.

Figure 10.1 shows where these problems are distributed. Chronic hunger is heavily concentrated in tropical Africa and in South Asia. More than one-third of the population in tropical Africa, especially central and southern Africa, is undernourished. In South Asia, between 20 and 33 percent of the population is chronically undernourished. When young children are undernourished, their physical development may be irreparably damaged, leading to adverse health consequences that last throughout their lives. Such consequences can include impaired brain development and vulnerability to various kinds of non-communicable diseases (such as cardiovascular disease or metabolic disorders) as adults.

Chronic undernourishment of young children is measured according to various indicators of severity. The first is *stunting*. Stunting means that a child has a very low height for his or her age. Specifically, children are assessed relative to a standard population distribution of height for age. Children who are more than two standard deviations below the norm are considered stunted. Stunting reflects the inadequacy of dietary intake but can result both from a poor diet and from

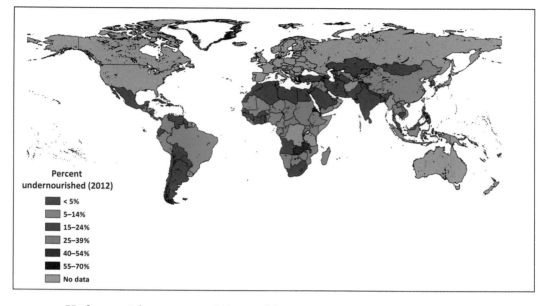

10.1 Undernourishment around the world

Source: World Bank. 2014. "World Development Indicators."

chronic infections, such as worm infections. As figure 10.1 shows, the most severe stunting is found in tropical Africa, and the highest stunting rates in the world are in South Asia, especially India. We noted earlier, in our discussion about poverty in South Asia, that stunting typically reflects a combination of factors: inadequate diets, chronic infections (such as worms), and lack of access to safe water and sanitation, making children vulnerable to new infections.

The second condition is even more urgent, and that is *wasting*. While stunting is a chronic condition in which the child does not grow, wasting is a low weight for height. Figure 10.2 illustrates the physical differences between the conditions. Wasting it is often a sign of acute, life-threatening undernutrition, of the kind one often sees in a famine. In those cases, children may require an urgent rescue through therapeutic foods (high-intensity nutritional foods designed to combat acute undernutrition) and emergency procedures to help keep the children alive.

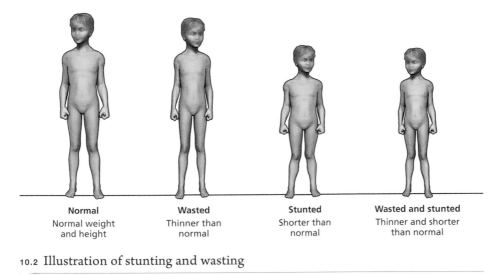

| **Normal** | **Wasted** | **Stunted** | **Wasted and stunted** |
| Normal weight and height | Thinner than normal | Shorter than normal | Thinner and shorter than normal |

10.2 Illustration of stunting and wasting

London School of Hygiene & Tropical Medicine.

There is a key distinction between chronic undernutrition (chronic insufficiency of calories and proteins) and acute undernutrition that may arise from wars, disasters, droughts, and displacement of populations. When those acute episodes occur, there is not only massive suffering but also the risk of a massive loss of life from starvation and disease.

Violence and conflict often break out in hungry regions. Figure 10.3 depicts food insecurity in August 2011. In West Africa there was drought and food crisis in the Sahel, covering Mali, Chad, and Niger; in East Africa, there was drought and food crisis in the Horn of Africa, covering Ethiopia, Somalia, northern Uganda, northeast Kenya, and Djibouti. In both cases, the drought and resulting famines led to large population movements and resurgent violence as migrants clashed with local populations. In Mali, regional conflicts and local conflicts combined to produce a massive and devastating civil war.

Hidden hunger afflicts not only those with chronic undernutrition but also another billion or so people who have an adequate caloric intake but an inadequate variety of nutrients in the diet. Figure 10.4 shows estimates of three particular micronutrient deficiencies (iron, vitamin A, and zinc; Muthayya et al. 2013).

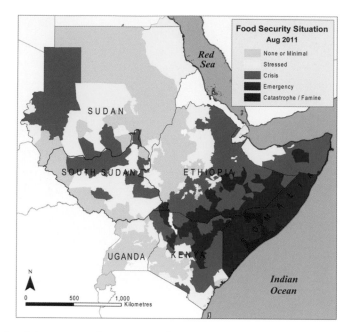

10.3 Acute food insecurity (August 2011)

FEWS NET (2012).

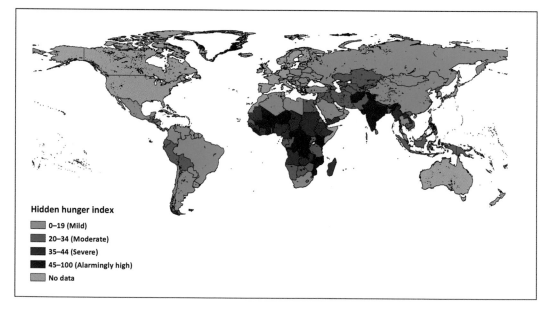

Hidden hunger index

- 0–19 (Mild)
- 20–34 (Moderate)
- 35–44 (Severe)
- 45–100 (Alarmingly high)
- No data

10.4 Hidden hunger index (zinc, iron, and vitamin A deficiencies)

Source: Muthayya, Sumithra, Jee Hyun Rah, Jonathan D. Sugimoto, Franz F. Roos, Klaus Kraemer, and Robert E. Black. 2013. "The Global Hidden Hunger Indices and Maps: An Advocacy Tool for Action." PLoS One 8(6): e67860. doi:10.1371/journal.pone.0067860.

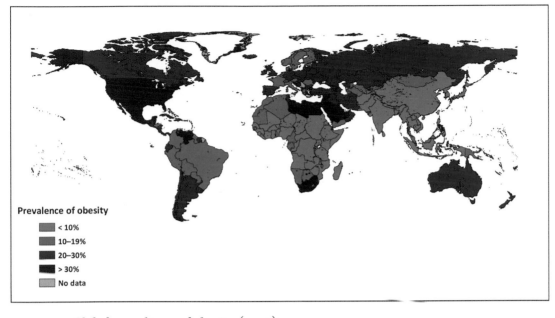

10.5 Global prevalence of obesity (2008)

Source: GHO data repository, World Health Organization.

We note the especially high rates of micronutrient deficiencies in South Asia, West and central Asia, much of tropical Africa, and the Andean region. Unfortunately, the data on hidden hunger are themselves hidden. There are no precise estimates of micronutrient deficiencies around the world.

Obesity marks the other end of the malnourishment spectrum and also causes a tremendous amount of disease and premature mortality. As we see in figure 10.5, the United States, Mexico, Venezuela, and several countries in the Middle East and North Africa (Libya, Egypt, Saudi Arabia), South Africa, and a few others, have an obesity rate above 30 percent. Several more countries, notably in Europe and the former Soviet Union, have an obesity rate between 20 and 30 percent. Why are we facing this obesity epidemic?

The fundamental causes of the obesity epidemic are still not fully clear. Part of the cause is total caloric intake and part is the result of relative inactivity in the urban environment. The high caloric intake may also result from the *kinds of highly processed foods* that people are eating, notably foods with a high "glycemic index." The glycemic index measures the rate at which a food raises the level of blood sugar. Foods with high glycemic indexes include soft drinks, potatoes, rice, and many baked goods. Foods with low glycemic indexes include whole grains, fruits, and vegetables. It is hypothesized that high–glycemic index foods give rise to a sharp rise in blood sugar followed by a sharp rise in insulin, which in turn lowers the blood sugar and raises the appetite. Satiety is therefore reduced, and overeating may result.

In short, the obesity epidemic most likely results from a combination of too many calories, the wrong kinds of calories, and the extreme physical inactivity of urban life. There is no doubt, as the map in figure 10.5 indicates, that a global epidemic is underway. It is already spreading from high-income countries to middle-income countries, and poses a rising threat to health and wellbeing.

To counteract this epidemic, dietary changes combined with more physical activity will be key. Breakthroughs in nutritional science are giving us guidance on improved diets. One of the leaders of modern nutrition is Walter Willett, chair of Harvard University's Department of Nutrition. He proposed a "healthy eating pyramid" (figure 10.6) that depicts the kinds of foods, and the relative frequency and amounts that should be eaten, in a healthy and well-balanced diet. A healthy diet includes whole grains (with low glycemic indexes), vegetables, fruits, and plant oils. Meats and foods with a high glycemic index (e.g., potatoes and rice) should be eaten sparingly. At the base of the pyramid is daily exercise. Unfortunately, actual diets in the United States and other countries with obesity epidemics are quite different, with very high intakes of processed grains, rice, potatoes, soft drinks, red meat, and unhealthy fats (trans fats) used in baked goods and fast foods.

Global food insecurity is already bad enough, but it is likely to get worse before it gets better. Not only is around 40 percent of the world malnourished, but the global food supply is also becoming destabilized by climate shocks and other environmental ills (e.g., freshwater depletion threatening irrigation of crops), even as the world population continues to increase. Moreover, the global demand for grains is rising even faster than population. Countries with rising incomes like China are also shifting to diets with more meat products. Since each kilogram of

White rice,
white bread,
white pasta;
potatoes,
soda, and
sweets

Red meat,
butter

↑ Use sparingly ↑

Dairy or calcium
supplement,
1-2 times/day

Multiple vitamins
for most

Fish, poultry, eggs, 0–2 times/day

Nuts, legumes,
1–3 times/day

Alcohol in
moderation
(if appropriate)

Vegetables (in abundance)

Fruits, 2–3 times/day

Plant oils (olive, canola, soy,
corn, sunflower, peanut,
and other vegetable oils)

Whole grain foods (at most meals)

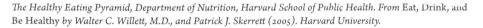

Daily exercise and weight control

10.6 The healthy eating pyramid

The Healthy Eating Pyramid, Department of Nutrition, Harvard School of Public Health. From Eat, Drink, and Be Healthy *by Walter C. Willett, M.D., and Patrick J. Skerrett (2005). Harvard University.*

beef requires 10–15 kilograms of feed grain to raise the cattle, there is a multiplier effect of higher incomes on the global demand for grains. The combination of unstable food production (due to climate change) and rising food demand (due to rising population and meat consumption) is resulting in upward pressure on global food prices.

Figure 10.7 shows food prices since the late 1970s (Rosegrant et al. 2012, 90). From 1970 to the early 2000s, real food prices were falling: the rise of food production was outpacing the growth of food demand. Yet since the early 2000s, food prices have been soaring in real terms (i.e., food prices have increased much faster than average inflation). Indeed, the rise of food prices during the period since 2000 marks a major reversal of an even longer trend of falling real food prices throughout the twentieth century.

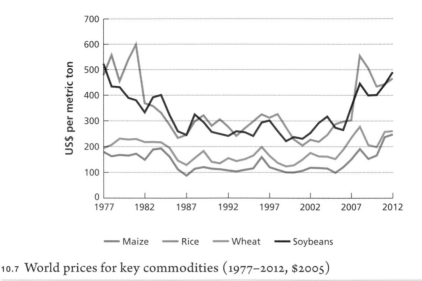

10.7 World prices for key commodities (1977–2012, $2005)

Source: Rosegrant, Mark W., Simla Tokgoz, Prapti Bhandary, and Siwa Msangi. 2012. "Scenarios for the Future of Food." In 2012 Global Food Policy Report, *89–101. Washington, DC: International Food Policy Research Institute.*

For wealthy people, the rise in food prices is an inconvenience. As we see in figure 10.8, only around 6 percent of U.S. household consumption is for foodstuffs. Yet for poor people, diets consume a large percentage of the family income, as much as 45 percent of the household budget in the case of Kenya, as shown in the figure. This inverse relationship of total consumption per person and the share spent on food is known as Engel's Law, and it is one of the most robust patterns in the economy. Because of Engel's Law, the recent rise in global grain prices is more than just an annoyance or a hindrance for the poor, especially the urban poor. It is often a profound threat to their wellbeing, one that pushes many people into abject poverty, hunger, and desperation. Of course, for farmers who are net sellers of grain on the market, the rise in food prices can be a blessing, not a curse, as it may raise the farmers' incomes by more than it raises their households' cost of food consumption.

In summary, we have around 40 percent of the world still not properly nourished, a food supply already under stress, rising food prices, and increasing demand for food production. What can be done? We now turn to the supply side:

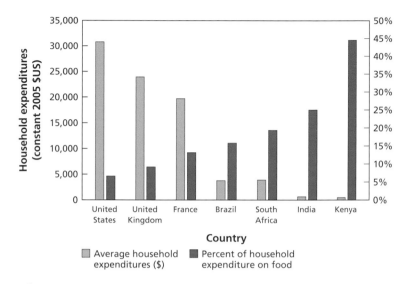

10.8 Engel's Law: Proportion of income spent on food falls as income rises

Source: World Bank. 2009. USDA.

how food is grown, where it is grown, and what might be the prospects for sustainable and nutritious food production in the future, especially in the era of climate change and water scarcity.

II. Farm Systems, Ecology, and Food Security

One of the challenges of addressing global food security is the remarkable variation in farm systems around the world. There is no "one size fits all" when it comes to farming or to methods to increase farm yields. This enormous diversity should not be surprising. Farmers differ incredibly in what they grow; how they grow it; and the challenges of climate, soil, water, topography, pests, biodiversity, and transport costs they face. These variations in turn have an enormous effect on farm systems and strategies. As a result there is no single or simple answer as to how farmers can become more productive and more resilient to environmental risks. Part of the proper diagnostics and solutions for a global sustainable food supply depends on thoroughly understanding how farm systems differ around the world.

There are about 130 million square kilometers of land on Earth and of that, a remarkably large proportion is already taken for human needs (United Nations Environment Programme 2014). Agriculture constitutes around 50 million square kilometers, roughly 40 percent of the world's total land area. Roughly 14 million square kilometers are arable land (land that can be used for agricultural crops), and roughly 34 million square kilometers are meadows and pasturelands. Farmland itself accounts for a little over 10 percent of the world's land area. Pastures are much bigger, around a quarter of the total land area. Another 39 million square kilometers are forests, covering about 30 percent of the Earth's land area. A modest proportion of the forests, perhaps around 15–20 percent of the total, are managed for pulp, paper, timber, logging, and other products. The remainder of the Earth's land is about 41 million square kilometers, roughly 30 percent of the total, much of which is uninhabitable land such as deserts and high mountains. Only a few percent of the world's land areas is in cities, where half the world's population lives. Recent estimates put the urban landscape at around 3 percent of the total, and rural households and businesses at another 3 percent.

Figure 10.9 maps where the agricultural land is, both for cropland and grazing land. The green shaded areas, where cropland is greater than 50 percent of the total, include the U.S. Midwest; parts of western, central, and eastern Europe; Russia; and much of China and India. In Africa and South America, the grazing land and cropland are mixed. The drier areas tend to be places where food crops cannot be grown with high productivity, and so are used more for animal grazing. In semi-arid environments, one finds nomadic populations that move herds of livestock across large areas in pursuit of the grasslands watered by the seasonal rains. In Africa, these pastoralist environments are found in the semidesert regions of the Sahel in West Africa, the Horn of Africa, and the deserts (like the Kalahari of Botswana) in Southern Africa.

There are two major forest areas to study in the map of the world's forests in figure 10.10. First are the rain forests around the equatorial belt. The Earth receives the highest solar radiation per square meter at the equator. The intense solar radiation warms the equatorial land and causes the humid equatorial air to rise and

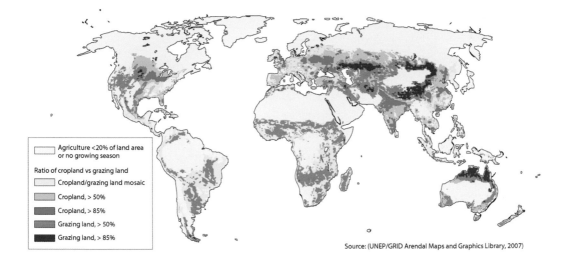

10.9 Global distribution of pastureland and cropland (2007)

Source: Ahlenius, Hugo. 2006. UNEP/GRID-Arendal. http://www.grida.no/graphicslib/detail/agriculture-land-use-distribution-croplands-and-pasture-land_aoab.

cool. The water vapor condenses and gives rise to massive precipitation at the equator (and a descending dry area at around 25 degrees north and south latitude, giving rise to deserts north and south of the equator). The equatorial rainfall and warm year-round temperatures produce the three great equatorial rain forests of the planet: the Amazon in South America, the Congo Basin in Africa, and the great rain forests of the Indonesian archipelago in Southeast Asia. This rain forest band circles the planet at the equator. The other major forest regions are in the high latitudes, like the boreal forest across the vast Eurasian land mass and Canada. Unlike the forests that once stood at midlatitudes in North America, India, China, and Europe, but have since been cut down, the boreal forests remain standing today in large part because the land under the forests cannot be used profitably for farming, as the temperatures are too low and the potential growing seasons are too short. Otherwise, human settlers would likely have deforested the high-latitude regions just as humans long ago deforested the temperate midlatitude regions.

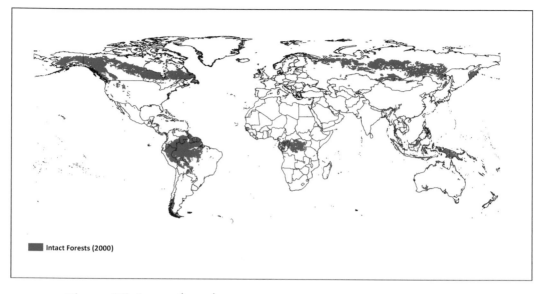

Intact Forests (2000)

10.10 The world's forests (2000)

Source: Potapov P., A. Yaroshenko, S. Turubanova, M. Dubinin, L. Laestadius, C. Thies, D. Aksenov, A. Egorov, Y. Yesipova, I. Glushkov, M. Karpachevskiy, A. Kostikova, A. Manisha, E. Tsybikova, I. Zhuravleva. 2008. "Mapping the World's Intact Forest Landscapes by Remote Sensing." Ecology and Society, 13(2): 51. http://www.ecologyandsociety.org/vol13/iss2/art51/.

We see therefore that the locations of cropland and forests are deeply rooted in the ecological conditions, including temperature, rainfall, topography (e.g., if the land is too steeply sloped it may be impossible to farm), whether irrigation is feasible, and more. All of these patterns shape the farm systems and shape the location of human populations. Population densities are nearly zero in the deserts and tundra and low in the near-desert areas and areas in high latitudes just lower than the tundra. Population densities and farmlands are prevalent in temperate midlatitude zones that are well watered, with good soils, moderate temperatures, and therefore good growing conditions for crops. These areas were heavily forested in prehistory, but humanity began deforesting them long ago to make way for croplands and pasturelands.

Many of the forests today are being threatened with deforestation, especially the equatorial rain forests. Populations are encroaching on these areas for a variety of reasons, including to make way for pastureland and cropland or to get fuel wood

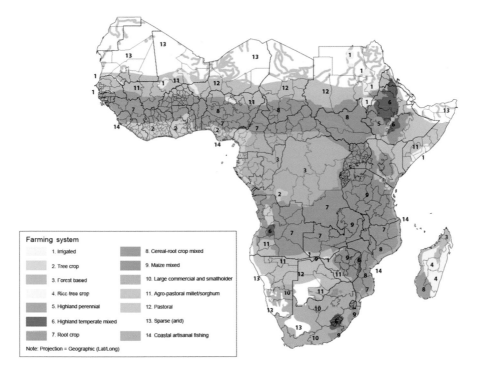

10.11 Major farm systems in sub-Saharan Africa

Source: Dixon, John et al. 2001. Farming Systems and Poverty. Food and Agriculture Organization of the United Nations. http://www.fao.org/farmingsystems/staticssa_en.htm. Reproduced with permission.

and other goods and services. The pace of deforestation is currently unsustainable in all of the great equatorial rain forests. Some of the forests are overlogged for tropical hardwoods, which are highly valued but used in an unsustainable manner around the world. Rain forests are also being cut down and replaced by massive tree plantations; for example, to grow high-demand products like palm oil, a problem that is particularly intense in Indonesia, Malaysia, and Papua New Guinea.

It is worthwhile to take a deeper look at one part of the world to illustrate how the geography shapes the farm systems and the society. In Africa, there are distinctive characteristics of climate that cause distinctive farm systems and distinctive economic results as well. Figure 10.11 maps the various farm systems in

Africa—the fourteen major agroecological zones, each with a specific kind of farm system adapted to the particular ecology.

Start at the equator. There the year-round high temperatures and high rainfall give rise to a rain forest ecology. The Congo Basin rain forest (shown as the bright green area 3) is not especially productive for annual crops (the soil nutrients are rapidly washed away by the intense rainfall), so the forest is mainly used for tropical forest products (tropical logs and tree crops such as rubber, palm oil, and cocoa). A similar situation applies in the tree-crop ecology of the coast of the Gulf of Guinea (green area 2) in West Africa, stretching from Liberia in the west to Cameroon next to the Congo Basin. This area too is given over to massive tree-crop plantations.

As one moves poleward (meaning toward the North Pole in the Northern Hemisphere and toward the South Pole in the Southern Hemisphere), the rainfall diminishes and becomes more seasonal. In general the rainfall is greatest in the summer months (e.g., in May–September in the Northern Hemisphere, examples being Nigeria and Kenya, and September–March in the Southern Hemisphere, examples being Malawi and Tanzania). Instead of growing tree crops, these regions grow annual crops. Maize growing is especially prevalent in East Africa (in purple zone 9). Root crops like cassava are grown in the brown zone 7. As one continues poleward in both hemispheres, the rainy season becomes shorter still. Only short-season crops that are well adapted to temporary dry spells, such as sorghum and millet, can safely be grown. These dryland crops are found in purple zone 11 in both hemispheres. Continue the poleward journey and one arrives at the arid regions, where the rainfall is too low to grow crops but just enough to water pasturelands for nomadic livestock. Thus ecological zone 12 in both hemispheres is home to the pastoralists of Africa, such as the Tuareg and Fulani of West Africa and the Khoikhoi of southern Africa. Finally, take one more step poleward, and we are in the deserts of the northern Sahara and the southern Kalahari.

Countries naturally straddle several of these zones. A country like Ghana has tree crops in the humid south near the Gulf of Guinea and maize and dryland crops in the north. Mali has irrigated rice in the south and pastoralism in the north. Kenya is a mosaic of farm systems, as is Ethiopia, with deserts, pastoralism, lowland and highland crops (crops are graded by elevation as well as by latitude). These distinct ecologies also are home to distinct ethnic, racial, and religious groups, leading to remarkable social and cultural diversity, as well as the potential for clashes, such as the age-old tensions that can arise between sedentary farm communities and migratory nomadic livestock herders.

People living in food-secure, well-watered (and often irrigated) croplands of the temperate regions often have little feel for the complexity of food production and the potential for food insecurity in seasonal tropical environments, especially those of low average rainfall and high vulnerability to drought. When the rains fail in dryland regions, populations may face hunger and may be forced to migrate in desperation, often bringing them into contact with other ethnic groups competing for scarce land and water. The result can be an explosion of violence, as in Darfur, Ethiopia, and northern Mali. (Ecological tensions are generally one factor among several that give rise to such violence, so complex phenomena such as interethnic violence typically have many drivers, not one factor alone.)

The world has especially failed to grasp the deep crises of the hyperarid regions, such as the Horn of Africa (including parts of Ethiopia, Kenya, Somalia, and other neighbors) and the Sahel (including parts of Senegal, Mali, Niger, Chad, and others). These are agropastoralist regions or in some cases solely pastoralist regions. They tend to be very poor, utterly dependent on rainfall, and suffering under the burdens of climate change, instability of rainfall, rising populations, falling trends in total precipitation, increasing hunger, and resulting instability and violence. However, the long-term drivers of crisis— including climate change and rising populations—have been too slow-moving to be recognized by most policy makers in Washington, London, and other rich-country capitals. When violence breaks out in these impoverished and vulnerable places, the rich countries have tended to respond with military approaches (e.g., to fight terrorist groups in the Sahel and coastal pirates in Somali) rather than to address the underlying problems of poverty, climate change, and unsustainable population increases (e.g., for lack of access to family planning). The military strategists in Washington or NATO fail to see the human and ecological dimensions of the crises.

III. How Environmental Change Threatens the Food System

Yet the problems are even deeper. Not only are massive numbers of people currently food insecure; farm systems almost everywhere are under tremendous stress, unable to ensure healthy diets and nutrition in economical and sustainable ways to meet the needs of the populations locally and globally.

There are some enormous challenges ahead that will make these problems even tougher than they are now. The most direct of these challenges is the fact that the world's population continues to grow relatively rapidly, even in absolute terms. Every year, another 75 million or so people are added to the world population. By 2025, the world will reach 8 billion people. The current medium-fertility variant of the UN Population Division puts the world's population at 10.9 billion people by 2100, although the number could be even larger than that (UN Department of Economic and Social Affairs Population Division 2013). (With a faster fertility decline than in the UN medium-fertility variant, the world population would stabilize at perhaps 8–9 billion.)

At the same time that the world will be grappling with the challenge of feeding more people, the current food supply, already under so much stress, is going to be even further stressed by another couple of features (SDSN 2013c). One that I noted earlier is the tendency of countries with rising incomes to add more meat to the diet, amplifying the demand for feed grains. The second major challenge is the environmental threats that will make it harder to grow food in many places in the world. These environmental threats and changes come in many shapes and forms. Climate change is the biggest of all. As the climate changes in complex ways under the force of human emissions of greenhouse gases (GHGs), for many parts of the world these changes will be highly adverse for food production.

Higher temperatures in general are going to be harmful for food production in today's warm environments. (The very high latitudes of the world, such as in Canada and Russia, could experience a rise in food productivity as very cold places become a bit less cold.) Especially in the poorest, tropical parts of the world, crops are likely to face temperature-related stresses. At high temperatures, crops may not develop at all, seeds may become infertile, and plant respiration may mean a net reduction of yields of farm crops. Higher temperatures mean faster evaporation of water in the soils and more transpiration of water through the stomata of the leaves of plants (the combination of evaporation and transpiration is called evapotranspiration). Climate change threatens the soil moistures, which in turn threatens the productivity of crops.

Warming also will be accompanied by changes in regional and global precipitation patterns. Many parts of the world will become drier, and many dry parts of the world will find it extraordinarily difficult, perhaps impossible, to continue to grow crops. It is a general principle that today's dry places in the tropics and subtropics will tend to get drier, while today's wet places closer to the equator will tend to get wetter, with more intense episodes of precipitation. Dryland places that today are on the very edge of crop growing may find themselves in a new climate that is too dry for food production. Wet places may find a great increase of flooding and extreme tropical storms.

Climate change will also mean rising sea levels. Coastal lowlands that are farmed right now will be threatened. Places on the deltas of the great rivers, like Bangladesh, may be inundated by floods or even permanently submerged.

In addition to climate change, carbon dioxide (CO_2) emissions are having a direct effect of acidifying the oceans. Ocean acidification has serious implications for another part of our food supply: marine life. Figure 10.12 illustrates the effect on shellfish of increasing acidification (from the top pictures to the bottom pictures in each column). A higher concentration of atmospheric CO_2 will lead to greater ocean acidity, which in turn will lead to smaller and damaged shellfish, as the acidity impedes the formation of the calcium carbonate shells. Many highly aquatic environments for marine life will be undermined, with a consequent threat not only to biodiversity but to human nutrition as well.

In addition to climate change and ocean acidification, many other environmental changes are already degrading farmlands and threatening agricultural productivity. Farmers use large amounts of pesticides and herbicides to grow crops, but these chemicals can poison the soils and the environment and take a major toll on biodiversity in farming regions. For example, pollinators like honeybees are vital for crop productivity, e.g., for growing fruits and other kinds of flowering crops. Yet the pollinator populations are plummeting. While the reasons are not clear, and perhaps include several factors, chemical pollutants are likely among the culprits.

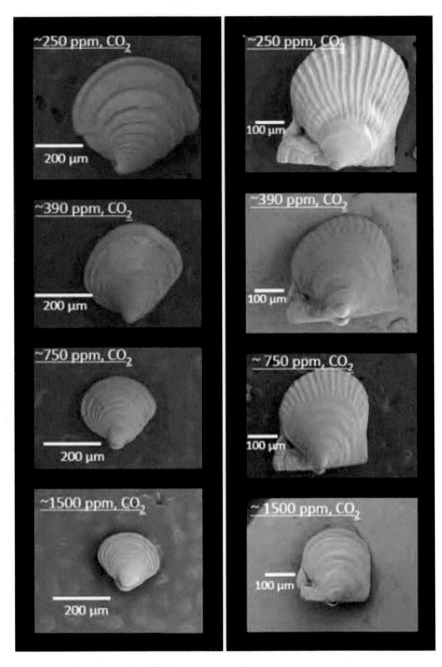

10.12 Impact of CO$_2$ on shellfish

Talmage, Stephanie C., and Christopher J. Gobler. "Effects of Past, Present, and Future Ocean Carbon Dioxide Concentrations on the Growth and Survival of Larval Shellfish." **Proceedings of the National Academy of Sciences.** *September 20, 2010. doi:10.1073/pnas.0913804107.*

Invasive species are another issue. This is when animal or plant species are deliberately or accidentally relocated from one environment to another environment, which can derange the entire ecology. A new species without any natural predators might be introduced in a new setting, and its population may run wild and overtake the native species. In this way there may be a rapid overgrowth of new superweeds, rodents, or other kinds of pests and pathogens that overtake the ecosystem where they have newly arrived.

Environmental stress also threatens the world's irrigated lands, which provide a disproportionate amount of the world's grain production. Crop production obviously depends on water availability, both rainfed and irrigated. Irrigation is the farm system of choice for farmers when they can afford it, because it offers the chance of a high degree of water control and even multiple crops during the year (including during the dry season). The problem is that our current global irrigation depends on freshwater sources—rivers, glaciers, and groundwater—that are all under threat from overuse and human-induced climate change. Glaciers are retreating as they melt under the warming climate. This melting can give rise to a temporary increase of river flow and crop production based on that flow. Yet when the glaciers finally disappear, the meltwater flows will swing from excess to zero. The result of the loss of glacial meltwater could prove to be a devastating and dramatic loss of food production.

Many major rivers have been so overused at this point that they are not even flowing to the sea. This, added to the pressures of climate change, will mean less river flow overall. The Nile, which hundreds of millions of people depend on, will most likely have a significant decline of river flow due to climate change. Those who depend on the vital Yellow River in northern China are experiencing the consequences of declining river flow and a river that no longer reaches the ocean. So too with the Rio Grande, shared by the United States and Mexico and now the cause of deep contention, as water supplies are facing grave stresses in the drought-prone regions of northern Mexico and the southwestern United States.

Groundwater pumping for irrigation, such as in the American Midwest, the Ganges plains, and the northern China plains, is now far faster than the natural recharge of those aquifers. The groundwater aquifers are therefore being depleted. The shocking reality is that hundreds of millions of people depend on irrigated crops where the underlying water sources—groundwater, rivers, glaciers—are already under tremendous stresses that are very likely to intensify in the future.

Rapid land degradation, soil loss, and depletion of soil nutrients are other results of intensive agriculture, when farms have encroached upon land areas with topographies not really suitable for farms, such as the steep slopes on mountain-sides. The consequences are very high. There is likely to be deforestation, the loss of habitat of other species, and a significant emission of CO_2 into the atmosphere as the forests are cut down and burned. There is also a habit of abandoning farm areas after the farm productivity quickly diminishes.

All of these environmental threats—climate change, ocean acidification, chemical pollutants, invasive species, retreating glaciers, excessive pumping of groundwater and river flows—emphasize the fact that farm systems, more than any other human activity, are dependent on the climate and environments we know and have had for the past centuries. Our food supplies are dependent on the known hydrologic patterns, ocean chemistry, and patterns of biodiversity, all of which are now undergoing enormous and rapid human-driven change. The human pressures on the planet are creating a new world—the Anthropocene—and it will be a dangerous one from the point of view of food security.

Of course there are possibilities for adaptation and far more efficient resource use. But the current inertia in habits, and the instability, crises, and conflicts that result from the collision of nature and our current systems, need to be addressed. We must realize how big the challenge will be. It has been hard enough feeding the planet—a challenge we have not even met for our current population in today's environment. When we consider the rising populations and the growing environmental stresses, we realize the magnitude of the food challenges that lie ahead.

IV. How the Food System Threatens the Environment

The problem with the food supply is further complicated by the fact that while the food supply is threatened by environmental change, today's agricultural systems are also the single largest source of human-induced environmental change! In other words, the agricultural systems themselves are a source of the threat to future food production. The arrows of causation run in two directions. On the one side is environmental change that threatens food production. Yet at the same time, agriculture as it is currently practiced gravely threatens the natural environment.

The damage agriculture does to the physical environment adds yet another dimension to the challenge of feeding the planet in a sustainable way. Our problem is not only about how to feed more people and how to feed the growing population more nutritiously than today. It is not only how to maintain farm yields in the face of environmental threats. It is also the challenge of changing current agricultural practices in order to stop inflicting so much environmental damage from the agricultural sector itself. Yet because farm systems differ so much around the world, there will have to be distinctive, localized problem solving in order to make local farm systems compatible with conservation of ecosystem functions, the preservation of biodiversity, and the reduction of human impacts on the climate system and freshwater supplies.

The agricultural sector is in fact the most important sector from the point of view of human-induced environmental change. Many people imagine the automobile or perhaps coal-fired power plants to be the biggest source of human-made environmental damage. And they are indeed major causes of global environmental unsustainability. Yet it is food production that takes the dubious prize as the most important single driver of environmental harms (SDSN 2013c).

What are the kinds of pressures generated by agriculture? The first is GHGs. The farm sector (including deforestation to make room for new farms and pasturelands) is a major emitter of all three of the major GHGs: CO_2, methane (CH_4), and nitrous oxide (N_2O). This means that farm practices will need to be redesigned to help the world move to lower GHG emissions.

The second major impact is on the nitrogen cycle. Our atmosphere is 79 percent nitrogen in the form of dinitrogen (N_2). That form of nitrogen is inert, odorless, without taste, and not very useful for us. However, nitrogen in the *reactive* forms of nitrates, nitrites, and ammonia is absolutely vital for living species, because nitrogen is the backbone of amino acids and proteins. Reactive nitrogen is absolutely core to our metabolism and to every aspect of our lives, including the ability to grow food. It is for that reason farmers put nitrogen on the soil in the form of chemical-based fertilizers and green manures. The nitrogen is a critical macronutrient for the crops. Nevertheless, the heavy use of nitrogen fertilizers (both green and chemical-based) causes major damage to ecosystems by changing the intensity of nitrogen fluxes in the environment.

The third major way that the farm system impacts the planet is the destruction of habitats for other species. This is not entirely surprising considering that an estimated 40 percent of the total land area of the planet is agricultural land. Humanity has already grabbed a huge amount of the land area, but it is still grabbing more. It is especially grabbing more in the forest areas, and the rain forests at risk right now are places of incredible and irreplaceable biodiversity. One of the major reasons that the Earth is vulnerable to a sixth great extinction of species is this process of habitat destruction by human hands.

There are many other ways in which the environment is damaged by farm activity. These include the pesticides (shown in figure 10.13), herbicides, and other chemicals that are used in farm production and that are a major threat to biodiversity. The overuse of freshwater for crop irrigation is another. Around 70 percent of the total human use of freshwater goes through agriculture, with only 10 percent going through household use and the remaining 20 percent or so for industrial processes. Agriculture is a voracious user of water, and that water itself is under threat.

For all of these reasons, the agriculture sector is a key driver of anthropogenic environmental loss. There is a strong need to change farm technologies and processes and patterns of land use to make the food system compatible with a sustainable planet.

The circle graph in figure 10.14 shows us the estimated total amount of GHGs that are emitted, according to the key sectors of the economy. Electricity and heat production, through the burning of coal, oil, and gas, is responsible for a massive amount of CO_2 emissions and for an estimated 25 percent of the total GHG emissions. The transport sector, with the internal combustion engine in automobiles, is responsible for an estimated 14 percent of total emissions. Industrial processes, such as steel production or petrochemical production, account for around 21 percent of total GHG emissions. In total, the direct and indirect use of fossil fuels accounts for around two-thirds of the GHG emissions.

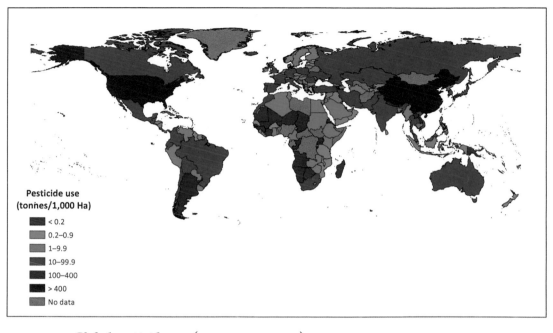

10.13 Global pesticide use (1992–2011 average)

Source: Food and Agriculture Organization of the United Nations. "Pesticide Use." Latest update: 7/18/2014.
http://faostat3.fao.org/faostat-gateway/go/to/download/R/RP/E.

The non-energy sphere is therefore responsible for around one-third of the total GHG emissions. These include CO_2 emissions, CH_4, N_2O, and chemical pollutants from specific chemicals like hydrofluorocarbons. Within the broad category of non-energy GHG emissions, agriculture plays by far the predominant role, both in the direct impacts of farming and the indirect impacts of deforestation and land-use change to make way for farming and livestock management. Of course, agriculture also emits CO_2 through energy use, for the planting, harvesting, storage, and transport of agricultural products.

Agriculture is a major source of CO_2 emissions through land use, but also a major source of the second- and third-ranking GHGs. Methane is emitted in the production of certain crops, notably paddy rice, and by livestock, through the natural processes of their digestion. Nitrous oxide is also emitted from agriculture,

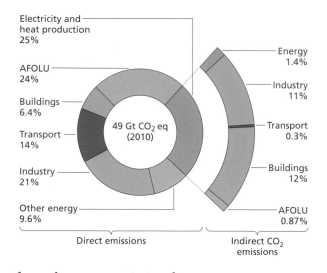

Electricity and heat production 25%

AFOLU 24%

Buildings 6.4%

Transport 14%

Industry 21%

Other energy 9.6%

49 Gt CO₂ eq (2010)

Energy 1.4%

Industry 11%

Transport 0.3%

Buildings 12%

AFOLU 0.87%

Direct emissions

Indirect CO₂ emissions

10.14 Sources of greenhouse gas emissions by economic sectors

Source: IPCC AR5 Summary for Policymakers. Total anthropogenic GHG emissions (GtCO₂eq/yr) by economic sectors. Inner circle shows direct GHG emission shares (in % of total anthropogenic GHG emissions) of five economic sectors in 2010. Pull-out shows how indirect CO₂ emission shares (in % of total anthropogenic GHG emissions) from electricity and heat production are attributed to sectors of final energy use. "Other Energy" refers to all GHG emission sources in the energy sector. . . . The emissions data from Agriculture, Forestry and Other Land Use (AFOLU) includes land-based CO₂ emissions from forest fires, peat fires, and peat decay that approximate to net CO₂ flux from the Forestry and Other Land Use (FOLU) sub-sector. . . . Emissions are converted into CO₂-equivalents based on GWP100 from the IPCC Second Assessment Report.

for example, through the chemical changes to nitrogen-based fertilizers. Instead of being taken up by the plants, the nitrogen in the fertilizer volatilizes (evaporates) and also goes into the water supply. Livestock and fertilizer use thereby emit N_2O in large amounts.

Indeed, the natural nitrogen cycle has now been overtaken in quantity by humanity. In nature, the N_2 molecules in the atmosphere are converted into reactive nitrogen through various biological processes of nitrogen-fixing bacteria, as well as by lightning. Yet now humanity is converting more N_2 into reactive nitrogen than even nature itself. Humanity does it through industrial processes that convert atmospheric nitrogen into ammonia and other forms of reactive nitrogen. Back in the early years of the twentieth century, two great chemical engineers, Fritz Haber and Carl Bosch, developed a process that some consider to be the twentieth century's single most consequential industrial innovation. The Haber-Bosch process is a way to break that N_2 triple bond through the application of high amounts of energy and the use of various identified catalysts, and create ammonia in a large-scale industrial process. This ammonia can then be used to provide the base stock for urea and other nitrogen-based fertilizers. This process, developed between 1908 and 1912, solved the scarcity problem of nitrogen nutrients needed to increase global food production. Up until the Haber-Bosch process, nitrogen deposited on soils came either from the manures of farm animals or from the mining of bird and bat excrement (guano), largely off the coasts of Peru and Chile. But those guano deposits were being quickly depleted, and there was a nitrogen crisis developing at the end of the nineteenth century. Along came the Haber-Bosch industrial process, which spurred production of nitrogen-based fertilizer. What was then a world population of fewer than 2 billion people could thereby become a population of more than 7 billion people one century later.

It was the advent of nitrogen-based fertilizer, along with high-yield seed varieties of the Green Revolution and other agronomic advances, that made it possible to produce enough food to support 7.2 billion people (recognizing that a large number of those 7.2 billion are not well nourished!). Yet with all that nitrogen now being converted from N_2 into reactive nitrogen, there is a huge problem, shown in the complicated flow chart in figure 10.15. What happens to all that chemical nitrogen when it is used in the farms? It runs off into the water supplies and volatilizes into the air, to be carried by the winds to land and water downwind. When reactive nitrogen enters the water supply as nitrates, it creates major dangers to the water supply and ecology. Some of the reactive nitrogen runs into the rivers and the sea, which leads to algal blooms and nitrification in downstream estuaries. Some of it enters the atmosphere, not as N_2, but as N_2O, a GHG. Some of it enters the atmosphere not as N_2O, but as NO_2 (nitrogen dioxide), which causes smog and local pollution.

The graphic from a European Union study in figure 10.16 shows the host of problems arising from the heavy use of fertilizers: more GHGs released; soil acidification; threats to water quality from nitrates and the nitrites in the water

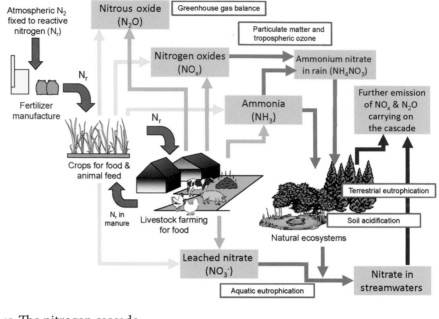

10.15 The nitrogen cascade

Sutton et al. 2011. Reprinted with permission from Cambridge University Press.

supply; eutrophication of the downstream estuaries; and fall in air quality as NO_2, NO_3, and other nitrogen-based molecules enter the urban atmosphere to create smog, tropospheric ozone, and massive health hazards in our cities (Sutton et al. 2011).

Here then is a major dilemma that is rarely discussed in our day-to-day life. We absolutely need the nitrogen, including the chemical fertilizers, for our global food production. Yet the multiple negative impacts of nitrogen on the physical environment, from climate change to eutrophication to urban smog, are serious and growing. The map in figure 10.17 comes from a study showing estuaries around the world suffering from eutrophication, particularly from nitrogen- and phosphorous-based fertilizers off the coasts of the economies with high rates of fertilizer use. The zones in red are "dead zones" in the coastal areas, where eutrophication (excess nutrient loading) has been followed by

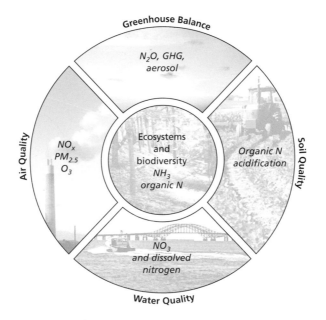

10.16 Key societal threats of excess reactive nitrogen

Sutton et al. 2011. Reprinted with permission from Cambridge University Press.

algal blooms and then by the bacterial decomposition of the algae leading to the depletion of oxygen in the water and a killing off of marine life. The problem is growing and is likely to get worse unless we address how to use the needed nitrogen in a more responsible way, for example, through far more precision in the use of chemical fertilizers to reduce the runoff and volatilization of the reactive nitrogen.

Another key area of heavy agricultural impact is on forests. Forest loss is occurring in all of the great rain forest regions, including the Amazon basin, the Congo Basin, and the Indonesian archipelago. In the Amazon, the clearing of the rain forest to make way for new pastureland, farmland, and construction of infrastructure accounts for most of the deforestation. Fortunately, there has been some reduction of Amazon deforestation in recent years. In Southeast Asia, tropical hardwoods are in huge demand for the booming economies of Asia,

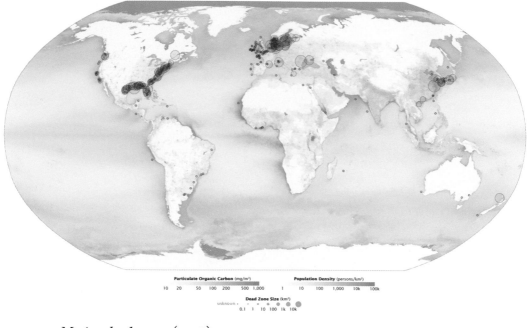

Particulate Organic Carbon (mg/m³)
10 20 50 100 200 500 1,000

Population Density (persons/km²)
1 10 100 1,000 10k 100k

Dead Zone Size (km²)
unknown ·   ·   · ● ● ●
0.1 1 10 100 1k 10k

10.17 Marine dead zones (2008)

NASA's Earth Observatory.

especially China. Logging and clearing of the rain forests to grow tree-crop plantations also account for a major part of the destruction of original rain forest.

In Africa, there is yet another driver: peasant smallholder agriculture. This is not large-scale clearing for logging but the spread of smallholder farmers into the forest margin. Often there is significant deforestation from the use of the forest for fuel wood (e.g., to make charcoal). In the wealthier rain forest regions of the ꞏꞏꞏon and Southeast Asia, the fuel wood problem is typically not as severe, ꞏꞏꞏse there are alternative energy sources. But in the Congo Basin and other ꞏꞏst areas of Africa where populations are very poor and alternative fuel sources ꞏ not available, charcoal is used in such large amounts that it is an important ꞏriver of the deforestation and habitat loss of other species.

In each of these areas, in order to preserve habitat, protect biodiversity, and reduce the GHG emission consequences of deforestation, actions that are responsive to the particular challenges in those areas and the particular needs of the local populations clearly need to be encouraged. This will not only play an enormously important role in helping to reduce the rate of climate change but will also be absolutely vital if we are to succeed in heading off the massive loss of biodiversity.

V. Toward a Sustainable Global Food Supply

Creating a sustainable farm system around the world is absolutely vital. The farm system must simultaneously feed a growing world population and reduce the tremendous pressures our current systems are placing on the Earth's key ecosystems. And the farm systems need to be made more resilient to climate change and other environmental changes that are already underway. We need to think about what will happen if we continue with business as usual (BAU), and contrast that with what we really need to do, which is reshape our own behavior with regard to food. We must reshape farm systems and create an alternative trajectory of sustainable development.

What are the threats of the BAU path? The chart in figure 10.18 attempts to summarize those risks and gives an expert assessment of how serious the particular risks are in particular regions of the world (SDSN 2013c, 10). The boxes that are shaded red are major alerts—these are places in the world where the challenges are of the first order of significance. The boxes that are yellow are like yellow warning signs—things could get bad in these areas with regard to these particular threats.

Business as usual will mean an increase in food insecurity in some parts of the world. The places under greatest threat are sub-Saharan Africa and South Asia, the two epicenters of undernutrition. But one should also note the growing threat to North Africa and the Middle East, because these are places where all of the climate evidence suggests there will be significant drying in the future, and therefore crop production will be under even greater threat than it is today.

Certainly the BAU course also means considerable dangers for East Asia and Southeast Asia, because these regions will be places of tremendous water stress and places where higher temperatures will wreak havoc on the food supply.

	North America	Latin America and Caribbean	Europe	Middle East and North Africa	Sub-Saharan Africa	South and Central Asia	Southeast Asia and Pacific	East Asia
Food insecurity				H	H	H	M	M
Malnutrition					H	H	M	M
Obesity, health	H	H	H			M	M	M
Poverty				M	H	H	M	M
Poor rural infrastructure		M		M	H	H	M	M
Land use change		H			H	M	M	M
Soil degradation				M	H	H	M	H
Water shortage	M			H	H	H	M	M
Water and air pollution	M		M	M		H	H	H
Biodiversity loss	M	H	M	M	M	M	H	H

10.18 The risks of business as usual

Source: Rockström, Johan, Jeffrey D. Sachs, Marcus C. Öhman, and Guido Schmidt-Traub. 2013. "Sustainable Development and Planetary Boundaries." Background Paper for the High-Level Panel of Eminent Persons on the Post-2015 Development Agenda. New York: Sustainable Development Solutions Network.

Malnutrition from deep undernourishment will continue to have its epicenter in sub-Saharan Africa and South Asia. On a BAU path, the other kind of malnutrition, obesity, is likely to worsen in North America and many parts of Latin America. It is already at epidemic proportions in Mexico. There are serious risks in Southeast Asia, East Asia, India, and other parts of South Asia as well.

Land-use change will have huge costs, especially in the rain forest regions. That means costs in Latin America, sub-Saharan Africa, and parts of Southeast Asia. Soil degradation is already a major crisis in parts of sub-Saharan Africa and South and East Asia, where the land has already eroded. The soils often have been swept away due to farmers trying to grow crops on steep slopes or in places where the winds blow the topsoil away, as is occurring in many parts of China. Biodiversity loss is also a moderately or acutely high threat in every region of the world.

While there are a few regions that will escape some of these risks, there are no regions that will escape all of them. Today's poor regions are in extreme peril, because they are already living on the edge. They already have people living in fragile environments, such as in tropical or dryland ecosystems. Part of their poverty has come from the fact that the natural environment is already fairly inhospitable. With environmental change, what is currently a difficult environment can quickly become an impossible environment for support of human life. When that happens, people will suffer, die, have conflict, and migrate. There will be environmental refugees created by the millions, possibly even hundreds of millions. In a world not often open to migration, poor newcomers forced to move can face a very hostile environment when they arrive. We are in for a lot of trouble if we maintain the BAU trajectory.

How do we move to a sustainable development (SD) trajectory? Because of the complexity of the food system; the interlinkage of land use, nitrogen use, and chemical pollutants; and the vulnerability of crops to higher temperatures, the kinds of responses that are needed will have to be varied, holistic in nature, and carefully tailored to local contexts. This is among the toughest sustainable development challenges that we face, because the world is in crisis and the problems will tend to get worse. It is not easy to say that one region will bail out the others, because all regions will have stresses. There will be no magic key that will suddenly make it possible to solve this problem. Each region is going to have to identify its own pathways to sustainable agriculture.

What are some of the things that can be done? The first is to improve the ability to grow food. We should be more productive in terms of higher yields per unit of land area and more resilient in terms of the ability of food crops to withstand the shocks that we already know are coming. Just as Norman Borlaug and his colleagues M. S. Swaminathan and Chidambaram Subramaniam made the Indian Green Revolution possible, we are going to need another Green Revolution of new crop varieties that will be especially propitious for the environmental challenges ahead.

For some regions this will mean a focus on drought-resistant varieties of crops, because the frequency of serious droughts is likely to become much higher. Certain plants in nature have a high level of drought tolerance. Plant scientists are now attempting to identify those genes and through various means, both conventional plant breeding or advanced genetic modification, to create new crop varieties that also share drought resistance. Natural breeding techniques have already helped to develop new seed varieties that are better able to withstand bouts of low rainfall during the growing season. The creation of genetically modified organisms

(GMOs), or GMO crops, has taken the experimental pathway of identifying the gene complexes in naturally occurring drought-tolerant plants and transplanting those gene complexes into crops.

Many people find this idea of genetic modification to be very threatening, risky to the environment and potentially to human health. Whatever is done in this research domain of cutting-edge genomics needs to be tightly monitored and regulated, but I would argue that we should certainly test these technologies to see what they have to offer. It seems very promising that by identifying genes for drought resistance or saline tolerance, we can get major advances in food security. We should therefore not dismiss a whole class of technology so quickly. While GMO technologies may pose certain risks, those risks are likely to be controllable and monitorable. The underlying technology itself may offer huge breakthroughs in food security in an age when we will need such breakthroughs.

The second step of what can be done is to make crop varieties more nutritious. Crops should not only grow better in harsh conditions, but they also should be more nutritious. Of course part of an improved diet involves choosing the right crops in the first place, with a well-balanced diet with fruits, vegetables, whole grains, and plant oils. Part of the solution is to make a particular crop (e.g., rice) more nutritious. This is the idea of the so-called golden rice, a crop that has been developed by the International Rice Research Institute (IRRI) in Los Baños, Philippines. The IRRI helped bring about the Green Revolution in rice. Now its scientists are reengineering the rice genome to express beta-carotene, a precursor for vitamin A, so that children who eat the golden rice will have the vitamin A they need, which will help to combat one of the key kinds of hidden hunger.

The third direction is absolutely essential and is known as "precision farming" or "information-rich farming." Such precision farming is already in widespread use in high-income countries. The point of precision farming is to economize on the use of water, nitrogen, and other inputs into production, so that more food can be produced with less environmental impact. In the coming years with the declining costs of information technology, poor farmers will use these techniques too. Precision farming involves, for example, a more precise application of fertilizer, so that there is less volatilization and runoff of fertilizer not taken up by the crops themselves.

Precision agriculture depends on information technologies, on detailed mapping of soil types, and often on global positioning systems that can tell a farmer exactly where that farmer is in the field and what is happening in the soil in that part of the farm. This kind of precision farming is on its way to the middle-income and poor countries, and it continues to be developed at lower costs. It is very promising, because it efficiently uses scarce resources and enables farmers to cut down significantly on the amount of fertilizer applied on the soil. Farmers can then economize, make a better income, not waste as many resources, and decrease their environmental impact.

More generally, better nutrient management can occur through better soil testing, soil mapping, and localized chemistry. Reading the qualities of the soil on handheld devices or from satellites makes it possible to get far more detailed resolutions of soil needs. This kind of soil nutrient management offers places with soil nutrient depletion a massive potential boost in yields. Africa is the first in line for this yield boost. It also offers places that use far too much fertilizer, such as China, a way to cut down very sharply on their fertilizer use.

Another breakthrough that lies ahead is improved water management. We need to apply less water to get more "crop per drop," because we are already depleting the scarce water we have and because the water challenges will get tougher in the future. That is not only because of water depletion but also because of the consequences of climate change. Solar-powered irrigation can play a very important role in microirrigation technologies, especially to help smallholder farmers.

Another major breakthrough that will offer us a tremendous opportunity will be better harvesting, storage, and transport of crops to avoid the very large losses of foodstuffs that now occur from farm to plate. Such food waste is often estimated to be around 30–40 percent of total food production. These large losses come from rodents and pests, food rotting, physical loss of the crops, exposure to rain, and so forth. Simple, low-cost means of more effective storage systems, better incentive systems for food handling, and the empowerment of local farmer cooperatives to invest in community-based storage facilities offer a tremendous hope for reducing losses in the agriculture value chain in low-income, hungry settings. These improvements will not only lead to large gains in farm incomes and more food security, but also to fewer human pressures on the environment.

Better business models for poor smallholder farmers are vital, not only for ending extreme poverty but also for empowering smallholders to make investments in improved crop varieties, irrigation, water management, and storage, all to raise farm yields and incomes. There are gains to be achieved through the aggregation of smallholders into farmer cooperatives and farmer-based organizations. These new business models can improve value chains and incomes.

Finally, we have to take responsibility ourselves for our personal health and for the way we approach the issues of food as individuals. Massive epidemics of obesity show that something is seriously wrong with prevailing diets. A lot of the problem comes from the fact that our governments have subsidized the wrong kinds of foods. Carbohydrates, trans fats, and other kinds of fast foods that are absolutely unhealthy are heavily subsidized and widely consumed. Our economic incentives have often been aligned against the very kinds of foods that are better for our health. We have subsidized, in effect, the feed grains that have led to massive overconsumption of beef in the United States and Europe, to the point where so much meat is eaten that it is detrimental to human health and is exceedingly damaging for the environment. Deploying 10 or 15 kilograms of feed grain for every kilogram of beef that is consumed tremendously multiplies the burdens on the land, on fertilizer use, and on water supplies, which aggravates all of the natural problems.

We have also been aggravating these problems more recently by diverting invaluable food production into the gas tank! In the United States, through unjustified subsidies driven more by politics than by any common sense or ecological insight, the U.S. government has turned a significant part of the annual corn (maize) production into ethanol for automobiles. This is a miserable deal because of the amount of vital resources needed to produce that ethanol. The maize-to-ethanol strategy creates no savings for the natural environment and diverts a tremendous amount of food production away from the food chain. These subsidies thereby push up food prices and place extra pressures on the natural environment. All this is being done at the behest of a few powerful companies with powerful lobbies in Washington (and comparably in European capitals). This is an example of where behavior and policy have gone awry.

The conclusion is as we have noted time and again: the pathway to sustainable development involves behavior change, public awareness, political and individual responsibility, and the mobilization of new systems and technologies that can dramatically reduce the pressures on the natural environment and help make our economy and way of life more resilient to the environmental changes already underway. Sustainable agriculture and food security remain a huge unsolved, yet solvable, problem. It is one of the areas that will require the most intensive problem solving at local levels all around the world. For these reasons, sustainable food systems and the fight against hunger will have a major place in the Sustainable Development Goals that lie ahead.

11

RESILIENT CITIES

I. The Patterns of Urbanization Around the World

We have discussed the world's great ecosystems, including rain forests, oceans, drylands, and polar ecosystems. It is time to focus on where more than half of the world's population now lives: the cities. Something remarkable happened in 2008 according to the UN official data: for the first time in all of human history, more than 50% of the human population lives in cities. Our species started out as hunters and gatherers, without settlements. Around 10,000 years ago, civilization began with the discovery and development of sedentary agriculture. Rather than hunting and gathering, humanity began to stay in one place. As a food surplus was gradually generated in the farm sector, an urban economy arose, with urban areas characterized mainly by the fact that the urban workforce is not (at least primarily) agriculturalists. Since their start 10,000 years ago, cities produced urban manufactures and services—food processing, light and heavy industry, public administration, religious rites, entertainment, finance, trade, banking—that it traded for food from the countryside.

Up to the time of the Industrial Revolution, the farm sector simply was not productive enough to support a large urban economy. Even though we think of the great historic monuments of Cairo, Rome, Beijing, Constantinople, Paris, London, and the other great cities of the world before the

Industrial Age, the share of the world's population actually living in cities was consistently around 10 percent or less. The vast majority of people in the vast majority of places and times lived in farm households in rural communities and engaged in local food production.

Then came the scientific, agricultural, and industrial revolutions beginning in the eighteenth century. With better farm practices (e.g., improved rotation of crops), better access to soil nutrients (e.g., the green and chemical fertilizers to boost soil nitrogen), and better transport conditions, the farm areas were able to produce more output per farmer and thereby support a much larger population in urban areas where food was not grown. The age of industrialization also coincided with the start of mass urbanization. As in other spheres of economic development, England and the Netherlands came first. According to one leading expert in this area, the late economic historian Paul Bairoch, Europe's average urbanization rate in 1800 was 10.9 percent (with urban areas defined by the threshold of 5,000 people or more in one aggregation; Bairoch and Goertz 1985, 289). Yet by that year, the British urbanization rate was 19.2 percent, and the Netherlands' urbanization rate was a record-setting 37.4 percent. By 1850, Britain had reached 39.6 percent and the Netherlands was at around 35.6 percent, while Europe as a whole was at 16.4 percent urban. Even by 1910, more than a century after the start of the Industrial Revolution, Europe's overall urbanization rate was no more than one-third (32.9 percent in Bairoch's estimate), though Britain, the Netherlands, and Belgium had passed the 50 percent mark.

The Industrial Revolution and its aftermath dramatically increased the output per farmer. This was the result of several factors: improved scientific knowledge; higher-yield crop varieties; scientific management of soil nutrients; and of course machinery, which enabled an individual farmer to manage a much greater land area in terms of clearing, preparation of soils, planting, harvesting, and transporting of output. The Haber-Bosch process to make nitrogen-based fertilizers at an industrial scale dramatically increased the yields of grain crops, as did a series of Green Revolutions in the twentieth century that centered on improved crop varieties. All of these advances mean that a smaller proportion of the population engaged in farming can grow the food for a rising share of the population living in cities.

Agriculture has two other properties that give it a distinctive role in the economy compared with industry and services. The first is that the demand for agricultural output does not increase in proportion with income. As gross domestic product (GDP) per capita rises, the consumption of food per capita does not rise at the same rate. A person who is ten times richer than another person will not eat ten times as much. We say that the income elasticity of food demand is less than one: food demand increases with income, but much less than proportionately. Food is a "necessity," not a luxury good. (A luxury good is one that rises in demand even faster than income, thereby assuming a growing share of consumption of richer households.) This means that as economic development takes place, agriculture will not keep pace as a share of the total economy.

The second property is that agriculture is land dependent, while industry and services are not. Farmers need land more than they need close neighbors. On the other hand, service providers such as barbers, doctors, lawyers, bankers, and movie theater operators need customers and neighbors more than they need large open spaces. Similarly, manufacturing companies need close access to both upstream suppliers (who provide semifinished goods to the factory) and downstream consumers and shippers. The result is that farmers need to live in sparsely settled areas, with lots of land per person, while industrial and service workers need to live in crowded areas, close to suppliers and buyers. Agriculture, in short, is intrinsically rural, while industry and services are intrinsically urban. (Of course, these statements apply more to services than to manufacturers, and at least some industrial activities are intrinsically rural, close to farms or mines rather than to customers and urban suppliers.)

The implication is straightforward and very important. Alongside the shift from agriculture to industry and services comes a parallel and fundamental shift from rural areas of dispersed populations to urban areas of densely settled populations. Moreover, one specific kind of activity is heavily concentrated in cities: research and development, built on scientific innovation and engineering breakthroughs. Cities are the major home of technological advances, even the technological advances (like the Haber-Bosch process) that greatly benefit farming. This leads to a dynamic symbiosis of farms and cities. More productive farms enable cities to grow, while cities in turn provide technological advances for farms that

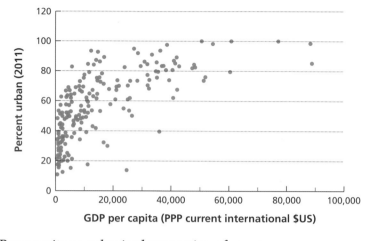

11.1 GDP per capita vs. urbanized proportion of country

Source: World Bank. 2014. "World Development Indicators."

lead to even greater farm productivity. Much of the 200-year dynamic of modern economic growth marks a constructive interplay in which advances on farms spur cities that in turn spur farms to further advances.

Figure 11.1 confirms that economic development is accompanied by urbanization. On the horizontal axis is GDP per capita, and on the vertical axis is the proportion of the country's population living in urban areas. This scatter plot is a worldwide snapshot for the year 2011. There is an upward sloping curve, where higher per capita incomes are associated with a higher proportion of urbanization. We expect that future global growth will therefore be accompanied by increasing urbanization.

Cities are also where most politics are settled. Capital cities are often places of great political contention and drama. There has been rising instability in recent years in the major cities around the world. Urban publics have protested some of the consequences of globalization itself: the rising inequalities and the rising unemployment that have resulted from shifting technologies and shifting trade patterns in many countries. The information age has also made people far more aware of political processes, and has increased the capacity of people to organize protests and even to overthrow governments. Protests and unrest, supported by social media such as Twitter and Facebook, are seemingly on the rise in the major cities. Yet the dynamics are complicated, since the information age has also enabled governments to use the new technologies to spy on their citizens and to crack down on protest movements.

Let us consider some of the factors that are distinctive about cities. First, cities have high concentrations of population. By definition, an urban area is a "densely settled area" of a certain minimum threshold, often taken to be 2,000 people or 5,000 people. (The definitions vary across countries.) Of course, the largest cities are not in the thousands but in the millions, and mega-cities have over 10 million people and significantly growing populations. The United Nations estimates there are around 30 urban agglomerations worldwide with 10 million or more people (29 make that threshold, and the 30th, Chennai, India, comes in at an estimated 9.9 million in 2015; United Nations Department of Economic and Social Affairs Population Division 2012). The list of these mega-cities is shown in figure 11.2. Note that these are not the populations of cities defined according to political boundaries, but rather of cities defined as a single concentration of population, often including dozens of individual legal entities that are grouped together in one contiguous geographical area termed an "urban agglomeration" by the UN Population Division.

Second, cities are distinguished by the kind of economic activity that they host. While cities are home to a small amount of farming, cities by and large are home to industry and services. In the high-income countries, services are the overwhelming activity. Retail and wholesale trade, education, finance, law, medicine, entertainment, public administration, and other service activities dominate the city economies.

Third, cities are relatively productive areas of the national economy. The average output per person in urban areas is often two or three times higher than in the rural areas of the same country. The migration of workers from rural to urban areas is often accompanied by a significant rise in national productivity, measured as total output per worker.

Fourth, as noted earlier, cities are the locus of a tremendous amount of innovative activities, whether it is universities, research laboratories, or major businesses introducing new products. Innovations then spread to outlying areas from the cities.

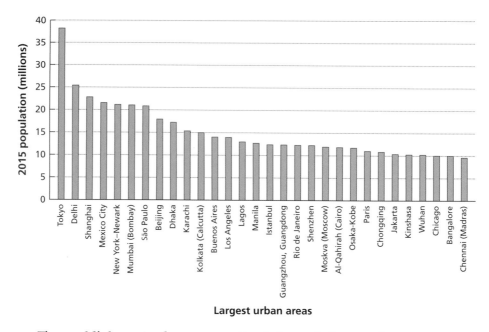

11.2 The world's largest urban areas, estimated population as of 2015

Source: United Nations Department of Economic and Social Affairs Population Division (DESA Population Division). 2013. "World Population Prospects: The 2012 Revision." New York.

Fifth, cities are trading centers, where a tremendous amount of activity involves the exchange of goods. Cities indeed exist as concentrated centers in large part to facilitate trade, exchange, and contracting, all greatly supported by proximity of buyers and sellers.

Sixth, major cities are generally coastal, to take advantage of the lower costs of sea-based shipping of goods. This recalls the observation of Adam Smith in 1776 that development normally starts at the coast and moves gradually to the interior. Most of the largest cities in the world are at or near the coast, where it is possible to move goods internationally at low cost. Inputs can be received from the rest of the world for the great cities, and goods can be moved along major riverways that connect the ocean ports with the country's interior. Great cities are often at the estuaries of great rivers. Cities like New York and Shanghai connect

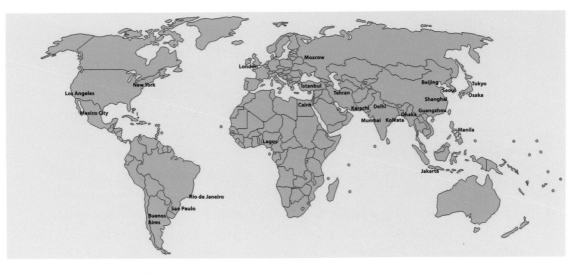

11.3 Major global cities

their respective countries with world markets and open up the interior of their countries to world markets via the Hudson River and the Yangtze River systems, respectively.

Have a look again at the great urban agglomerations in figure 11.2. Most of the cities are within 100 kilometers of a major port if not directly on the coast themselves, as we see in figure 11.3 for some of the world's mega-cities in 2012. Most of the truly interior cities are on major inland waterways that connect them with the oceans via rivers (e.g., Chicago and Chongqing). Only a few like Mexico City, Moscow, and Bangalore are interior cities without major rivers connecting them to the coast. And consider New York City, my hometown. It is not only a great trading city, but it is also the terminus of a major sea-based network. On the one hand, goods come from the Atlantic Ocean trade, and on the other, the interior of the United States is connected to New York through a waterway system that started operating at the beginning of the nineteenth century. Goods could come from Chicago, an inland city, then be carried by ship through the Great Lakes, the Erie Canal, and down to New York City via the Hudson River, which enters into the

Atlantic Ocean. New York's marvelous location on the Eastern Seaboard allowed New York not only to connect the United States with the world, but also to connect the interior of the United States with the coast. That is one of the reasons why the Chicago-New York linkage was so essential. A close counterpart is Shanghai, China's greatest industrial city, which is also the terminus of the major Yangtze River. The Yangtze connects Shanghai with great interior cities in China, notably Chengdu and Chongqing, and subsequently from Shanghai to world markets.

Seventh, cities are places of rapid population growth. They are the only places in the world that are growing right now, because rural areas have peaked in population.

Eighth, cities are often places of glaring inequality. (Rural areas of course can be as well, between large landowners and the landless.) Cities can put the rich and the poor next to each other, often in shocking proximity, as seen in figure 1.11, which shows Rio de Janeiro's grand, towering modern buildings right next door to the favelas. New York City is not very different in this regard. It puts some of the richest parts of the United States (e.g., the Upper East Side of Manhattan) just next to some of the poorest areas, such as Harlem to the north of the Upper East Side.

Ninth, cities enjoy enormous advantages of economies of scope and scale, meaning that their productivity is enhanced by the large markets they offer, making possible an enormous range of activities and depth of specialization (economies of scope) and enormous scale of production.

Tenth, and finally, cities face major challenges of "urban externalities," resulting from the high density of population and economic activities. Cities must cope with intense pollution of air and water, massive traffic congestion, and the rapid transmission of diseases if left uncontrolled. Many cities must cope with the potential for massive crime and violence resulting from the high concentrations of human interactions. Yet we should also mention that the concentration of population and economic activity may work to the advantage of the provision of public services (e.g., access to vaccines to stop disease transmission), pollution control (through infrastructure), and policing against crime.

As the world economy continues to grow and develop in the twenty-first century, and as rural productivity increases (assuming that the agricultural gains are

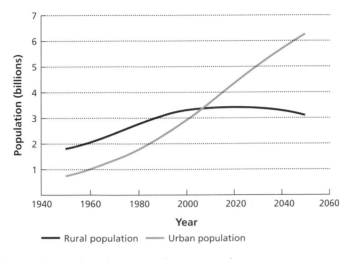

11.4 Global population distribution, urban vs. rural

Source: United Nations, Department of Economic and Social Affairs, Population Division (DESA Population Division). 2012. "World Urbanization Prospects: The 2011 Revision." New York.

not choked off by climate change), the world's urban areas are expected to continue growing, as seen in figure 11.4. In 2008, the rural-urban populations crossed, and for the first time half the world became urban. There is no looking back. The UN Population Division forecasts that by 2030, urban areas will be an estimated 60 percent of the world's population; and by 2050, 67 percent of the world's population will live in urban areas (UNFPA 2012). In other words, all of the expected increase in population, going from 7.2 billion to 8 billion and beyond, will be associated with a rising urban population and a stable or even declining absolute rural population.

As with GDP per capita and health, there is also now a tendency toward convergence of urbanization rates across regions of the world. Just as poor countries now tend to grow faster than rich countries, poor countries also tend to urbanize more rapidly than rich countries, which are already nearly entirely urban. Figure 11.5 shows the percentage of the population residing in urban areas for different regions (UNFPA 2012). Asia and Africa are the two dynamic

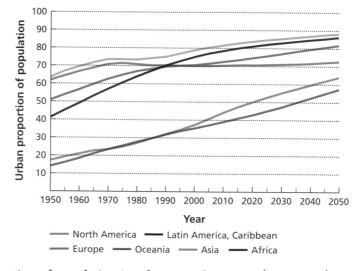

11.5 Proportion of population in urban areas by region (1950–2050)

Source: United Nations, Department of Economic and Social Affairs, Population Division (DESA Population Division). 2012. "World Urbanization Prospects: The 2011 Revision." New York.

urbanizing regions of the world now—they are becoming urban societies after a long history of being village-based rural societies.

When we examine the share of the world population in different regions, we see that something quite remarkable is happening. The global distribution of urbanization, just like the global distribution of the world economy, is shifting in a fundamental way. According to figure 11.6, in 1950, 38 percent of the world's urban population was in Europe. Europe was the site of the imperial powers and dominated the rest of the world economically and politically. The European and North American populations in 1950 constituted an amazing 53 percent of the world's urban population, compared with around 29 percent of the global population. Yet by 2050, a time in which Asia and Africa will have substantially urbanized, the United Nations forecasts that Europe will only be 9 percent of the world's urban population. This is because Europe's share of total population is falling, while the rest of the world is urbanizing. North America will be 6 percent. Rather than 53 percent of the world's urban areas

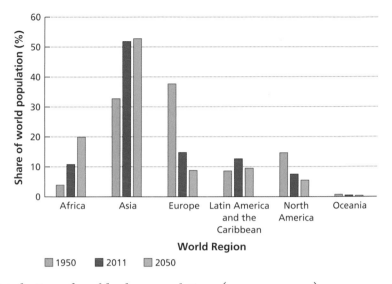

11.6 Distribution of world urban populations (1950, 2011, 2050)

Source: United Nations, Department of Economic and Social Affairs, Population Division (DESA Population Division). 2012. "World Urbanization Prospects: The 2011 Revision." New York.

as in 1950, together Europe and North America will constitute just 15 percent. The era when the West, and Western cities in particular, led the world is coming to an end.

This is also borne out by the dynamics of the world's largest cities. Back in 1950, there were just 2 mega-cities with more than 10 million people, both in the developed world: Tokyo and the New York area. In 1990, there were 10 mega-cities, 4 from high-income countries (Tokyo, Osaka, New York, Los Angeles), and 6 from developing countries (Mexico City, São Paulo, Mumbai, Calcutta, Seoul, and Buenos Aires). By 2011, there were 23 such mega-cities, with only 5 in the high-income countries, and the remaining 17 in today's developing world. As of 2015, as shown in figure 11.2, of the 29 urban agglomerations with more than 10 million, only 5 are from today's high-income world: Tokyo, New York, Los Angeles, Paris, and Chicago. According to the UN forecast for 2025, there will be an astounding 36 mega-cities. Only 7 of them will be in the high-income world, roughly 20 percent of the total number.

II. What Makes a City Sustainable, Green, and Resilient?

Since most of the world will live in cities, it is important to ask what makes a city sustainable. The answer is threefold (according to the three dimensions of sustainable development). Sustainable cities are economically productive, socially (and politically) inclusive, and environmentally sustainable. In other words, they must promote efficient economic activities, ensure that all citizens can benefit from them, and must do so in a way to preserve the biodiversity, safe air and water, and physical health and safety of the citizens, especially in an age of climate change and increasing vulnerability to extreme climate catastrophes.

In the age of the Anthropocene, cities will be buffeted by environmental shocks. When mega-cities are on the coasts and sea levels are rising, the cities will become far more vulnerable to intense storms and storm surges, as New York City learned painfully with Hurricane Sandy in October 2012. Cities need to prepare for those shocks, not as disasters that seemingly come out of the blue, but rather as events that are rising in frequency even if specific occurrences cannot be well predicted in advance.

Urban sustainability is therefore an enormous task. We can summarize some of the major features of sustainability in the following schematic manner.

- *Urban productivity*. Cities need to be places where individuals can find decent, productive work, and businesses can produce and trade efficiently. The basis for success is a productive infrastructure: the networks of roads, public transport, power, connectivity, water and sewerage, waste flows, and other "connective tissue" that enable the urban economy to operate with low transaction costs. Infrastructure also includes "software," like an effective court system to enforce contracts. When the urban infrastructure fails, the city is overwhelmed by congestion, crime, pollution, and broken contracts that impede business, job creation, and forward-looking investment.
- *Social inclusion*. Cities can be places that create high social mobility, or places that widen the divides between the rich and the poor. Neighborhoods can be mixed by income and ethnicity, or divided by class and race. Schools can be unified in a strong public system or divided between strong private schools for

the rich and weak public schools for the rest. The social stability, trust, and harmony in the society (including political stability and level of violence) will be affected by the extent of social mobility. When it is low and falling, protest, unrest, and even conflict are more likely to ensue. Effective urban planning and politics can lead to cities in which people of different races, classes, and ethnicities interact productively, peacefully, and with a high degree of social mobility and trust. With ineffective planning, lack of civic participation, and neglect of social equity, cities can become deeply divided between rich neighborhoods facing off against slums.

- *Environmental sustainability*. By definition, cities are places of high population density. They are therefore highly vulnerable to environmental ills: pollution of the air and water, despoliation of the land, the rapid spread of communicable diseases, and climate shocks and other catastrophes such as droughts, floods, extreme storms (e.g., tropical cyclones), and seismological disasters such as volcanic eruptions and earthquakes (in certain geological zones). Cities need to make two kinds of environmental efforts. The first, *mitigation*, is to reduce their own "ecological footprint," for example, the greenhouse gas (GHG) emissions caused by urban activities. The second, broadly speaking, is *adaptation*, meaning preparedness and resilience to changing environmental conditions, for example, rising temperatures and sea levels (for coastal cities).

How effectively a city plans and prepares for the future is decisive in determining its prospects for sustainable development. None of these core issues—infrastructure, social mobility, quality education, and environmental preparedness—can be solved by market forces alone. Urban productivity, social inclusion, and environmental sustainability require considerable brainstorming, planning, deliberation, and political engagement of stakeholders.

Urban Sprawl Versus High Density

One key determinant of a city's productivity and environmental footprint is its density, the concentration of population per square kilometer. Densely settled cities, if properly designed, tend to be more productive and to emit fewer GHGs

Table 11.1 Urban density and commutes in major U.S. cities							
Urbanized area	Perceived density (people per square mile) (rank)		Density gradient index (rank)		Percentage of commutes by public transit (rank)		Percentage of commutes by public transit or walking (rank)
New York–Newark, NY–NJ–CT	33,029	(1)	6.2	(1)	30.6%	(1)	36.5% (1)
San Francisco–Oakland, CA	15,032	(2)	2.2	(5)	15.9%	(2)	20.5% (2)
Los Angeles–Long Beach–Santa Ana, CA	12,557	(3)	1.8	(8)	5.8%	(8)	8.2% (8)
Chicago, IL–IN	10,270	(4)	2.6	(4)	11.9%	(4)	14.7% (5)
Philadelphia, PA–NJ–DE–MD	8,457	(5)	3.0	(3)	9.7%	(6)	13.3% (6)
Boston, MA–NH–RI	7,711	(6)	3.3	(2)	11.6%	(5)	16.1% (4)
San Diego, CA	7,186	(7)	2.1	(6)	3.1%	(12)	5.0% (11)
Washington, DC–VA–MD	6,835	(8)	2.0	(7)	15.7%	(3)	18.6% (3)
Miami, FL	6,810	(9)	1.6	(12)	3.6%	(10)	5.3% (9)
Phoenix–Mesa, AZ	5,238	(10)	1.4	(14)	2.5%	(13)	4.1% (13)
Detroit, MI	4,955	(11)	1.6	(10)	1.7%	(15)	3.0% (15)
Seattle, WA	4,747	(12)	1.7	(9)	7.6%	(7)	10.3% (7)
Dallas–Fort Worth–Arlington, TX	4,641	(13)	1.6	(11)	1.9%	(14)	3.2% (14)
Houston, TX	4,514	(14)	1.5	(13)	3.2%	(11)	4.6% (12)
Atlanta, GA	2,362	(15)	1.3	(15)	4.0%	(9)	5.1% (10)

Source: Eidlin, Eric. 2010. "What Density Doesn't Tell Us About Sprawl." *ACCESS* 37: 2–9.

than sprawling, low-density settlements. This may seem surprising. High density implies that a lot of people are jammed together. Yet in high-density areas, it is also possible to achieve lower emissions in transportation (including via walking and public transportation) and more efficient trade (with smaller distances to cover). So places of high population density tend to be places with lower ecological impacts, including lower carbon emissions per person of the population. The overall urban agglomeration of New York City is about 33,000 people per square mile, as seen in table 11.1 (Eidlin 2010, 8). Compare that with Los Angeles, for example, at about 12,000 people per square mile. It is not surprising that Los Angeles is an automobile city, whereas New York is not. Density makes a huge difference. Atlanta, which is decidedly an automobile city, is just one-fifteenth of the density of New York, at about 2,000 people per square mile. It is estimated that about 36 percent of all commutes or transit in New York are by walking or public transportation. Compare that with an automobile city. For Los Angeles, the estimate is 8 percent, and in Atlanta, only about 5 percent of travel is by public transit or walking. It is important to keep in mind that how one defines the geography of these urban agglomerations makes a difference in the precise comparisons. Yet the general patterns hold: higher density means more walking and low-emission public transit.

New York City's density and transport infrastructure means it is doing quite well compared with the rest of the United States in carbon emissions. On average, around 2008, Americans were emitting about 20 tons of CO_2 per person per year (more recently, emissions are around 17 tons per person). New York City, however, was roughly one-third of that, at around 6 tons of CO_2 per capita (PlaNYC 2009, 6). Figure 11.7 shows New York City at the low end of CO_2 emissions. The benefits are not only in transport but also the heating and cooling of buildings. Emissions per person in apartment buildings and row houses tend to be lower than in large stand-alone houses found in most other cities and suburbs.

If cities are smart in the kinds of energy and transport systems they build, and in the encouragement of high-density settlements with close proximity of people to shops, offices, amenities, and, of course, other people, then urbanization offers a real chance for a lower ecological footprint and lower carbon emission levels per capita. To get the best of what cities can offer in terms of combined low ecological footprint, high productivity, ease of movement, low congestion, and low level of time wasted, we have to look at how cities invest in the infrastructure and what choices they make.

III. Smart Infrastructure

Cities need to make choices about infrastructure. How will they handle energy, transport, water, and waste? Cities are complex systems, with millions of people interacting with each other and with industrial processes, and complex

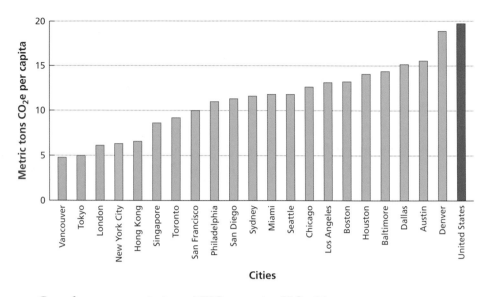

11.7 Greenhouse gas emissions, NYC vs. major U.S. cities

Source: Dickinson, Jonathan, and Andrea Tenorio. 2011. Inventory of New York City Greenhouse Gas Emissions, September 2011. *New York: Mayor's Office of Long-Term Planning and Sustainability.*

transportation, communication, water, sewerage, and waste systems. What kind of infrastructure is best for such systems? Cities that plan and design infrastructure well are able to maximize economic opportunities, improve quality of life, promote public health, and minimize the impact of the population on the natural environment, including through a relatively low-carbon economy.

One aspect of this core infrastructure includes transportation in densely settled areas. Relying on the automobile is a recipe for massive congestion, air pollution, and high GHG emissions from the tailpipes of vehicles. There are much better options of public transport, such as buses, metro systems, and properly managed biking and walking routes. Some cities manage their transport well, and other places are chaotic beyond imagination. Figure 11.8 shows commuters in Indonesia, packed to a life-threatening extent, and crowds entering a train station in the Philippines.

11.8 (*left*) Commuters in Indonesia   *Reuters/Crack Palinggi.*

(*right*) Line to enter a train station in the Philippines   *Reuters/John Javellana.*

Compare that with the picture in figure 11.9 of one of the world's most spar-kling, efficient, and dynamic metro systems in Seoul, South Korea. They have a huge, highly efficient system of hundreds of kilometers of rail lines and modern stations, shown in figure 11.10. Many Chinese cities, including Beijing, have also built metro systems in recent years that serve them very well. This is a wise invest-ment in mass public transportation, because the alternative for China would be a huge explosion of automobile use, with all of the accompanying ecological and economic ills.

Bogotá, Colombia offers a well-known example of another kind of public trans-portation: bus rapid transit. It, in turn, followed another city, Curitiba, Brazil, which pioneered bus rapid transit in the 1970s. The idea is to encourage people to shift from automobiles to buses by giving buses favorable conditions of access on dedicated lanes. There is very frequent service with comfortable ways for people to get in and out of buses and convenient waiting areas in stations. Many people are learning the lessons of Curitiba's rapid transit system, and it consequently has emulators around the world.

11.9 Seoul subway station

"Train Waiting for Start," Doo Ho Kim, Flickr, CC BY-SA 2.0.

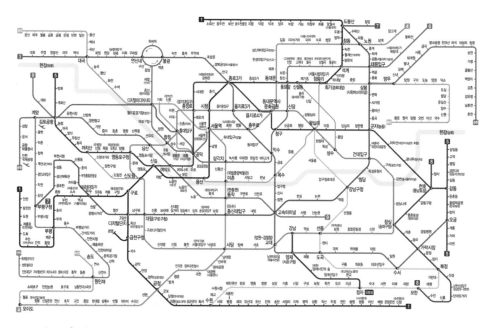

11.10 Seoul Metro map

Courtesy of the Seoul Metropolitan Rapid Transit Corporation.

Recently cities have started to make a full arc to their earlier patterns by making sure there are safe open areas for bicycling, and sidewalks and footpaths for walking. Europe has taken the lead with bicycle lanes and with innovative smart card–based shared-bicycle systems. An example is the Vélib bike-sharing system in Paris, which is leading to a surge in cycling. Bicycles were once discarded in favor of the car, but because of congestion, the expenses of managing automobiles, difficulties in parking, traffic jams, and individuals' concerns with their contributions to climate change and their personal health, people are seeking healthier ways to travel. Bicycling is returning to New York City, London, and other cities, and walking through Manhattan's dazzling shopping streets keeps the city vibrant and the population in better health.

Urbanites in countless cities around the world have recently come to appreciate the need for walking and cycling. The health costs of urban, sedentary lifestyles, including obesity, diabetes, and heart disease, are very high. Increasing numbers of urbanites are therefore eager to find ways to get back out onto the sidewalks and bike paths.

Public policies make a huge difference. In the 1950s, America doubled down on the automobile with the Federal-Aid Highway Act of 1956 (Pub. L. No. 84-627), which created a national interstate highway system, and with the encouragement of new roads and land development in the suburbs. This new road network had a very large economic impact, helping to move goods and people and greatly facilitating suburban sprawl. The exodus from the densely populated central cities to the more dispersed suburbs contributed to America's remarkably high ecological footprint and its extraordinarily high level of CO_2 emissions per capita.

Now it is China's turn to choose, as rapid urbanization is bringing hundreds of millions of people into China's cities. China has the world's largest network of large cities, with more than 100 urban agglomerations, each with more than 1 million people. China's record of encouraging automobiles versus public transit and walking is so far a mixed one. The cities themselves are increasingly automobile dependent. China is already by far the world's largest automobile market for new car sales, well over 20 million units per year. By the end of the decade, if not sooner, China will equal the United States in the number of personal vehicles, around 250 million. That number could soon double or go even higher if China tries to emulate the U.S. pattern of very high personal car ownership. (The United States has around 250 million vehicles for 316 million people; the equivalent for China would be 1 billion vehicles for 1.3 billion Chinese people!)

Yet Chinese leaders also appreciate the dangers of such high car dependency in an already very crowded country. China is already courting massive air pollution, among the world's highest, as well as massive dependency on imported oil, unrelieved traffic congestion, and huge emissions of CO_2 from petroleum use. This is why China is emphasizing public underground metro systems in all major cities, as well as fast rail rather than car-based highway traffic for intercity transport.

Still, China's main urban transport choices still lie ahead. Will China emphasize walking, rapid bus transit, and densely settled cities, or sprawling automobile cities with an American-style ownership of personal vehicles? Will China continue to produce and use automobiles powered by internal combustion engines and imported oil, or will China aim to deploy electric or fuel-cell vehicles powered by low-carbon electricity? These choices lie ahead, and they will do much to determine China's own environmental sustainability as well as China's emissions of GHGs, already the highest in the world for any country (though not the highest in per capita terms).

Water Supplies for Future Cities

Another key aspect of infrastructure is water supply. Every big city has to provide drinking water for its population, as well as water for other uses, including periurban agriculture, healthy ecosystems, and industrial processes. How to provide that water safely, fairly, reliably, sustainably and at reasonable cost is a massive challenge. In New York City, this issue has been addressed for more than a century by tapping into two watersheds outside the city, the Catskill and Croton watersheds, and then carrying the water from these watersheds to the city via huge underground pipes, as seen in figure 11.11.

About fifteen years ago New York faced the problem that water coming from both of these areas was becoming polluted and increasingly contained chemical products from the outflow of industrial and farm activities in the areas nearby. The proposal at the time was to build new, multibillion-dollar water treatment sites

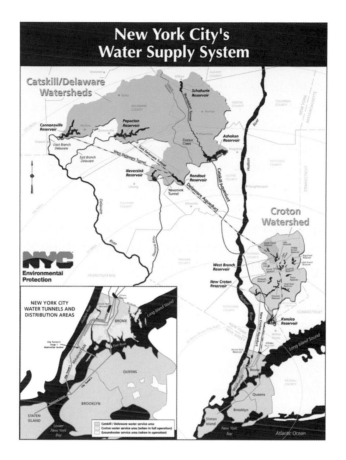

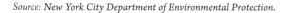

11.11 New York City's water supply system

Source: New York City Department of Environmental Protection.

to keep New York City water safe. This was an intuitive step; when the water is becoming less safe, the city should treat and clean it. Giant water treatment sites for a city of 8 million people (and a metropolitan area with around 20 million people) seemed inevitable, yet the New York City planners at the time had a clever realization. It would be safer and smarter to encourage the areas outside the city and near the watersheds to engage in fewer activities that endangered the water, so that these multibillion-dollar investments in water treatment would not have to be made.

Of course, those outlying areas had their own economic interests, so New York City realized that it would have to provide a *financial incentive* to both the Catskill and Croton watershed areas so they would agree to desist from the kinds of farm and industrial activities (e.g., the use of fertilizers, pesticides, and chemical industry effluents) that were endangering New York City's drinking water. The city therefore negotiated a financial transfer to these outlying areas to compensate them for agreeing to cut back on the activities that threatened the New York City water supply. The arrangement proved to be ingenious. New York City preserved the safety and quality of its water supply. The watershed areas were compensated for the lost economic opportunities. And both parties came out ahead, since what New York City paid the outlying areas was considerably less than what would have been the cost of building and operating giant water treatment plants.

This solution was of course specific for New York City. Every major urban agglomeration has to solve problems like this. New York's solution was a very interesting one, because it is very unusual and it had to be put together with a lot of creativity, insight, and good political management. It also highlights something very important: sustainable development is inherently an exercise in *problem solving*. It is about being creative and creating new models to combine economic, social, and environmental concerns.

The challenges of urban water supplies will become far more important, and even dire, in many regions. The continued massive growth in urban populations and industry will raise the demand for water just as climate change, groundwater depletion, and the melting of glaciers due to global warming will lead to reduced availability of fresh water and greater competition between cities and farmers for scarce water supplies. Cities will also have to invest massively in urban water-related infrastructure.

Urban Waste Management

There is yet another aspect of urban infrastructure that is absolutely essential: waste management. Some waste, like paper and plastic, may be recyclable. Some scrap metal can be reprocessed. A lot of urban waste is organic, such as rotted food. Some is highly toxic. How should cities handle all of this waste?

11.12 Ankara landfill gas to energy facility

Source: ITC Invest Trading & Consulting AG.

The typical way was to put it someplace, such as landfills. Many large cities around the world have built massive human waste dumps that are eyesores, nose sores, carriers of pollution, and sites of massive methane emissions as the organic wastes are decomposed by bacteria. Engineers have increasingly understood that simply filling land with this kind of waste is unwise and actually dangerous for the land, water supply, residents living nearby, and the GHG emissions of methane. Landfills also represent a huge missed economic opportunity, because a lot of what is placed into the landfills are potential inputs for recycling, industrial processing, and energy.

During the past twenty years, cities have experimented with many different kinds of recycling programs and innovative waste-to-energy facilities. For example, a facility in Ankara, Turkey (shown in figure 11.12), sorts waste into plastics, organics, and metals. Some of the waste is composted for fertilizer, and some is recycled through industrial processes. The organic waste is fed into a large biodigester unit where it is decomposed through bacterial action, releasing methane that is collected into large storage tanks and then used for electricity generation. The electricity is sold back into the urban grid, and the heat from the methane combustion is then fed into two pipes, one that runs to a series of greenhouses that grow food products and the other to heat a new retail mall. Indeed the nearby mall could only be built after the landfill had been converted into the waste-processing plant, thereby eliminating the noxious odors and gases that had previously made the area too polluted for other use. In short, what was once a polluting, stinking landfill is now a profitable recycling operation and source of waste-to-energy power. This kind of clever waste management is now being considered and adopted in many parts of the world.

The Ankara facility is an example of "smart infrastructure" that reconfigures the urban metabolism to give off less pollution (including GHGs) and to use materials and energy flows far more efficiently. Smart transport systems will more seamlessly connect public transport, car sharing, and walking into an integrated system. Smart power grids will enable electric vehicles and buildings with photovoltaic panels to sell power to the grid as well as to draw power from it. Smart power grids will have time-of-day pricing and other management tools to enable the urban area to economize on power use and smooth the peaks of power generation. One underlying commonality of these smart systems will be the incorporation of extensive information technology—meters, monitors, machine-to-machine communications—to enhance efficiency and convenience.

IV. Urban Resilience

Cities can choose to be efficient and have low carbon emissions per capita through the right kind of infrastructure. Cities must also plan for a future of rising ecological shocks resulting from human-induced climate change and other environmental change. Even if humanity does everything it ought to do to head off the worst of climate change, we are still going to be living through many decades in which average temperatures will be increasing, extreme climate-related events (including heat waves, storms, floods, and droughts) will increase in frequency, and sea levels will be rising. We will examine some of these threats in more detail in later chapters.

Infrastructure must be designed or refitted to be ready to face these threats. The massive flooding, loss of life, and horrendous damage in New Orleans caused by powerful Hurricane Katrina in 2005 was greatly exacerbated by the failure of

11.13 Smog in Beijing

Beijing Air Pollution, Kentaro IEMOTO, Flickr, CC BY-SA 2.0.

infrastructure: the breaking of the levees that protected the low-lying city from flooding. Engineers had warned for years that the levees of New Orleans were under threat, yet the city put off the necessary reinforcements to the levee system because of budget constraints. The levees burst, New Orleans flooded, and death and disaster ensued. Many cities around the world face this kind of risk, are being warned of it, and yet are still not taking it seriously or face financial constraints that make them unable to respond.

Pollution is likely to worsen as well, through combinations of changing weather patterns, higher proportions of automobiles per household, and an increase in industrial activity and other smog-related sources. Beijing residents were shocked in January 2013 when a major smog descended on Beijing that was so heavy it exceeded the global guidelines for particulate pollution by 20 or 30 times. The smog brought much of the city to a halt. It was a wake-up call for China to take on its challenge of massive air pollution. Another intense smog crisis hit Beijing in January 2014 (figure 11.13).

11.14 Smog attack in New York City (1966)

"Midtown and Lower Manhattan covered in smog. 1966. New York," Andy Blair, Flickr.

The good news is that it is possible for cities to overcome such crises. New York City and London had similar smog attacks in the 1950s and 1960s (figure 11.14). Since then, pollution regulations and a shift of energy use from coal to gas and other cleaner sources have cleared the air. Economists speak of the *co-benefits* of moving from coal to low-carbon energy: cleaner, safer air as well as the reduction of CO_2 emissions.

Earthquakes are another profound threat to the many cities that lie in seismically active zones, typically at the boundaries of tectonic plates. The high-earthquake zones are shown in figure 11.15, which records earthquakes during the period 1900–2012. Many populous cities are located in the danger zones. Large cities at high risk include Los Angeles, Manila, Istanbul, Lima, Tehran, Santiago de Chile, San Francisco, Kunming, Nagoya, and Izmir.

Cities in earthquake zones can make preparations. Buildings can be reinforced at relatively low cost. When this is not done, tragedies ensue like the earthquake

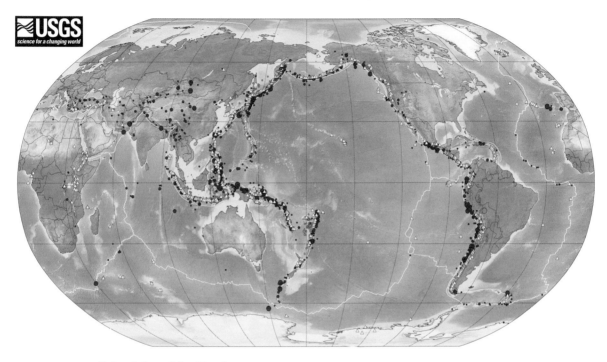

11.16 Seismicity of the Earth 1900–2012

Source: U.S. Geological Survey, Department of the Interior/USGS.

that hit Port-au-Prince, Haiti, in January 2010, causing more than 100,000 deaths (figure 11.16). Yet when a similar magnitude earthquake hit Kobe, Japan, a few years earlier, the death toll was high at about 5,000, yet nothing akin to the deaths and devastation in Haiti. Haiti, a very poor country, had not taken earthquake precautions. The buildings were constructed of rock or brick and were not fortified against earthquake risk. Moreover, many settlements were on steep hillsides that suffered calamitous landslides following the earthquake. Better zoning can also help to prevent such massive deaths by preventing building and settlements in dangerous areas.

Floods are another risk for cities along coasts, which are most of the large cities. Tokyo, Delhi, Mexico City, New York, Shanghai, São Paulo, Dhaka,

11.16 Port au Prince, Haiti after 2010 earthquake

"haiti_postearthquake13," Colin Crowley, Flickr, CC BY 2.0.

Calcutta, Buenos Aires, and Rio de Janeiro are all cities at risk of flash floods as a result of their topography, climate, and coastal proximity. Even the most sophisticated cities are not necessarily very well prepared. Figure 11.17 shows downtown Manhattan, one of the richest places on the planet, in darkness following the storm and flooding caused by Superstorm Sandy in October 2012. It turned out that New York City was not well prepared for such a major storm. Power stations were flooded; backup generators for hospitals were in the basements and were quickly submerged in water. Major hospitals required emergency evacuations. Many New York City scientists and engineers had been predicting for years that New York would be highly vulnerable to a major storm surge and ensuing flood. The warnings had been given but far too few preparations had actually been made.

Cities are now finding that they must protect themselves against a rising sea level. The Netherlands is probably the world's most experienced country when

11.17 Blackout in downtown Manhattan after Hurricane Sandy (October 2012)

"Hurricane Sandy Blackout New York Skyline," David Shankbone, Wikimedia Commons, CC BY 3.0.

it comes to battling the sea level. Much of the country is significantly below sea level, so throughout its history, it has had to create special fortifications, dykes and other barriers, to protect the land and habitation from floods. When the Netherlands experienced horrible episodes of flooding and loss of life in the 1950s, its engineers once again became world leaders in new creative solutions. This time they aimed not only to protect the land from the sea but also to do so in a way that would protect the fragile ecology of the coastal ecosystems. In particular, the engineers realized that simply blocking the ocean from the land with traditional barriers would damage estuaries where the rise or fall of the tides is key to the ecology. The engineers therefore sought solutions that would allow for normal ecosystem functioning but also protect against major storms. Figure 11.18 shows the Eastern Scheldt barrier, an ingenious multibillion-dollar creation of Dutch engineering that allows for protection against storm surges using gates that close

11.18 Eastern Scheldt storm-surge barrier, the Netherlands

"The Oosterscheldekering seen from the sky," Bryan Tong Minh, Wikimedia Commons, CC BY-SA 3.0.

during storms but remain open at other times, thereby allowing for the normal ocean flow and tidal fluctuations.

There are many ways in which ongoing climate change and other environmental changes are interacting with rising populations and more crowded cities to create new risks. It is important for every city to assess those changing risks in a detailed and rigorous way. There is not a fixed blueprint. Each city has distinctive topography, population density, and vulnerabilities. Each city needs to assess its particular challenges. The UN Population Division has made a valuable classification of the hazards facing the world's major cities; including tropical cyclones, droughts, earthquakes, floods, landslides, and volcanoes. These major hazards need to be modeled, understood, and anticipated by each city.

These hazards are on the rise. This is consistent with the idea that we have entered a new era, the Anthropocene, or the Age of Sustainable Development. We need forward-looking planning that combines ecology, engineering, and public policy to keep our cities resilient and desirable places to live in the twenty-first century.

V. Planning Sustainable Development

Sustainable cities are green and resilient. They are green in the sense that they have a low ecological impact, low GHG emissions per capita, and a pleasant and healthful environment for people to live and work in, including safe and clean air, accessible parks, and ways for people to remain active and healthy through walking, bicycling, and other means. Sustainable cities are resilient because they recognize and plan ahead for the shocks they may experience in the future.

My own city, New York City, has been making a major effort to become green and resilient, one that has become even more urgent in the wake of Superstorm Sandy, which exposed the extent of the risks that the city faces. That single storm caused an estimated $60 billion or so in damages! Earlier and more effective planning might have saved lives and tens of billions of dollars.

Two very green U.S. cities, Portland and Seattle, were perhaps the first in the United States to adopt comprehensive sustainability plans, and they have a rightly deserved reputation for great environmental sensitivity and foresight (PlaNYC 2013, 10). (See the timeline in figure 11.19.) Back in 1994, they looked at how they could reduce GHG emissions and become more resilient to climate change. Other cities began this kind of serious planning, and New York City was also relatively early in adopting its PlaNYC in 2007. Copenhagen followed with a major program in 2009, Rotterdam in 2010, and various cities around the world are now adopting similar sustainable development plans. As the United Nations itself adopts Sustainable Development Goals, this hopefully will spur thousands of cities around the world to do the same. Indeed, there is a powerful argument that sustainable cities should be one of the major Sustainable Development Goals, thereby sending a clear signal to mayors and city governments around the world that sustainable development is a subject for their focused attention, not one that can or should be left just to the national government!

What does New York City's PlaNYC call for? It has 10 Sustainable Development Goals shown in figure 11.20 (PlaNYC 2013, 11). Goal 1 is to anticipate a rise of roughly 1 million in New York City's population by 2030 and to therefore plan for more affordable housing and neighborhoods. Goal 2 is to create more parks and public spaces, which are vital for the quality of life and public health. Goal 3 is to

Timeline of Municipal Sustainability Plans

1994
Portland — Global Warming Reduction Strategy
Seattle — Toward a Sustainable Seattle

2002
San Diego — Sustainable Community Program

2004
San Fransico — Climate Action Plan

2005
Columbus — Get Green Columbus Action Plan
San Diego — Climate Protection Action Plan

2006
Denver — GreenPrint Denver

2007
Boston — Climate Action Plan
Los Angeles — Executive Directive No. 10
New York — PlaNYC

2008
Austin — Climate Action Plan
Charlotte — Environmental Sustainability Program
Chicago — Climate Action Plan

2009
Baltimore — The Baltimore Sustainability Plan
Boston — Renew Boston
Cleveland — Sustainable Cleveland
Copenhagen — CPH 2025 Climate Plan
D.C. — Green Agenda
Minneapolis — Plan for Sustainable Growth
Philadelphia — Greenworks
Portland — Climate Action Plan
— Mission Verde

2010
Miami — GreenPrint: Our Design for a Sustainable Future
Rotterdam — Rotterdam Programme on Sustainability and Climate Change

2012
Austin — Imagine Austin
Charlotte — Mecklenburg County: Environmental Sustainability Plan
Chicago — Sustainabile Chicago 2015

2013
D.C. — Sustainable D.C. Plan

11.19 Timeline of municipal sustainability plans

City of New York, PlaNYC Progress Report 2013. Mayor's Office of Long-Term Planning and Sustainability, New York, 2013.

Progress

Our goals for achieving a greener, greater New York

 Housing and Neighborhoods
Create homes for almost a million more New Yorkers while making housing and neighborhoods more affordable and sustainable

 Parks and Public Space
Ensure all New Yorkers live within a 10-minute walk of a park

 Brownfields
Clean up all contaminated land in New York City

 Waterways
Improve the quality of our waterways to increase opportunities for recreation and restore coastal ecosystems

 Water Supply
Ensure the high quality and reliability of our water supply system

 **Transportation**
Expand sustainable transportation choices and ensure the reliability and high quality of our transportation network

 Energy
Reduce energy consumption and make our energy systems cleaner and more reliable

 Air Quality
Achieve the cleanest air quality of any big U.S. city

 Solid Waste
Divert 75% of our solid waste from landfills

Climate Change
Reduce greenhouse gas emissions by over 30%

Increase the resiliency of our communities, natural systems, and infrastructure to climate risks

11.20 PlaNYC's Sustainable Development Goals for New York City

City of New York, PlaNYC Progress Report 2013. Mayor's Office of Long-Term Planning and Sustainability, New York, 2013.

clean up polluted areas, the so-called brownfields. Goal 4 is to improve the quality of waterways for transport, recreation, water safety, and coastal ecosystems. Goal 5 is to ensure the safety and adequacy of New York City's water supply. Goal 6 is a robust, resilient, efficient, low-cost, and ecologically sound public transportation system. Goal 7 is energy efficiency and reliability. Goal 8 is to improve air quality. Goal 9 is to shift the management of solid wastes from the traditional landfill model to new approaches such as waste-to-energy systems and more recycling. Goal 10 is to reduce the city's GHG emissions.

New York's per capita CO_2 emissions are already less than one-third of the U.S. national average, but at around 6 tons per capita, the city's emissions are still far higher than the very low level of CO_2 emissions the world will need to achieve by midcentury. As we shall note later, global emissions should probably not exceed 1.7 tons per person as of 2050, if the world is to keep the rise of global temperatures below 2° Celsius, as the world's governments have agreed. Thus, New York City is going to have to make major efforts in reducing CO_2 emissions, and the rest of the United States will have to do even more.

As part of PlaNYC, New York City has adopted a goal of reducing its carbon dioxide emissions by 30 percent by the year 2030. As shown in figure 11.21, PlaNYC anticipates several means to achieve this goal, including more energy-efficient buildings, cleaner sources of energy, more sustainable transportation systems, and improved waste management (PlaNYC 2011, 152). New York City's emissions goal is notable for a number of reasons, including the specifics of the plan as well as the fact that New York City is moving forward on emissions reduction even while the U.S. federal government remains gridlocked on climate policy.

The New York City action plan identifies four major categories for emissions reduction. The first is a reduction of CO_2 emissions in the heating, cooling, and ventilation of buildings. This can be achieved through a variety of means, including better air seals and insulation, natural heating and cooling, use of heat pumps in place of boilers and furnaces, and the cogeneration of electricity and heat by power stations.

The second is a cleaner energy supply, one that moves away from fossil fuels and toward renewable energy such as solar arrays on New York's roofs. New York also has plans to bring in more hydroelectric power from Canada through

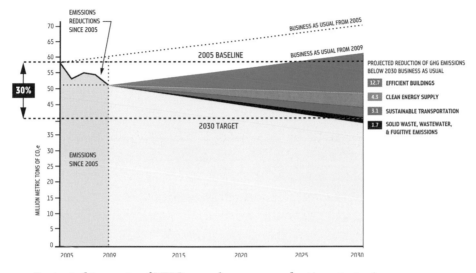

11.21 Projected impacts of NYC greenhouse gas reduction strategies

City of New York, PlaNYC Full Report, April 2011. Mayor's Office of Long-Term Planning and Sustainability, New York, 2011.

long-distance transmission lines as part of its strategy for reducing CO_2 emissions. Offshore wind power is a third large-scale possibility.

The third component of emissions reduction is sustainable transportation. Vehicle fleets should be transitioned to electrification (e.g., battery-powered or fuel cell–powered vehicles), and transportation should be increasingly directed toward mass transit, walking, and bicycling.

The fourth category is improved management of solid wastes. With a move from landfills to recycling and waste-to-energy systems as in Ankara, CO_2 emissions are reduced through lower methane emissions at landfills and the harnessing of the energy in the organic waste.

A last and extremely important part of PlaNYC is the use of new metrics. The dashboard in figure 11.22 describes for each category of goal a specific quantification of what the goal is supposed to accomplish, the most recent data, and whether the city is making progress or falling further behind (PlaNYC 2011, 179).

CATEGORY	METRIC	2030 TARGET	FIGURE FOR MOST RECENT YEAR	TREND SINCE BASE YEAR
HOUSING AND NEIGHBORHOODS	Create homes for almost a million more New Yorkers while making housing and neighborhoods more affordable and sustainable			
	Increase in new housing units since January, 2007	314,000	125,837[1]	↗
	Total units of housing in NYC	INCREASE	3,415,500[1]	↗
	% of housing affordable to median-income NYC household	INCREASE	60.0%[1]	↘
	Vacancy rate of least expensive rental apartments	INCREASE	1.0%[1]	↗
	% of new units within a 1/2 mile of transit	> 70%	93.9%[1]	↗
	Residential building energy use per capita (source MMBTU) (3 yr rolling avg)	DECREASE	47.28[2]	↘
PARKS AND PUBLIC SPACE	Ensure all New Yorkers live within a 10-minute walk of a park			
	% of New Yorkers that live within a 1/4 mile of a park	85%	76.3%[1]	↗
BROWNFIELDS	Clean up all contamined land in New York City			
	Number of vacant tax lots presumed to be contaminated	DECREASE	1,500 – 2,000[1]	NEUTRAL
	Number of tax lots remediated in NYC annually through the Brownfield Cleanup Program	INCREASE	11[1]	↗
WATERWAYS	Improve the quality of our waterways to increase opportunities for recreation and restore coastal ecosystems			
	Fecal coliform rates in New York Harbor (Cells/100mL) (5 yr rolling avg)	Decrease	42.97[1]	↗
	Dissolved oxygen rates in New York Harbor (mg/L) (5 yr rolling avg)	INCREASE	6.5[1]	NEUTRAL
WATER SUPPLY	Ensure the high quality and reliability of our water supply system			
	Number of drinking water analyses below maximum contaminant level	100%	99.9%[1]	↗
	Water usage per capita (gallons per day) (3 yr rolling avg)	DECREASE	124.46[1]	↘
TRANSPORTATION	Expand sustainable transportation choices and ensure the reliability and high quality of our transportation network			
	Sustainable transportation mode share (Manhattan CBD bound commute)	INCREASE	74%[2]	↗
	Change in transit volume minus change in auto traffic volume since 2007	POSITIVE	0.9%[1]	↗
	Vehicle revenue miles (Miles transit vehicles travel in revenue service)	INCREASE	915,096,265[1]	↗
	% of bridges meeting a state of good repair (FY)	100%	41.4%[1]	NEUTRAL
	% of roads meeting a state of good repair (FY)	100%	73%[1]	↗
	% of transit station components meeting a state of good repair	100%	71%[1]	
ENERGY	Reduce energy consumption and make our energy systems cleaner and more reliable			
	Greenhouse gas emissions per unit of electrical power (lbs C/MWh)O	DECREASE	657.69[2]	↘
	System reliability: CAIDI (Customer Average Interruption Duration Index)	DECREASE	2.39[1]	↗
	System reliability: SAIFI (System Average Interruption Frequency Index)	DECREASE	104[1]	↘
	Energy use per capita (source MMBTU) (3 yr rolling avg)	DECREASE	123.20[2]	↘
AIR QUALITY	Achieve the cleanest air quality of any big U.S. city			
	City ranking in average P(3M yr rolling avg)	#1 (LEAST)	5.67[2]	↘
	Change in average PM year-on-year % change in 3 yr rolling avg)	DECREASE	-9.4%[1]	↘
SOLID WASTE	Divert 75% of our solid waste from landfills			
	Percentage of waste diverted from landfills (includes fill)	75%	54%[1]	↘
CLIMATE CHANGE	Reduce greenhouse gas emissions by over 30%			
	Increase the resilience of our communities, natural systems, and infrastructure to climate risks			
	Greenhouse gas emissions (MTCe) O	DECREASE 30%	53,358,868[2]	↘
	Greenhouse gas emissions (100% = 2005 GHG emissions)	70%	84%[2]	↘
	Greenhouse gas emissions (MTCe) peO r GCP ($M)	DECREASE	93.82[2]	↘
	Greenhouse gas emissions (MTCe) per capita	DECREASE 30%	6.47[2]	↘

1   Results are for FY or CY 2011
2   Results are for FY or CY 2010; data is only available with a lag
3   Data updated every three years
4   Updated data not available

11.22 PlaNYC sustainable development indicator dashboard

City of New York, PlaNYC Full Report, April 2011. Mayor's Office of Long-Term Planning and Sustainability, New York, 2011.

The fact there is a lot of green on the dashboard in the final column is an indication of some progress. Yet there are also many areas that require enhanced policy efforts. Dashboards like this for water supplies, transport, CO_2 emissions, energy, and other dimensions of sustainable development will help governments to remain focused on Sustainable Development Goals and will help civil society keep their governments accountable for their promises. The indicators help to create a feedback from measurement to policy action or policy correction, thereby accelerating the transition to sustainable development.

12

CLIMATE CHANGE

I. The Basic Science of Climate Change

Roughly forty years ago, a small group of scientists and policy makers began to realize that humanity was on a dramatic collision course, as the rapidly growing world economy and population threatened to collide with the planet's finite resources and fragile ecosystems. The danger was first highlighted globally at the 1972 UN Conference on the Human Environment (UNCHE) in Stockholm. A famous and influential book that same year, *Limits to Growth*, warned that business as usual could lead to an economic collapse in the twenty-first century.

Back in 1972, as the core idea of planetary boundaries was first being understood, the kinds of boundaries that would turn out to be the most important were not yet very clear to the scientific community. The big concern in 1972 was that humanity would run out of certain key minerals or ores and that the resulting scarcity would make it difficult to maintain the level of economic activity, much less to continue to achieve economic growth.

What was not so clearly appreciated back in 1972 was that the real limits were not the minerals, but rather the functioning of the Earth's ecosystems, the biodiversity, and the ability of the atmosphere to absorb greenhouse gases (GHGs) emitted by humanity from fossil fuels and other agricultural and industrial processes. It is only now that we are beginning to see that the real planetary boundaries are mainly ecological rather than limits of mineral ores. There is no doubt that the greatest of all of these threats is human-induced climate change, coming from the buildup of GHGs including carbon dioxide, methane, nitrous oxide, and some other industrial chemicals.

There has never been a global economic problem as complicated as climate change. It is simply the toughest public policy problem that humanity has ever faced. First, it is an absolutely *global crisis*. Climate change affects every part of the planet, and there is no escaping from its severity and threat. Humanity in the modern period has faced some pretty terrible threats, including nuclear annihilation along with mass pandemic diseases. Climate change ranks right up there on the scale of risks, especially for future generations.

Every part of the world is contributing to the problem, though on a per capita basis, some places like the United States are causing far more damage and risk than other parts of the world. Roughly speaking, emissions are in proportion to income levels. High-income countries tend to have the largest GHG emissions per capita, while poor countries are often great victims of human-induced climate change without themselves having contributed much to the crisis.

Second, when crises are global, as this one is, there are huge challenges in getting the world mobilized to take corrective actions. The UN Framework Convention on Climate Change (UNFCCC), signed at the Rio Earth Summit in 1992, has 195 signatory governments plus one regional organization, the European Union (EU). The 195 nations in the UNFCCC have vastly different perspectives. Some are exporters of fossil fuels; others are importers. Some use massive amounts of renewable energy (such as hydroelectric power); others use very little. Some are rich; others are poor. Some are highly vulnerable to climate change (such as small island economies or tropical countries); others believe themselves to be less vulnerable (such as countries in cold climates in high latitudes). Some countries are democracies; others are not. All of these differences give rise to sharp differences of opinion and interests on the proper way forward.

Third, the problem crosses not only countries but also generations. The people who are going to be most profoundly affected by human-induced climate change have not yet been born. They are not voting, writing op-eds, publishing papers, or giving speeches right now. They are not even on the planet yet. Humanity is surely not very good at considering, much less solving, such a multigenerational crisis. Who represents the future generations? Is it the politicians facing election next year? Is it the business people worrying about the next quarterly report? Is it any of us as we focus on today, tomorrow, or the next day? It is no doubt very difficult for a political system, or any of us, to keep in mind and fairly represent the interests of generations yet to come.

Fourth, the challenge is also complicated because the problem of GHG emissions goes to the core of a modern economy. The success of modern economic growth arose from the ability to tap into fossil fuel energy. First came the steam engine and its ability to harness coal; then the internal combustion engine and its ability to use petroleum; and then the invention of the gas turbine with its ability to use natural gas. The entire world economy has grown up as a fossil fuel–based economy, and yet fossil fuels are at the core of the climate change crisis. The number one human contributor to climate change is the burning of fossil fuels that emit carbon dioxide into the atmosphere and thereby change the planet. We must undertake a kind of "heart transplant," replacing the beating heart of fossil fuel energy with an alternative based on low-carbon energy!

Fifth, climate change is a slow-moving crisis. To be more precise, it is a very fast-moving crisis from the perspective of geological epochs, but very slow from the point of view of daily events and the political calendar. If the climate change crisis were going to culminate in a single event in a year's time, there could be little doubt that humanity would get itself organized to prevent or adapt to the crisis. Yet the climate changes underway will play out over decades, not months.

Our situation is a bit like the proverbial frog that is put in water that is very slowly heated. The story has it that a frog in gradually warming water will never jump out and will eventually be boiled alive. Perhaps humanity will be the same. The changes year to year may be too gradual to provoke large-scale political actions, yet the cumulative effects could prove devastating; or, we may wake up to reality when it is simply too late to change course decisively.

Sixth, the solutions to climate change are inherently complex. If there were one action, one magic bullet, one new technology that would do the trick, the problem would be solved by now. The kinds of changes that are needed in response to human-induced climate change involve every sector of the economy, including buildings, transportation, food production, power generation, urban design, and industrial processes. With such operational complexity on the pathway to deep decarbonization, it is no surprise that very few governments have been able to establish workable plans or pathways.

Seventh, the energy sector is home to the world's most powerful companies. The large oil and gas companies are generally among the world's largest companies by revenues. A remarkable seven of the ten largest companies in the world in 2013, as ranked by Global Fortune 500 (with the rank shown in parentheses), are in the energy sector:

Royal Dutch Shell (1)
Exxon Mobil (3)
Sinopec Group (4)
China National Petroleum (5)
BP (6)
China State Grid (7)
Total (10)

Incidentally, companies ranked 8 and 9 are Toyota and Volkswagen, which both produce petroleum-based vehicles. The lobbying clout of the oil, gas, and automobile industries is therefore staggering.

In short, we are dealing with the heaviest of heavyweights of the world economy and of global politics. By and large these companies hope, plan, and lobby for the world to remain heavily dependent on oil and gas, despite the risks to ourselves and to future generations. These companies are able to win political support to stall the conversion to low-carbon energy through many tools: campaign financing, lobbying, and other means of persuasion. Some companies have gone so far as to promote antiscientific propaganda and to sow doubt in the public mind regarding well-known and mainstream science. With enough money, any big lie can be defended, at least for a while. In the United States, the wealthy Koch brothers, who own a major U.S. oil company among other interests, have financed an aggressive campaign against climate science and against measures to convert to low-carbon energy.

Altogether, climate change is therefore one very tough issue, and time is running out! The emissions of the main GHGs that lead to human-induced climate change are increasing each year, and the threats to the planet are growing as well. We are losing time even though the stakes for the planet are incredibly high.

The Basics of Climate Science

When seeking a true solution to the problem, the best place to start is with the science itself. The science is not new. The basics of human-induced climate change were already worked out by scientists in the nineteenth century. One great scientific genius, Svante Arrhenius, a Swedish Nobel laureate in chemistry, calculated accurately by hand, without a computer, the effects of doubling the atmospheric concentration of carbon dioxide (CO_2) (Arrhenius 1896). And he did so back in 1896! He correctly calculated that a doubling of the CO_2 in the atmosphere would cause a rise in the mean temperature of the planet of around 5° Celsius, an estimate that is within the likely range today based on advanced computer models and vastly more extensive data than Arrhenius had at his disposal.

Yet Arrhenius was a better scientist than an economic forecaster. He was not accurate in his guess about the timescale in which the CO_2 concentration would double. Arrhenius expected that human use of coal and oil and other fossil fuels would cause the atmospheric CO_2 to double in around 750 years. In fact, because of the remarkable geometric growth and energy use of the world economy since Arrhenius's time, the doubling of CO_2 is likely to occur roughly 150 years after Arrhenius's study, that is, around 2050.

The basic reason the likely doubling of CO_2 is so frightening can be understood with the schematic diagram of the GHG effect in figure 12.1. As the diagram explains, the sun's radiation reaches the Earth as ultraviolet radiation. A large part of the ultraviolet radiation passes through the atmosphere and arrives on the planet. A small part of the incoming solar radiation is reflected by clouds and goes back into space; and some of the solar radiation that lands on the surface of the Earth, for instance on the ice, is also reflected directly back into space.

The Earth warms as a result of the solar radiation that reaches the surface and is not immediately reflected back to space. By how much does the Earth warm?

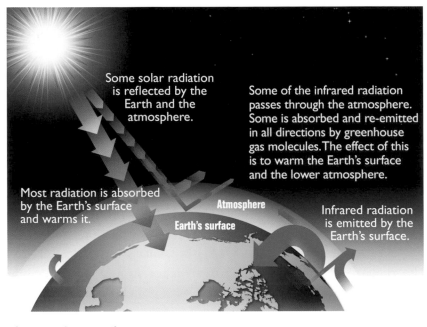

Some solar radiation is reflected by the Earth and the atmosphere.

Some of the infrared radiation passes through the atmosphere. Some is absorbed and re-emitted in all directions by greenhouse gas molecules. The effect of this is to warm the Earth's surface and the lower atmosphere.

Most radiation is absorbed by the Earth's surface and warms it.

Atmosphere

Earth's surface

Infrared radiation is emitted by the Earth's surface.

12.1 The greenhouse effect

Source: U.S. Environmental Protection Agency, 2012.

By just enough that it reaches a temperature at which the Earth radiates energy to space at the same rate that the sun transmits energy to Earth. The key to understanding this energy balance is a concept known as "black-body radiation." Any warm body, including the Earth itself, radiates electromagnetic energy. The warmer the body, the greater the radiation. When the sun radiates energy to Earth, the Earth warms to just the temperature at which the Earth radiates energy to the sun equal to the sun's radiation reaching the Earth. An energy balance is thereby struck. (This basic concept of how the Earth's temperature is determined was discovered by the great French scientist Joseph Fourier in 1824.)

While the sun radiates ultraviolet radiation to the Earth's surface, the Earth radiates infrared (long-wave) radiation back to space. In energy balance, the incoming ultraviolet radiation must equal the outgoing infrared radiation. But here is the kicker at the heart of the entire climate change problem. The Earth's atmosphere contains some special molecules, like CO_2, that trap part of the infrared radiation heading out to space. These gases, called GHGs, thereby change the energy balance: more ultraviolet hits the Earth than infrared radiation reaches space. On net, the Earth absorbs net radiation, and the planet begins to warm. (Note that the GHGs do not absorb the incoming ultraviolet radiation, only the outgoing infrared radiation.)

Yet by how much does the Earth warm as a result of the GHGs? The Earth warms by just enough so that at the higher temperature the Earth radiates *extra infrared radiation*, by just enough that even after some is trapped by the GHGs, the remainder that leaves the Earth and goes into space just balances the amount of solar radiation that reaches the Earth's surface from the sun. Now indeed we can see by how much the GHGs will warm the planet. If we know by how much the CO_2 traps infrared radiation, we also know by how much the Earth must rise in temperature to restore a net energy balance with the sun.

There are several major GHGs: CO_2, methane (CH_4), nitrous oxide (N_2o), and some industrial chemicals called hydrofluorocarbons (HFCs), perfluorocarbons (PFCs), and sulfur hexafluoride (SF_6). Another major GHG is water vapor (H_2O), which, like CO_2, traps infrared radiation and thereby warms the planet. The first kind of GHGs (CO_2, CH4, N_2O, and HFCs) are all directly emitted by human activity. Water is only indirectly affected by human activity. As the planet warms, the water vapor in the atmosphere tends to increase, and this increase causes an additional greenhouse effect, meaning an additional rise in temperature.

The basic greenhouse effect is a lifesaver for us. If the Earth, like the moon, had no GHGs, then the Earth would be a much colder place and would not support life as we know it. Without the greenhouse effect, the average Earth temperature would be around $-14°C$ (about 6.8°F), well below the freezing point of water. With the greenhouse effect, the average temperature on Earth is around 18°C (around 64°F). For this much we must be grateful!

Yet as we put more GHGs into the atmosphere, we warm the planet from the range we have known throughout human history to a much warmer and essentially unfamiliar planet Earth. Our food crops and farm systems, the locations of plants and animals, the location of cities, key infrastructure (roads, bridges, ports, buildings), and public health have all been shaped by a planet with a fairly stable temperature range during the period of civilization, roughly the past 10,000 years. This modern period, known as the Holocene (preceded by the epoch known as the Pleistocene, which was characterized by periodic ice ages), is the period of civilization. It has been remarkably stable in temperature and benign in overall average climate. It is that period of stability that we are now threatening to over-turn by our massive production of GHGs.

There are a number of points about the various GHGs. Perhaps most impor-tant, CO_2 stays up in the atmosphere for a long time. We speak of a long "resi-dence time," in the case of CO_2, lasting for centuries. When it comes to CO_2, what goes up does not come down, at least not anytime soon. The CO_2 is not washed back to Earth by rainfall, for example. Other GHGs differ from CO_2 in their heat-trapping capacity (what is called their "radiative forcing") and in their residence time. Methane, for example, traps roughly 23 times more heat than CO_2, counting each molecule of CH_4 compared with each molecule of CO_2. Yet the residence time of methane is much shorter, around ten years rather than hundreds of years in the case of CO_2.

The total warming effect of all of the anthropogenic (human-caused) GHGs is determined by adding up the separate radiative forcings of each of the six GHGs. For each GHG, we measure its radiative forcing in units of CO_2 equiva-lent (CO_2E). For example, since CH_4 has a radiative forcing equal to 23 times that of CO_2, we say that each single molecule of CH_4 in the atmosphere should be counted as equivalent in warming potential to 23 molecules of CO_2. Similarly, each molecule of N_2O counts as equivalent to 296 molecules of CO_2. In this way, we are able to take any combination of CO_2, CH_4, N_2O, HFC, PFC, and SF6 and express the total radiative forcing in units of CO_2 equivalents, as if there were only one GHG, CO_2, with a radiative forcing equivalent to the actual forcing caused by the presence of six distinct GHGs.

We can then ask the share of each of the GHGs in the total warming effect. Carbon dioxide takes the prize. As we see in the final column of table 12.1, CO_2 accounted for fully 77 percent of the total greenhouse effect of the six molecules. Taken together, the top three GHGs (CO_2, CH_4, and N_2O) account for the lion's share of the total warming effect, around 99 percent of the total greenhouse effect.

Table 12.1 Greenhouse gas characteristics			
	Lifetime in the atmosphere (years)	100-year global warming potential (GWP)	Percentage of 2000 emissions in CO_2E
Carbon dioxide (CO_2)	5–200	1	77%
Methane (CH_4)	10	23	14%
Nitrous oxide (N_2O)	115	296	8%
Hydrofluorocarbons (HFCs)	1–250	10,000–12,000	0.50%
Perfluorocarbons (PFCs)	> 2,500	> 5,500	0.20%
Sulfur hexafluoride (SF6)	3,200	22,200	1%

Source: The Stern Review Report © Crown copyright 2006.

We don't actually count the number of CO_2 molecules that we add to the atmosphere. Instead, we measure the total number of tons of CO_2 that humans emit into the atmosphere (mainly by burning coal, oil, and gas). From there, we are able to convert the tons emitted into the atmosphere into a change in the CO_2 concentration in the atmosphere, measured not in tons but in molecules of CO_2 per million molecules in the atmosphere. Here is the rough calculation. Each 1 billion tons of CO_2 added to the atmosphere amounts to an additional 127 molecules of CO_2 for each 1 billion molecules of the atmosphere. Thus, an extra 16 billion tons of CO_2 in the atmosphere equals 2 extra molecules of CO_2 per million overall molecules. When the world burns around 35 billion tons of CO_2 each year in energy use, around 46 percent of that, equal to 16 billion tons, stays in the atmosphere. The other 54 percent of the CO_2 is absorbed into the forests, soils, and oceans. The part that stays in the atmosphere results in a rise in CO_2 concentration by roughly 2 parts of CO_2 per million atmospheric molecules.

In total, the world is emitting around 55 billion tons of CO_2E (meaning the CO_2 equivalent tons, counting all six GHGs). The CO_2 part of that total is about 35 billion tons, which comes from burning coal, oil, and gas. An additional amount, perhaps 3.5 billion tons of CO_2 per year, results from cutting down trees and clearing land for farms and pasturelands. There is more uncertainty about the net CO_2 emissions due to land-use changes than due to energy changes, since the land is both a source of emissions (e.g., through deforestation) and also a CO_2 "sink," meaning that increases in soil carbon and aboveground plant matter capture some of the CO_2 in the atmosphere. The net effect year to year is hard to measure with precision.

How big is an annual CO_2 emission of 35 billion tons due to fossil fuel use and other industrial processes? It is certainly big enough to cause huge planetary-scale dangers. About 50 years ago a very far-sighted scientist, Charles Keeling, put monitors on a mountaintop in Hawaii and started to measure the amount of CO_2 in the atmosphere. Thanks to those measurements from 1958 till now, we have the annual and even seasonal levels of CO_2 (Scripps 2014). The resulting instrument record is known as the Keeling Curve (figure 12.2), showing that the amount of CO_2 in the atmosphere has been rising significantly over the years. As usual, the CO_2 is measured as parts per million (ppm) of CO_2, or the number of CO_2 molecules per million of total molecules in the air.

Starting back in 1958 when that machine first went up on the top of Mauna Loa in Hawaii, the CO_2 was 320 molecules for every 1 million molecules in the air (320 ppm). By now, CO_2 has reached 400 ppm. Before James Watt came along with his brilliant steam engine, the atmosphere contained about 280 ppm. In the geologic history of the last 3 million years, the CO_2 varied roughly between 150 and 300 ppm. Then came humanity and the Industrial Revolution, and we have since been burning so much oil, gas, and coal, and deforesting so many regions, that we have sent the CO_2 levels soaring, reaching 400 ppm in the spring of 2013. This is a concentration of CO_2 not seen on the planet for 3 million years. In other words, human activity is pushing the planet into a climate zone completely unknown in both human history and Earth's recent history.

Notice the within-year ups and downs of the CO_2 in the Keeling Curve. Atmospheric CO_2 is high in the winter and spring months, reaching a maximum in May, and is low in summer and fall, reaching a minimum in October. We are watching the planet breathe. During the winter months in the Northern Hemisphere (where most land and vegetation is located), the trees reduce their photosynthesis and shed their leaves, thereby releasing CO_2 into the atmosphere. During the

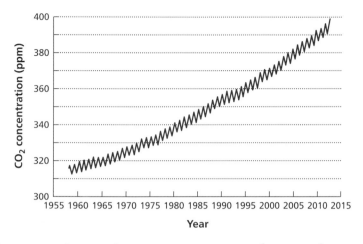

12.2 Keeling curve of atmospheric CO2 concentration (1958–2013)

Source: Scripps 2014.

summer months in the Northern Hemisphere, the trees build up their carbon content, thereby withdrawing atmospheric CO_2 and building up the terrestrial plant mass.

Some great scientists, like James Hansen of Columbia University, are able to use various techniques, such as measuring the isotopic properties of CO_2 in ice cores, to look at the long history of CO_2 and temperatures on the planet. Figure 12.3 is a kind of open manuscript of the Earth's climate history, showing a long reconstruction of CO_2 and temperatures over the past 450,000 years. We see that atmospheric CO_2 fluctuated in natural cycles not caused by humanity. These were natural fluctuations of CO_2 driven by natural processes of volcanoes, the fluxes of CO_2 between the ocean and the atmosphere, and changes in the Earth's orbital cycle with a periodicity of tens of thousands of years.

This paleoclimate (ancient climate) record shows that when CO_2 concentrations were high as a result of natural processes, the Earth's temperature was also high. This is the basic greenhouse effect at work: *raise the CO_2 in the atmosphere (by natural or human means), and the result is a warmer planet.* This relationship has been true throughout history, and it is true now.

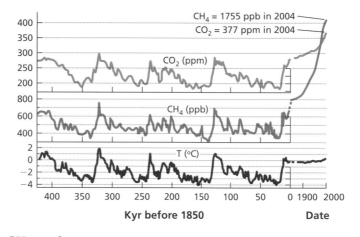

12.3 CO2, CH4, and temperature, 450,000 years ago–present

Source: Hansen, James E. 2005. "A Slippery Slope: How Much Global Warming Constitutes 'Dangerous Anthropogenic Interference'?" Climatic Change *68(3): 269–279.*

If we look at the temperature from the start of the Industrial Revolution until now (figure 12.4), the Earth has warmed by about 0.9 of 1°C, and it has not finished its warming in response to the GHG increases that have already taken place. Even if we were to put no further GHGs into the atmosphere, Earth would continue to warm by perhaps another 0.6 of 1°C, because the oceans take a long time to warm up in response to the GHGs that have already risen in the atmosphere. (There is some evidence that the year-to-year warming slowed a bit after 1998, with growing evidence suggesting that changes in Pacific Ocean patterns, with more La Niña conditions, contributed to this slowing; a swing back to El Niño conditions would, in that case, lead to a return to higher year-to-year warming.)

Yet we are certainly not done emitting GHGs. As the world economy has grown in recent years, the total emissions per year have also increased significantly. Even though the world's governments promised to curb emissions of CO_2 when they signed the UNFCCC at the Earth Summit in Rio de Janeiro in 1992, the actual emissions per year have continued to soar, not least because of China's

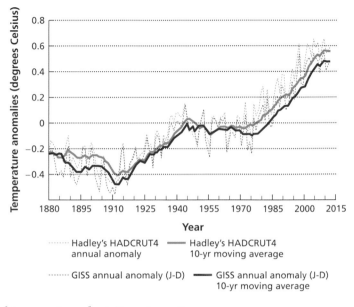

12.4 Global temperature deviation since 1850

Source: GISS/NASA.

remarkable economic growth combined with China's dependence on coal as its major energy source. As emissions rise, the CO_2 concentration in the atmosphere continues to rise (remember that the CO_2 residence time is a matter of centuries, not years), so that we can expect the Keeling Curve to continue to increase for decades to come.

It has been more than twenty years since the Rio Earth Summit where the world's governments agreed that we have an urgent challenge in heading off the human-induced GHGs; but we have still not reduced emissions. In fact, the rate of emissions has been increasing year to year as the world economy increases in scale, as figure 12.5 shows. With the growth of China, there has been an enormous increase of GHG emissions in recent years. China, by virtue of its huge size and use of coal as its primary energy source, has become the world's largest emitter of CO_2.

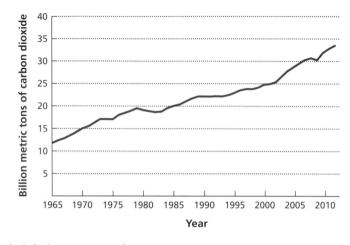

12.5 Annual global emissions of CO2, 1965–2012

Source: BP Statistical Review of World Energy June 2014.

II. The Consequences of Human-Induced Climate Change

Why should we care about human-induced climate change? The fact is that we should be truly scared, and not just scared, but scared into action—both to mitigate climate change by reducing GHG emissions and to adapt to climate change by raising the preparedness and resilience of our economies and societies. The consequences of a business as usual (BAU) trajectory for this planet could be absolutely dire. The temperature increase by the end of the century compared with the preindustrial average temperature could be as much as 4–7°C. Such an increase in temperature would be very likely to have devastating effects in many ways.

There is no absolute precision on how big the average temperature increase might be. It is very difficult to determine how much GHGs humanity will emit on the BAU path with a growing world economy. There are also uncertainties about the Earth's physical processes and the precise feedbacks from CO_2 to temperature increases. Climate models cannot precisely get the exact decimal points of the likely increases of temperature. Yet there is overwhelming evidence coming from many different directions—the instrument record, the paleoclimate, the statistical models used by climate scientists, the direct measurements of energy fluxes in space and the oceans, and the overwhelming evidence of changes already underway in physical and human systems—to tell us that we are on a dangerous path of rising temperatures with dangerous potential consequences.

An important report on climate change produced by Lord Nicholas Stern, known as the Stern Review of Climate Change, offered a graphical representation of the potential dangers (Stern 2006). In figure 12.6, the top of the chart shows the various possible concentrations of CO_2 depending on the policies we follow. The higher the CO_2 concentrations, the higher the temperature increases will be. Then, along the left-hand side of the chart are the various sectors that will be impacted by the temperature increases. These include: food, water, ecosystems, extreme weather events, and major irreversible changes to the Earth's physical

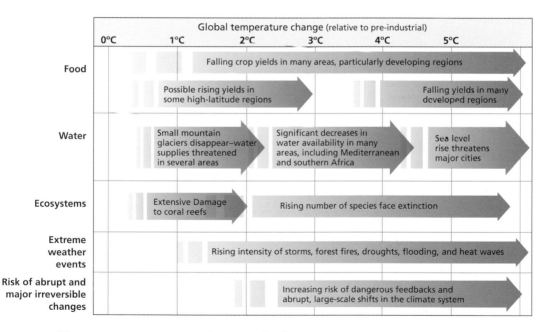

12.6 Temperature increases and potential risks

Source: Stern, Nicholas. 2006. The Stern Review Report: The Economics of Climate Change. © Crown copyright 2006.

systems (such as the melting of the great ice sheets on Greenland and Antarctica, which would raise the ocean level by tens of meters).

The graph makes clear that the danger in each of these areas (shown by the intensity of the color red in the diagram) rises markedly as the mean global temperature increases. By the time the world average temperature is raised by around 3°C, the danger in every area—food supply, water supply, hazards, and so forth—is already in the bright-red danger zone. By a 4°C increase, we are contemplating truly catastrophic potential changes. And yet that is the trajectory of BAU. We simply need to change course.

Consider food, for example. At just 1°C above the preindustrial temperature (basically what has already occurred), one of the consequences is likely to be severe impacts on food production in the Sahel region. The Sahel is the part of West Africa just below the Sahara Desert. It is a very dry region already (caught graphically in the photo in figure 12.7), so the consequences for the Sahel of even a 1°C increase in temperature are quite serious. What would happen at a 4°C

12.7 A dry region of the Sahel Desert

EC/ECHO/Anouk Delafortrie, Flickr, CC BY-ND 4.0.

increase? According to the evidence, entire regions of the world would experience major declines in crop yields, with up to a 50 percent decline of crop yields in Africa. Such a catastrophic decline in food production would likely result in mass hunger. If temperatures rise by more than 4°C, the consequences are absolutely terrifying. Glaciers will disappear, soil moisture will be lost (as water evaporates at a much higher rate), rainfall will decline in many regions (notably, today's sub-humid and arid regions in the subtropics, like the Mediterranean basin countries), and extreme events such as massive heat waves, droughts, floods, and extreme tropical cyclones, will all become far more frequent.

With temperature increases of 5°C or more, the ensuing sea level rise would likely threaten major world cities, including London, Shanghai, New York, Tokyo, and Hong Kong. Calamitous events are possible with a mega-rise in sea levels. If the big ice sheets in west Antarctica and Greenland melt sufficiently or even partially break up into the ocean, the sea level will rise by many meters (in addition to the sea level increase directly resulting from the expansion of ocean water at higher temperatures). Whenever the Earth was a few degrees Celsius warmer than now, the ice sheets and glaciers retreated, and sea levels were indeed several meters higher than today. Yet in those episodes tens of thousands and hundreds of thousands of years ago, we didn't have mega-cities of millions of people dotting the Earth's coastline!

On the northeast coast of the United States, the sea level has already risen by around one foot or roughly one-third of a meter. Worldwide, the average sea level has increased by roughly one-quarter of a meter since the late nineteenth century, as shown in figure 12.8. (One might have thought that the sea level rise around the world would be fairly uniform, as in a bathtub filling with water, yet in fact differences in the Earth's topography and other geologic features of Earth mean that the sea level will rise at somewhat different rates across the planet.) More storm surges and coastal erosion are already occurring as sea levels are rising. New evidence suggests that by the end of this century, on a BAU path, sea levels could be a meter higher than now; in the worst case, the rise could be several meters.

There is no precise estimate of how and when the great ice sheets of Greenland and Antarctica might melt or break apart, but the human impact is large enough to cause a massive loss of those ice sheets and a massive rise in sea level. The ice

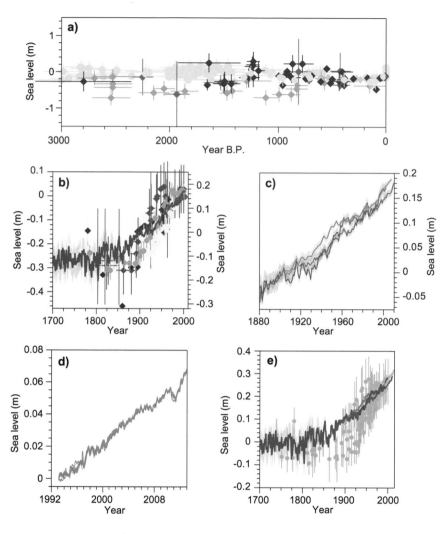

12.8 Sea level rises over time

Church, J.A., P.U. Clark, A. Cazenave, J.M. Gregory, S. Jevrejeva, A. Levermann, M.A. Merrifield, G.A. Milne, R.S. Nerem, P.D. Nunn, A.J. Payne, W.T. Pfeffer, D. Stammer, and A.S. Unnikrishnan, 2013: Sea Level Change. In Climate Change 2013: The Physical Science Basis. Contribution of Working Group I to the Fifth Assessment Report of the Intergovernmental Panel on Climate Change [Stocker, T.F., D. Qin, G.-K. Plattner, M. Tignor, S.K. Allen, J. Boschung, A. Nauels, Y. Xia, V. Bex and P.M. Midgley (eds.)]. Cambridge University Press, Cambridge, United Kingdom and New York.

sheets are already under stress, as figure 12.9 shows (Hansen and Sato 2012, 41). Together, the consequences for the urban areas hugging the oceans and for our food supplies around the world are extraordinary.

Certain regions of the world are extraordinarily vulnerable to higher temperatures and the loss of soil moisture needed for agriculture. I have already noted the vulnerability of the Sahel. Yet the problems are not just in the poor, dry parts of the developing world. The U.S. Southwest (Texas, New Mexico, Arizona, and southern California) is also extraordinarily vulnerable to drying. The Mediterranean basin, including the countries of southern Europe (Spain, Italy, and Greece), North Africa (Morocco, Algeria, Libya, Tunisia, and Egypt), and the eastern Mediterranean (Turkey, Syria, Israel, and Jordan), could also be devastated by drying.

Note the changes of rainfall in the Mediterranean basin over the last century in figure 12.10. The Mediterranean basin has experienced a significant trend of drying. The record clearly indicates that if we continue with BAU, this region could experience further dramatic drying with quite devastating consequences to economies, nature, ecosystems, and food security. This is a region of potentially great instability, because higher food prices combined with politics have already created a tremendous amount of unrest in places like North Africa and the eastern Mediterranean (Syria and Palestine) in recent years.

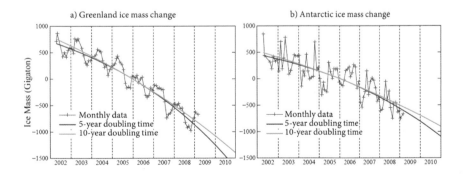

12.9 Ice mass changes in Greenland and the Antarctic (2002–2010)

Source: Hansen, James, and Makiko Sato. 2012. "Paleoclimate Implications for Human-Made Climate Change." In Climate Change: Inferences from Paleoclimate and Regional Aspects, *ed. André Berger, Fedor Mesinger, and Djordjie Šijački, 21–48. Heidelberg: Springer.*

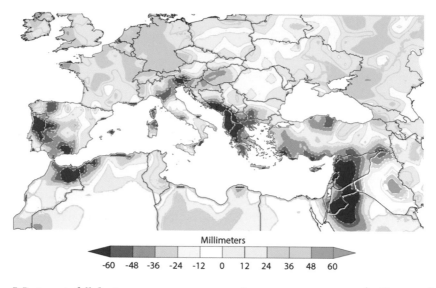

Millimeters

-60 -48 -36 -24 -12 0 12 24 36 48 60

12.10 Winter rainfall during 1970–2010 compared to 1900–2010 average (millimeters)

Source: NOAA.

Recent studies show that many populous parts of the world are likely to experience significant declines of soil moisture needed to grow food. One recent study, summarized in figure 12.11, estimates the increase of drought risk (using two technical indicators, called PDSI and SPEI) around the world for the period 2080–2099 using a series of climate models that incorporate the

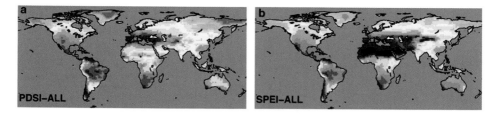

12.11 Prospects for drought 2080–2099

Cook et al. 2014. With kind permission from Springer Science and Business Media.

implications of temperature and precipitation on soil moisture. Most of the world near the equator to the midlatitudes is shaded brown, meaning a tendency toward drought! Only the higher latitudes are found to have more, rather than less, soil moisture.

The map in figure 12.12 comes from a study asking what might happen to food production if the combination of warmer temperatures and more drying were to take place. While the net effects of food production are rather uncertain, the evidence suggests the possibility of massive losses of food productivity in many parts of the world, especially in the tropics and subtropics (i.e., the equatorial region to the midlatitudes). In South Asia and tropical Africa, the map is filled with red zones, meaning the likelihood of major loss of agricultural production. This is the same in the southern part of the United States and much of Latin America and Australia. The only areas of consistent increase in food productivity are likely to

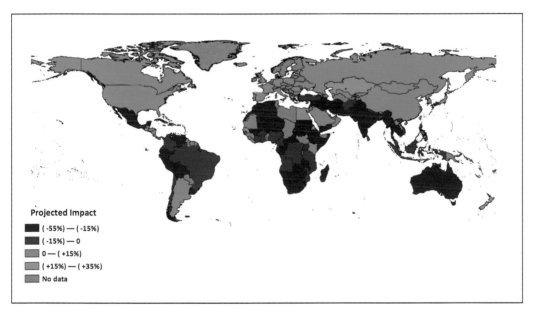

Projected Impact

■ (-55%) — (-15%)
■ (-15%) — 0
■ 0 — (+15%)
■ (+15%) — (+35%)
■ No data

12.12 Projected changes in agricultural productivity (2080)

Source: Hugo Ahlenius, UNEP/GRID-Arendal. Source: Cline, W. R. 2007. Global Warming and Agriculture: Impact Estimates by Country. Washington D.C., USA: Peterson Institute. http://www.grida.no/graphicslib /detail/projected-agriculture-in-2080-due-to-climate-change_141b

be the high latitudes. In short, the world's food supply will be in increasing peril on the BAU path.

Even if one put aside all of the climate-induced changes from the rise of CO_2 concentrations—such as all of the major storm events, the rising sea levels, the rising temperatures, the increased floods and droughts, and the loss of soil moisture needed to grow food—the basic physical fact is that a higher CO_2 concentration in the atmosphere will also lead to more CO_2 dissolving into the oceans, which in turn will raise the acidity of the oceans (as shown by the already-occurring decline in ocean pH depicted in figure 6.2). As the ocean becomes more acidic, major classes of animal life, including shellfish, animals with exoskeletons like lobsters and crabs, certain microscopic plankton (a vital part of the major food chains), and the coral reefs that are so vital for marine ecosystems, are all likely to experience a massive dying-off.

These multiple threats are beyond our easy imagination; and unfortunately there is another kind of climate denial that has been promoted by systematic propaganda from major vested interest groups, including some of the big oil companies. There is every reason to change the game, every reason to mitigate the human-induced climate change for our own safety and for the safety of the planet and future generations. Yet how can human-induced climate change best be brought under control? How can we mitigate human-induced climate change?

III. Mitigation of Greenhouse Gas Emissions to Limit Global Warming to Two Degrees Celsius

There needs to be a strong global response to the climate change challenge. There are two terms to reflect the two different ways of responding, both of which are important. One term, *mitigation*, means to reduce the GHGs causing human-induced climate change. The world has agreed on several occasions to try to limit the increase in average global temperature to no more than 2°C above the pre-industrial mean temperature. The other term used is *adaptation*, which means preparing to live more safely and effectively with the consequences of climate change. Adaptation includes steps like safeguarding cities against storm surges; protecting crops from high temperatures and droughts; and redesigning agricultural technologies to promote more drought resistance, heat tolerance, and flood tolerance in our crops and production systems.

There is a limit to how much we can adapt, because if the changes are so dramatic that sea levels rise several meters or the global food supply is profoundly threatened by higher temperatures and drier conditions, then we are unlikely to be able to control the consequences of massive and worldwide crises. Mitigation is essential. At the same time, it is important to adapt, because climate change is happening and will continue to happen, even if mitigation is highly successful. There is inertia in the warming, as already noted, and it will take us some considerable time at the global scale to bring GHG emissions under control.

In short, mitigation is therefore an enormous priority and requires a careful diagnosis and prescription. Measures must be taken to head off further increases of GHG concentrations. Since about three-fourths of the increased radiative forcing of anthropogenic GHGs is due to CO_2, our highest mitigation priority should be to reduce the emissions of CO_2. Since most of the CO_2 emissions come from the burning of fossil fuels, the reduction of energy-related CO_2 emissions is the number one item on the mitigation agenda. The second way that CO_2 concentrations are increasing is land-use change, so next on the list (actually to be pursued simultaneously with energy-sector reform) is to head off the deforestation that is causing the emissions of CO_2 from land-use change. The third priority is to reduce the emission of CH_4, which results from several processes, both agricultural and nonagricultural. Our fourth priority is the reduction of emissions of N_2O.

For each of these human-induced emissions of GHGs, feasible and economical reductions in emissions must be sought. How long will it take to shift to a low-carbon energy system? What are the technological alternatives available for low-carbon energy? What are the most cost-effective ways to substantially reduce GHG emissions?

The right place to start is with CO_2. Scientists have usefully posed the mitigation question as follows: What would it take to reduce CO_2 emissions (mainly from fossil fuels, but also from land use) to keep the total increase of the Earth's temperature below the limit of 2°C? The basic answer is that since temperature has already increased by almost 1°C, we would need to dramatically reduce CO_2 emissions in the coming decades.

One recent scientific study of what this would require is shown in figure 12.13. Note that the measure of emissions is PgC yr^{-1}. PgC is petagrams (10^{15} grams) of carbon (C) per year, which translates to billions (= 10^9) of tons (= 10^6 grams) of carbon per year. To translate into billions of tons of CO_2 per year, we must multiply by a factor 3.667 (= 44/12), which is the atomic weight of CO_2 relative to the

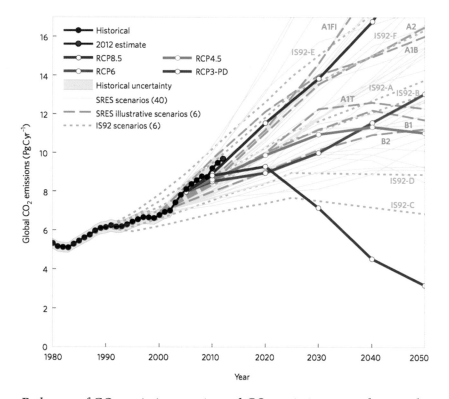

12.13 Pathways of CO2 emissions: estimated CO2 emissions over the past three decades compared with the IS92, SRES, and the RCPs.

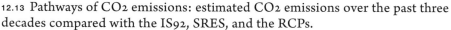

atomic weight of carbon. Thus, the current emissions level in 2014 of around 9.5 petagrams of carbon is equivalent to around 35 billion tons of CO_2.

There are many possible trajectories for future CO_2 emissions. On the horizontal axis are the years to 2050. In the various pathways in the figure, two are most important. The red path is the BAU trajectory, which assumes continued rapid growth of the world economy and few gains in energy efficiency. Global emissions reach around 17 billion tons of carbon by 2040, or as much as 60 billion tons of CO_2. In this scenario, the world economy is growing rapidly, and it uses more and more fossil fuels as it grows. Such a trajectory would take us to massive increases in global temperatures by 2100, probably to between 4°C and 7°C above the preindustrial level.

What trajectory of CO_2 is needed to avoid a 2°C increase? One trajectory that would most likely succeed is shown by the blue curve that bends down sharply after 2020. The blue trajectory holds CO_2 levels to around 450 ppm, and would be likely (but not certain) to contain the rise in temperature below the 2°C limit.

Yet such a trajectory will be very tricky to accomplish, especially with a growing world economy. We basically need a trajectory in which the world economy grows by a factor of perhaps 3 times by 2050 (reaching \$250–300 trillion in today's prices), yet emissions fall by half or more as of 2050 compared with today. A frequent assumption for a 2°C limit is that 2050 emissions should be somewhere between 10 and 15 billion tons of CO_2 (2.7 and 4.1 billion tons of carbon) compared with 35 billion tons in 2014. That would mean that emissions per dollar of gross world product (GWP) would need to decline by a factor of 6 or even more!

The term *decarbonization* is used to mean a sharp reduction of CO_2 per dollar of GWP. A deep decarbonization of the world economy is necessary to remain within the 2°C limit. Since most of the CO_2 comes from burning fossil fuels, we therefore need a sharp reduction in the use of fossil fuels or a large-scale system to capture and sequester the CO_2 that is used.

One major economy, the state of California, is committed by law to reducing its emissions by 80 percent per by the year 2050. This is no small step, given California's importance in the U.S. economy and in the world economy. Indeed, if California were an independent country, its gross domestic product (GDP) would rank twelfth in the world (as of 2012).

A fascinating recent study has examined California's pathway to this goal (Williams et al. 2012). The pathway found in the study is quite important, because it sets certain general principles of deep decarbonization that will be widely applicable. There are three key steps of deep decarbonization, shown in figure 12.14. The first is *energy efficiency*, to achieve much greater output per unit of energy input. Much can be saved in heating, cooling, and ventilation of buildings; electricity use by appliances; and energy directed toward transportation.

The second necessary step is to *reduce the emissions of CO_2 per megawatt-hours of electricity*. This involves, first and foremost, dramatically increasing the amount of electricity generated by zero-emission energy such as wind, solar, geothermal, hydroelectric, and nuclear power while cutting the production of power based

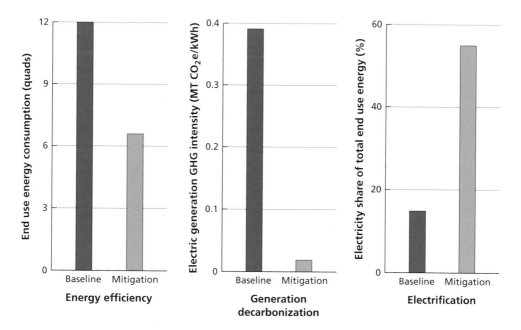

12.14 Three energy transformations to reduce GHGs in California by 2050

Source: Williams, James H., Andrew DeBenedictis, Rebecca Ghanadan, Amber Mahone, Jack Moore, William R. Morrow III, Snuller Price et al. 2012. "The Technology Path to Deep Greenhouse Gas Emissions Cuts by 2050: The Pivotal Role of Electricity." Science 335(6064): 53–59. Reprinted with permission from AAAS.

on fossil fuels. It may also utilize carbon capture and sequestration as an adjunct or fallback technology, depending on the eventual costs of capturing and storing CO_2 from fossil fuels.

The third step is a *fuel shift,* from direct use of fossil fuels to electricity based on clean primary-energy sources. This kind of substitution of fossil fuel by clean energy can happen in many sectors. Internal combustion engines in automobiles can be replaced by electric motors. Furnaces and boilers to heat buildings can be replaced by heat pumps run on electricity. Open furnaces in industry can be replaced by fuel cells run on hydrogen, with the hydrogen produced by electricity. And so on. There are innumerable ways in every sector to shift from direct burning of fossil fuels to reliance on electricity. The trick then is to generate the electricity with low or zero carbon.

Regarding energy efficiency, one policy that has been quite successful is to put appliance standards into effect through regulation. Some economists do not like this approach, but markets are often not very effective in spurring transformations in energy efficiency at the necessary speed. Basic standards can be placed on automobile mileage per gallon or energy use in refrigerators and air conditioners. Building codes, which are part of the normal policy framework of any well-run city, can make a big difference. Building material quality, the insulation and ventilation properties, the choice of heating and cooling systems, and, of course, the types of power sources, all make a huge difference in the energy efficiency of buildings.

There are also several scalable approaches to low-carbon energy. One key option is photovoltaic (PV) cells. Photovoltaic cells have the ability to convert the energy in light rays (photons) into electrical energy. Albert Einstein first explained the underlying physical phenomenon, the photoelectric effect, in 1905. Photovoltaic systems can be the basis for large-scale power generation in much of the world. Figure 12.15 is a map of the solar energy potential across the planet, determined mainly by latitude and by average cloud cover. Note, for example, that solar potential is very high over the midlatitude deserts (such as the Mojave in California and the Sahara in Africa), but actually a bit lower at the equator, where the solar rays are more direct (i.e., overhead) but cloud cover is high.

12.15 World solar energy potential

Credit: Hugo Ahlenius, UNEP/GRID-Arendal; Source NASA. 2008. "NASA Surface Meteorology and Solar Energy (SSE) Release 6.0 Data Set, Clear Sky Insolation Incident On A Horizontal Surface." http://www.grida.no/graphicslib/detail/natural-resource-solar-power-potential_b1d5#.

Figure 12.16 shows another potentially scalable approach to zero-carbon electricity: wind power. The wind turbine uses electromagnetic induction (rotating a coil of conducting material such as copper through a magnetic field) to generate electricity. Wind power is already cost competitive with fossil fuels in many windy places. Figure 12.17 is a map of average wind speeds around the world measured at 80 meters above the surface, showing land regions of high wind potential in orange and red areas. We can see many high-potential areas, including the U.S. Midwest and Northeast, the southern tip of South America, several desert regions of Africa (including Morocco, Sudan, Ethiopia, and Somalia), northern Europe along the North Sea, and parts of central and western China, among others.

Another zero-carbon alternative is geothermal energy. In favorable locations, for example, along the boundaries of tectonic plates, it is possible to tap large-scale heat energy in the Earth's mantle. The geothermal energy is used to boil water to turn steam turbines for electricity generation. Geothermal energy already powers much of Iceland (which uses the energy both to produce electricity and to heat water that is then piped to homes and offices) and is being deployed at an

12.16 Wind turbines

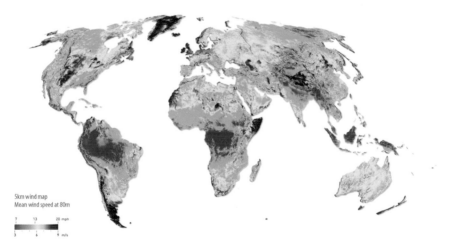

5km wind map
Mean wind speed at 80m

| 7 | 13 | 20 mph |
| 3 | 6 | 9 m/s |

12.17 Average global wind speeds

3TIER by Vaisala.

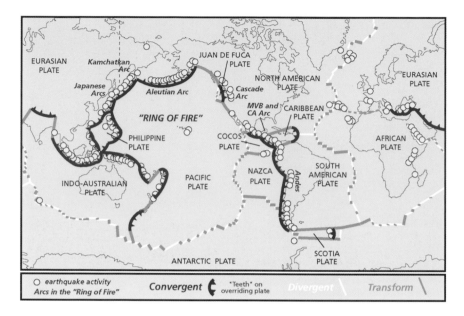

12.18 World geothermal provinces

"What Is Geothermal Energy?" by Mary H. Dickson and Mario Fanelli. Pisa: Istituto di Geoscienze e Georisorse, CNR, 2004. International Geothermal Association http://www.geothermal-energy.org.

increasing scale in the Rift Valley of East Africa and other geothermal sites. Figure 12.18 offers an estimate of geothermal potential in different parts of the world. Notice for example the geothermal zone along the Rift Valley of East Africa.

Nuclear power, such as the British nuclear plant shown in figure 12.19, also offers zero-carbon energy at a relatively low cost, and currently accounts for around 12 percent of global electricity generation. Yet nuclear plants are controversial because of non-climate risks, such as the secret diversion of nuclear fuel and waste for nuclear weapons use, and accidents that cause the release of nuclear radiation into the surroundings, as occurred in the 2011 Fukushima disaster in Japan (when the power plant was hit by a tsunami) and the 1986 Chernobyl disaster in the Ukraine (when nuclear fuel rods were accidentally allowed to overheat as the result of inappropriate procedures). Another challenge is the long-term disposal of nuclear waste materials. Nuclear power is set

12.19 Nuclear power plant in England

to grow markedly in East Asia, notably in China and Korea, while other countries, including Germany, have decided to abandon nuclear power. Still others, such as the United States, are on the policy fence, with society deeply divided between supporters and opponents.

When electricity is produced with low-carbon or zero-carbon technologies, electricity offers an indirect means to reduce carbon emissions from other sectors of the economy that now directly burn fossil fuels. Rather than running vehicles on internal combustion engines, vehicles can be powered by electric motors run on low-carbon electricity (figure 12.20). There are many ways to do this, including battery-powered vehicles that are recharged on the power grid or fuel-cell vehicles in which the fuel cell uses an energy source such as hydrogen that is produced with low-carbon electricity. (The electricity can be used to split water molecules, H_2O, into hydrogen and oxygen.) Synthetic liquid biofuels such as methanol can also be produced through industrial processes using low-carbon energy.

12.20 Electric vehicle at a charging point, London

"Electric car charging," Alan Trotter, Flickr, CC BY 2.0.

Similarly, buildings that are now heated by burning coal, oil, or natural gas on the premises can instead be heated with an electric-powered heat pump, in which the electricity is generated with a low-carbon source. In this way, the direct emissions of CO_2 from the building are eliminated. A heat pump is like a refrigerator run in reverse, pumping heat from a relatively cold to a relatively warm reservoir. In this case, the pump takes heat from outside the building (e.g., heat from underground in the wintertime), and pumps it inside the building. Since the heat is transferred from a relatively cold exterior reservoir (in the ground) to a relatively warm reservoir (the building interior), it must be "pumped" against the natural flow (like pumping water uphill).

There are also many industrial processes that can be converted from the direct burning of fuels (e.g., in furnaces) to heat provided by hydrogen fuel cells and other sources produced by electricity. As with vehicles and buildings, low-carbon electricity offers a way to eliminate the reliance on fossil fuels and thereby to reduce CO_2 emissions from industry. Some of the highest-emitting industrial processes today, such as steel production, can be reengineered to be part of a low-carbon economy.

When the California study added up the numbers, the engineers found a pathway to reach California's bold target of an 80 percent reduction of CO_2 by 2050. That path is illustrated in figure 12.21 (Williams et al. 2012, 54). The baseline emissions are the line at the top, which shows that CO_2 emissions with BAU are on the rise in California because of long-term economic growth. The preferred mitigation trajectory is the downward-sloping line at the bottom of the curve. The gap is explained in the list of the ways of reducing CO_2 emissions. The light blue zone, for example, shows the reduction of emissions coming from energy efficiency. The purple zone shows the reduction of emissions coming from decarbonizing electricity generation. The yellow zone shows the reduction from fuel shifting (electrification), such as the transition from vehicles with internal combustion engines to electric vehicles.

There are also other smaller categories of low-carbon energy, such as the deployment of biofuels. Biofuels use biomass to produce a liquid fuel that is a substitute for fossil fuel. Figure 12.22 shows one example of an advanced biofuel. These panels look like PV solar cells, but they are in fact filled with genetically modified bacteria engineered to use solar energy to synthesize liquid hydrocarbons. There are many biological pathways by which biomass can be grown and converted into fuels. The problem with biofuels, however, is that in many cases the production of the biomass feedstock competes with food production. This is very much the problem with the large U.S. program to convert corn (maize) to ethanol through the anaerobic respiration of yeast. The diversion of maize production for this program has driven up food prices (by shifting maize out of the supplies of food and feed) while doing little to reduce net CO_2 emissions.

Regional Solutions for Renewable Energy

There are two further crucial aspects to tapping renewable energy sources like wind and solar power. First, the greatest potential for renewable energy is often located far from population centers. Solar energy, for example, is highest in the

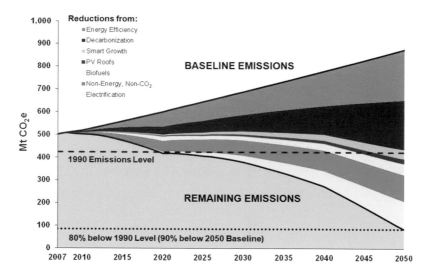

Wedge Category:	Emissions Reduction Mt CO₂e (% of Total)		Types (and Numbers) of Measures Used	Key Attributes in 2050
	2030	2050		
Energy Efficiency	102 (33%)	223 (28%)	Building EE (18); Vehicle EE (9); Other EE (6)	Improve energy efficiency 1.3% per year on average for 40 years
Electricity Decarbonization	72 (23%)	217 (27%)	High renewables, high nuclear, high CCS, and mixture of the three	Meet 90% of generation requirement with CO₂-free sources. Equivalent decarbonization in each scenario.
Smart Growth	13 (4%)	41 (5%)	Reductions in vehicle miles traveled (VMT) (6)	VMT reduced in light duty vehicles (LDV) by 10%; freight trucks 20%; other transportation 20%
Rooftop PV	8 (3%)	21 (3%)	Residential and commercial PV roofs (2)	Rooftop PV displaces 10% of electricity demand by 2050.
Biofuels	18 (6%)	49 (6%)	Transportation biofuels: ethanol, biodiesel, biojet fuel (9); Residential, commercial, industrial biomethane (3)	By 2050, biomethane displaces 2% of natural gas use in buildings, and biofuels displace 10-20% of petroleum-based fuels for vehicles
Non-Energy, Non-CO₂	67 (22%)	116 (15%)	Cement, agriculture, and other (3)	Non-fuel, non-CO₂ GHG emissions reduced 80% below baseline
Electrification	29 (9%)	124 (16%)	Transportation electrification (9); Other end-use electrification (5)	75% of LDV gasoline use displaced by PHEVs & electric vehicles; 30% of fuel use in other transport sectors electrified; 65% electrification of non-heating/cooling fuel use in buildings; 50% electrification of industrial fuel uses
Baseline Case Emissions	688	875		
Mitigation Case Emissions	380	85		
Total Reduction	308	791		

12.21 Emission reduction wedges for California in 2050

Source: From Williams, James H., Andrew DeBenedictis, Rebecca Ghanadan, Amber Mahone, Jack Moore, William R. Morrow III, Snuller Price et al. 2012. "The Technology Path to Deep Greenhouse Gas Emissions Cuts by 2050: The Pivotal Role of Electricity." Science 335(6064): 53–59. Reprinted with permission from AAAS.

12.22 Biofuel plant: making ethanol from sunlight and CO_2

Credit: Joule.

desert regions. Second, both wind and solar power are intermittent energy. Solar power obviously varies predictably by time of day, but it also depends on the random fluctuations of cloud cover. Winds also fluctuate unpredictably. Even very windy locations occasionally experience hours or days of becalmed conditions with little power generation, and in many places winds are highly seasonal.

There are three main implications. First, tapping renewable energy on a large scale will generally require building new transmission lines to carry the power from remote locations to the major population centers. Second, the storage of renewable energy—for hours, days, or longer—makes them far more attractive as energy sources. There are many proven and emerging technologies for storing intermittent power sources. Third, there is a strong case for joining disparate renewable energy sources into a shared transmission grid. When it is cloudy in some part of the network, it is likely to be sunny in other parts of the network, thereby helping to smooth out the fluctuations in any single location.

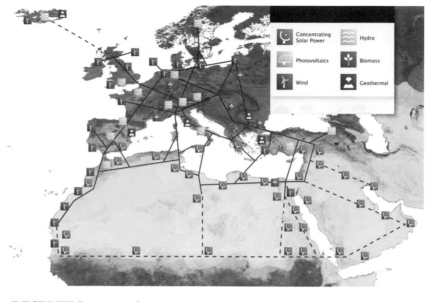

12.23 DESERTEC energy plan

Source: DESERTEC.

Consider three examples of potential large-scale power generation and distribution based on renewable energy. None has yet been developed, yet each is under consideration by governments and private investors. The first project is known as DESERTEC, and is designed to link North Africa, the Middle East, and Europe into a single grid (shown in figure 12.23). This system would tap the strong solar and wind potential of North Africa and the Arabian Peninsula, both to supply energy for these economies and to export the surplus to Europe. The challenges to realizing this concept are enormous, beginning with an estimated price tag of several hundred billion dollars and technical challenges of managing a far-flung grid based heavily on renewable, intermittent energy. Yet the concept is potentially a key solution to Europe's unsolved challenge of deep decarbonization and an enormous boost to the economies of North Africa and the Middle East.

A second major concept is to tap the enormous offshore wind potential of the United States, illustrated in figure 12.24. Proponents of wind power have argued

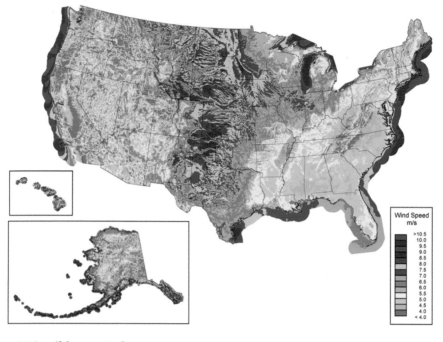

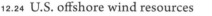

12.24 U.S. offshore wind resources

Wind resource map developed by NREL with data from AWS Truepower.

persuasively that the wind off the shore of the Eastern Seaboard could potentially meet most of the electricity needs of the U.S. Northeast, from Virginia to
Maine. Yet despite many proposals and business plans, there is still no offshore
wind power tapped in the United States, due to regulatory, political, and environmental challenges and debates. There are also unsolved technological challenges
that seem to be within reach of solution yet have not been explored with adequate public or private research and development (R&D) funding. Of course, the
United States has vast untapped large-scale renewable energy potential, including
solar energy in the Mojave, onshore wind in the Dakotas, and the offshore wind
shown in figure 12.24.

A third renewable energy project with the potential to transform its region with
zero-carbon energy is the vast hydropower potential of Inga Falls in the Congo River

12.25 Map of Inga Falls dam site

Source: International Rivers, Congo's Energy Divide (2013).

basin. The Grand Inga Dam Project, discussed for half a century, could produce up to 40 gigawatts of hydroelectric power, more than one-third of the total electricity currently produced in Africa. However Inga Falls is in one the least bankable places in the world, the Democratic Republic of Congo. Yet many close observers now feel that an arrangement is now within reach in which the nations of the region, including the Democratic Republic of Congo, the Congo (Brazzaville), Burundi, Rwanda, and perhaps South Africa, join together to back a multilateral project. Potential funders of the project, which is estimated to cost around $50 billion in total, might include the African Development Bank, the Chinese Development Bank, and the World Bank Group (including the International Finance Corporation).

All three of these projects illustrate a basic reality of deep decarbonization. Large-scale, zero-carbon projects are within reach. Yet they are politically complex, require massive upfront investments, and need further R&D to bring them to fruition. In short, massive renewable energy is possible but far from assured. A serious global commitment to low-carbon energy will be required.

Carbon Capture and Sequestration

In addition to energy efficiency, low-carbon electricity, and fuel switching, there is one more potential way to reduce the CO_2 emissions from fossil fuel use. Currently, when fossil fuels are burned, the CO_2 enters the atmosphere, where it may reside for decades or centuries. A potential solution is to capture the CO_2 instead of allowing it to accumulate in the atmosphere. Two main ways to do this have been proposed. The first is to capture the CO_2 at the site where it is produced (e.g., the power plant), and then to store it underground in a geologic deposit (e.g., an abandoned oil reservoir). The second is to allow the CO_2 to enter the atmosphere but then to remove the CO_2 directly from the atmosphere using specially designed removal processes (e.g., collecting the CO_2 with special chemical sorbents that attract the CO_2). This latter approach is called "direct air capture" of CO_2. Figure 12.26 is a mock-up of a direct-air-capture facility as proposed by Professor Klaus Lackner, one of the world leaders in the engineering of direct air capture of CO_2 (Lackner et al. 2012).

If carbon capture and sequestration (abbreviated as CCS) proves to be successful, then there is a wonderful way to reduce CO_2 emissions without having to change our current technologies or energy mix! Rather than shifting to new sources of noncarbon energy, we could continue to use fossil fuels but then remove the CO_2 that is produced, either at the power plant or via direct air capture. Some oil companies, for example, have presented climate change scenarios in which CO_2 mitigation is achieved largely through the scaling up of CCS.

There are vigorous technical and policy debates about the feasibility and cost-effectiveness of large-scale CCS technologies. There are, indeed, many questions. First, how costly will it be to capture CO_2 on a large scale (through either method)? Second, how costly will it be to ship the CO_2 by a new pipeline network and then store the CO_2 in some safe, underground geologic deposit? And third, if the CO_2 is put underground (e.g., in an abandoned oil reservoir or perhaps a saline aquifer that can hold the CO_2), how sure are we that the CO_2 will stay where it is put, rather than

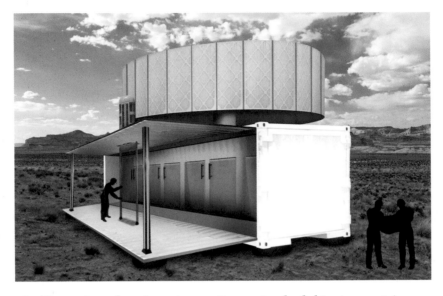

12.26 An illustration of an air capture unit on a standard shipping container (by Prof. Klaus Lackner)

Credit: GRT, 2009.

returning to the surface and then into the atmosphere? Leakage rates of CO_2 would have to be very low to make this technology feasible on a large scale.

Governments, including the United States, European Union, Australia, and China, have been talking about the large-scale use of CCS for at least a decade, but there is still far too little R&D underway to test the economic and geologic potential for large-scale CCS. Remember that tens of billions of tons of CO_2 would have to be captured and stored each year for CCS to play the leading role in addressing CO_2 emissions. Perhaps it will prove feasible and economical at a smaller scale, where the location of power plants and suitable geological storage sites make CCS an especially low-cost option.

Geoengineering as a Final (Desperate?) Option

There is one more idea around, called geoengineering. The basic idea is that if carbon emissions cannot be stopped at a reasonable cost or on a reasonable timeline, then there may be other ways to compensate for or counteract the effects of the rising CO_2. For example, if CO_2 continues to rise and dangerously warm the planet, some scientists have suggested that we should deliberately add sulfate aerosol particles into the air to dim the incoming sunlight and thereby cool the planet in order to offset the warming effects of the CO_2. Another idea is to place giant mirrors in space in order to deflect some amount of incoming solar radiation. These are, evidently, very radical, and perhaps completely unworkable ideas.

Another huge problem with such suggestions is that the compensatory action (in this case, the deliberate emission of sulfate aerosols) may have hugely deleterious effects (e.g., air pollution or dimmer sunlight), so that they "solve" the CO_2 problem only by introducing an even greater or more unpredictable problem. Remember that if we actually try to offset the CO_2 warming by adding sulfate aerosols, the CO_2 concentrations in the atmosphere would continue to rise. This continued increase would have two huge implications. First, it would mean that if we ever stop adding sulfate aerosols into the atmosphere, the warming effect of the CO_2 would quickly be exposed. Temperatures would surge as the sulfate aerosols are washed back to Earth (e.g., in rainfall). Second, the high concentrations of atmospheric CO_2 would continue to acidify the oceans, even though the aerosols temporarily offset the warming effect of the CO_2.

For these reasons it seems unlikely that offsetting geoengineering could ever make it safe for humanity to continue to increase the atmospheric concentration of CO_2. Humanity most likely has no good alternative other than to keep the carbon emissions below the trajectory associated with a 2°C increase in temperature.

IV. Adaptation

It is possible to reduce human emissions of GHGs substantially. The technologies are within reach. Energy efficiency, low-carbon electricity, and fuel switching (e.g., electrification of buildings and vehicles) are all needed. Carbon capture and sequestration may play some role. Yet even hugely successful efforts in these directions are bound to involve an ongoing buildup of atmospheric

CO_2 for years to come, and with it, continued climate change and global warming. In other words, it is too late to prevent at least some further increase of climate damage.

In fact, the situation is even grimmer than that. Suppose (unrealistically!) we could immediately stop all new net emissions entirely, and thereby maintain the atmospheric levels of CO_2 and other GHGs as they are in 2014. This would not be enough to stop global warming. The Earth's average temperature has so far increased by 0.9°C compared with the preindustrial temperature, yet the oceans have not yet warmed as much as the land (given that oceans have an enormous capacity to absorb heat). When the oceans finally warm in line with the GHG concentrations, the Earth's average temperature is likely to be an additional 0.6°C warmer than now (or a total warming of 1.5°C). Thus, further warming is in store for two reasons: (1) "thermal inertia" (the delay in ocean warming); and (2) the inevitability of a further buildup of greenhouses gases in the short term.

For these reasons, we will need not only to prevent future climate changes by decarbonizing the energy system (and taking actions vis-à-vis the other GHGs), but also learn to live with at least some climate change as well. With great diligence and global cooperation it may be possible to keep the global average temperature from rising by 2°C above the preindustrial level, yet even so, a 2°C rise will imply massive changes to the climate system, including more droughts, floods, heat waves, and extreme storms. We need to get ready for such eventualities.

Adaptation will require adjustments in many sectors. In agriculture, crop varieties must be made more resilient to higher temperatures and more frequent floods and droughts (depending on location). Cities need to be protected against rising ocean levels and greater likelihood of storm surges and flooding. The geographic range of some diseases, such as malaria, will spread as temperatures rise. Biodiversity will suffer as some animals and plants are unable to adjust to the changing climate conditions; special efforts will be needed to ensure that particular species are not thereby driven toward extinction. The list, in short, is very long and location-specific.

Policy Instruments for Deep Decarbonization

Economists rightly emphasize the need for corrective pricing to provide proper incentives for producers and consumers to reduce CO_2 emissions and avoid the "externalities" associated with CO_2-induced climate change. Carbon dioxide imposes high costs on society (including future generations), but those who emit the CO_2 do not pay for the social costs that they impose. The result is the lack of a market incentive to shift from fossil fuels to the alternatives. Ideally, producers and consumers would choose among alternative energy technologies in order to minimize the true social costs of energy use, including the costs of climate change and the costs of adverse health consequences of polluting energy sources. On both counts—climate and health—users of fossil fuels should be required to pay a higher price than users of clean energy, in order to shift the incentives to a low-carbon economy.

There are several ways to overcome part or all of the current incorrect pricing of fossil fuel use. The most straightforward is that all users of fossil fuel should bear an extra "carbon tax" equal to the social cost of the CO_2 emitted by the fuel. This would raise the costs of coal, oil, and gas compared with wind, solar, nuclear, and other low-carbon energy sources, shifting the energy use toward the low-carbon options. (Of course if these alternatives also impose social costs, such as the risks of nuclear accidents, those alternatives should also bear the true social costs inclusive of those risks.) Economists have proposed a carbon tax on the order of $25–100 per ton, on the grounds that the social cost of an extra 1 ton emission of CO_2 is estimated to be in the range of $25–100 per ton. Over time, as climate change intensifies, the social cost of CO_2 emissions, and hence the carbon tax, would most likely increase.

A related alternative approach, in use in Europe and in some U.S. states, is a permit system, in which emitters of CO_2 must buy a permit to do so. This is closely akin to the carbon tax, except that emitters buy a permit on the open market (or receive it from government) instead of paying a tax. If an emitter would like to increase emissions of CO_2 (perhaps because the business is expanding so that energy use is rising), the emitter can buy an extra emissions permit on the market from another firm that is successfully reducing its carbon emissions.

There have been heated debates for two decades about whether carbon taxes or emissions permits are the appropriate policies. Carbon taxes are likely to give more predictability as to the future price of carbon. Emissions permits may (or may not) give more predictability to the future quantities of emissions. They would seem to give more predictability about the total emissions (since emissions are limited in theory by the availability of the permits), but in fact permit systems are often not very credible, since an expected scarcity of permits (driving up their price) frequently leads governments to increase the allotment of permits. Taxes in general are much easier to administer, while permit systems are in principle easier to configure to meet special interests (e.g., specific favored industries can be given permits for free in order to delay their adjustment to alternative energy sources). In practice, both types of systems are likely to be used in future years, though the tax-based systems are likely to be significantly more robust, predictable, and easier to administer.

A third way to adjust market prices is through "feed-in tariffs." The government tells a utility company or a power generator, "We will buy electricity from you, and will pay an extra high price if the electricity that you are bringing into this system is from a low-carbon source such as solar power." Rather than taxing the CO_2, the government instead gives an added boost to the alternative sources. These positive incentives can be quite powerful in inducing companies to shift to low-carbon energy generation. The main problem of feed-in tariffs compared with a carbon tax is that the government may not have the budget revenues available to pay the subsidy for low-carbon energy. Indeed, several countries that promised such feed-in tariffs pulled back their commitments after the 2008 financial crisis.

The Double-Edged Sword of Technological Advance

It is heartening to realize that advances in technological know-how can enable humanity to find a safe, efficient, and relatively low-cost transition from fossil fuels to a low-carbon economy based on greater energy efficiency, low-carbon electricity, and fuel switching. Recent technological advances include sharp reductions in the cost of wind and solar energy; improved geothermal energy; improved batteries for electric vehicles; smarter power grids; improved building materials; better waste management; new building design requiring less energy for heating, cooling, and ventilation; and much more. And there are significant advances ahead, such as the potential for direct air capture of CO_2, storage of intermittent renewable energy, highly efficient long-distance power transmission, advanced biofuels, and new nanotechnologies for strong, lightweight construction materials, among others. Technological advances can save the day.

Yet we should not be overly simplistic about the saving grace of technological advances. Ironically, in a world of externalities (such as CO_2 emissions), technological advances can worsen rather than improve the situation, since they can exacerbate the tendencies toward the exploitation of high-carbon energy sources. The simple fact is that the oil and gas sector has been quite technologically sophisticated in recent years, dramatically improving the capacity to find, develop, produce, and transport fossil fuel–based energy! Here are a few pertinent examples.

The first advance, shown in figure 12.27, is a true technological marvel: a floating liquefied natural gas (LNG) plant, designed and built by Royal Dutch Shell, and soon to be introduced into service (Shell 2014). This facility, described as the largest structure ever sent to sea, will cool offshore natural gas into LNG for onward ocean transport. Before the advent of this new facility, offshore gas must be transferred by pipeline to a land-based LNG plant. Ocean deposits of methane too far from the land are not economical when they must be transferred by pipeline, but now will be economical to produce. Moreover, pipelines are not only expensive but are also vulnerable to storms, leaks, ruptures, and other accidents in the open seas. A technological marvel, yes—and one that will accelerate the production and use of natural gas.

A second example of a technological breakthrough is the capacity to develop Canada's vast reserves of oil sands, which are sand and rock deposits that contain bitumen, a highly viscous form of petroleum. One of the development sites is shown in figure 12.28. Canada's oil sands (and also the oil sands of Venezuela) are vast deposits that would substantially raise the quantity of petroleum available to world markets. They have been too expensive to produce until recently, when the combination of improved mining and processing technologies and higher world oil prices have made these deposits profitable. The proposed Keystone XL

12.27 Model of floating liquefied natural gas factory (Shell Prelude)

Photographic Services, Shell International Ltd.

12.28 Canadian oil sands, Alberta

"Tar sands, Alberta," Dru Oja Jay, Howl Arts Collective, Flickr, CC BY 2.0.

Pipeline, a highly controversial new pipeline development, would carry the Canadian oil to refineries in the Gulf of Mexico, and (mainly) on to global markets. A technological breakthrough: yes, but one associated with massive pollution on site (as evident in the figure) and a vast increase of fossil fuel resources that will tend to push the world even faster over the 2°C carbon budget.

A third remarkable technological breakthrough is shown in the illustration in figure 12.29. The figure illustrates the breakthroughs of horizontal drilling and hydraulic fracturing (hydrofracking) of natural gas caught in shale rock. In this process, the drilling is first down and then horizontal (as shown) into shale rock containing methane in the rock pores. To release the methane, a high-pressure mix of fluids and drilling materials are blasted into the rock, thereby fracturing

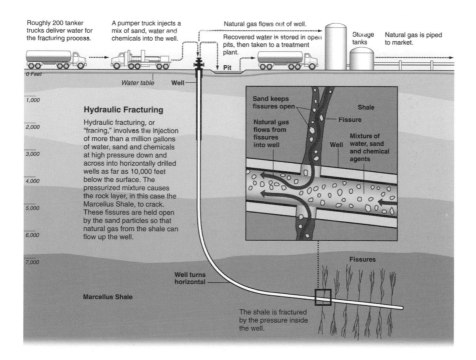

12.29 Hydrofracking diagram

Al Granberg/ProPublica.

the rock and freeing the methane, which rises to the surface, where it is collected. The shale gas boom (and a similar shale oil boom) has been transforming the U.S. energy landscape and rural landscape in recent years. The process is highly contentious. On the one hand it is leading to an oil and gas boom in the United States. On the other hand, it is leading to massive local pollution and a boom in fossil fuels that is at least delaying, if not blocking, an eventual shift to low-carbon energy.

All three advances have greatly expanded the world's capacity to tap fossil fuel reserves, but we must pause to ask ourselves if we are really doing ourselves a favor by slowing down the transition to urgently needed low-carbon energy. These advances are making it harder, not easier, to live within the carbon budget. They have made the politics around climate change even more difficult, since the fossil fuel lobby has something important to show for itself: real resources earning real profits (and large profits at that). Yet none of this changes the basic truth: we are on a path of grave long-term planetary danger at the price of short-term market returns.

V. The Politics of Carbon Dioxide Mitigation

There are many obstacles to a low-carbon world: technological, economic, engineering, and organizational. Getting to a low-carbon economy will not be easy. Indeed, it will require serious planning alongside market forces. It will require global cooperation to invest in the improvement of low-carbon technologies. It will require a commitment to much deeper decarbonization than most governments are now considering. The right approach is to recognize that by 2050 we must have cut emissions by more than 50% of today's levels, and then to "back-cast" (that is, work backwards from 2050) to the present period in order to chart out the timing of deep changes in the energy system. None of this is easy—far from it.

Yet perhaps no obstacle is as important as politics, at least in countries with large domestic supplies of coal, oil, and gas. The fossil fuel industry is probably the most powerful lobby in the United States and in most other major fossil fuel countries. The biggest obstacle to a strong global agreement on climate change remains the bargaining positions of the major fossil fuel countries: the United

States, Canada, China, Russia, and the Persian Gulf economies. These positions, in turn, reflect mainly domestic political considerations.

Figure 12.30 shows two maps. The shaded areas in brown on the top map are states that produce coal, about half of the U.S. states. The bottom map shows in red the states where the senators voted against the Climate Stewardship Acts (also known as the McCain-Lieberman Acts), which would have introduced a cap-and-trade system for GHGs. It is an almost-perfect fit. Coal, oil, and gas interests finance the politicians in the "brown states" and have so far been able to maintain a veto on federal climate control legislation. This is the case all over the world, which makes it extremely difficult to make progress. Interestingly, many of the "green states" in the voting map have implemented state-level mitigation programs, such as California's decision to reduce CO_2 emissions by 80 percent by 2050.

The global politics of climate change have been largely stuck since 1992, when the world's governments adopted the UNFCCC at the Rio Earth Summit. It is a well-reasoned, well-balanced document that points the way forward on global mitigation. The main objective of the treaty is described clearly in article 2, which states that:

> the ultimate objective of this Convention and any related legal instruments that the Conference of the Parties may adopt is to achieve, in accordance with the relevant provisions of the Convention, *stabilization of greenhouse gas concentrations in the atmosphere at a level that would prevent dangerous anthropogenic interference with the climate system* [emphasis added]. (UN 1992b)

This objective makes perfect sense and has been made more precise and operational in recent years by associating "dangerous anthropogenic interference" with a rise in the mean global temperature of 2°C. Yet since the UNFCCC went into effect in 1994 (upon ratification by enough countries), the world has failed to implement it properly. The treaty parties have met year after year, and have just finished the COP20 (Conference of the Parties, 20th session) in Lima in 2014. Yet the treaty has not even succeeded in slowing the year-to-year of increase of GHG emissions, much less forced the emissions curve to turn downward.

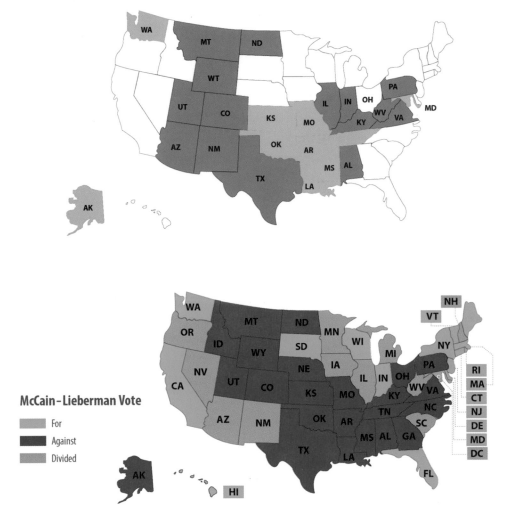

McCain-Lieberman Vote

- For
- Against
- Divided

12.30 Coal producing states vs. Climate Stewardship Acts voting patterns

Source: U.S. Department of Energy & U.S. Senate.

The first major attempt to implement the treaty came with the Kyoto Protocol, signed in 1997 (UN 1998). This was an agreement by the high-income countries to reduce their emissions by an average of 20 percent by 2012 compared with 1990. The developing countries, including the fast-growing emerging economies such as China, were not obligated to meet specific emissions targets. The treaty did not work. On the one hand, the United States never signed and Australia and Canada did not implement the treaty despite having signed it. (Notice the pattern of major fossil fuel–producing countries!) On the other hand, the emissions by China and other emerging economies soared, thereby keeping the global emissions levels on a steeply rising course.

Since 1992, the U.S. Senate (which must ratify all treaties) has been in the grips not only of the coal, oil, and gas lobbies but also of a perception that the United States should do nothing if China will not do as much or more. The U.S. rationale has held that it is "unfair" to expect the U.S. to act in advance of China, as that would leave China in an advantaged competitive position in world trade. This is an odd sense of "fairness," because the U.S. for decades has been changing the climate of the entire world without any sense of fairness about the huge costs it has imposed on the rest of the world. Though President Clinton's administration actually signed the Kyoto Protocol in 1997, the president never sent it to the Senate for ratification, as its defeat in the Senate was assured.

The UNFCCC actually assigns the initial mitigation responsibilities to the high-income countries (known as Annex I countries under the treaty). The high-income countries are assigned this responsibility for a few reasons: (1) they are better able to bear the extra costs of low-carbon energy; (2) they are disproportionately responsible for the rise in CO_2 in the past; and (3) the poorer countries need time and help to catch up with the richer countries. China has long insisted that the United States and Europe should lead the way and that it would follow some years later as its economy gained strength.

Since 1992, however, much has changed. China has now become the world's second-largest economy and has actually become by far the world's largest GHG emitter. Even though the Chinese economy is not as large as the United States, it emits far more CO_2 for three reasons: (1) it is less energy efficient (higher energy input per unit of GDP); (2) it relies more on coal, the most CO_2-intensive of all fossil fuels (higher CO_2 per unit of energy); and (3) it is more industrial, so that the economy has several large, energy-intensive sectors such as steel production. Indeed, one of the reasons that the United States and Europe emit less CO_2 than China is that they are net importers of energy-intensive products in their trade with China.

Still, twenty-two years after the UNFCCC was agreed upon, the global politics are shifting, and China is now being called upon by countries around the world to take up more global leadership on climate mitigation. China today is a far richer country than it was in 1992. It has had another two decades of very rapid economic growth. As just noted and shown in figure 12.31, China is now the world's largest GHG emitter, having overtaken the United States around 2007. China notes in its own "defense," however, that in per capita terms it still emits much less CO_2 than does the United States. The United States emits 17.6 tons of CO_2 per person, while China emits about 6.2 tons of CO_2 per person. Still, the Chinese leadership clearly acknowledges that China must do far more in order for the world to achieve the 2°C target.

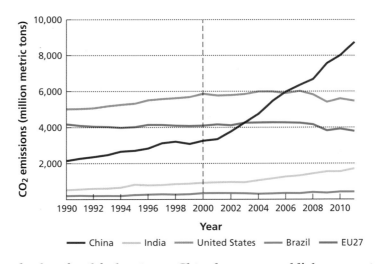

12.31 Top absolute fossil fuel emitters: China becomes world's largest emitter

Source: The Policy Climate. 2013. San Francisco: Climate Policy Initiative.

There are internal pressures as well. For one thing, China itself is highly vulnerable to climate change. A significant part of China is already very dry and is likely to get drier as a result of climate change. China is highly vulnerable to extreme storms, extreme events, and massive flooding. China is deeply vulnerable to climate change and so has a real reason to participate in a global mitigation effort.

Heavy smog pollution is becoming more frequent in major Chinese cities. This smog arises from a mix of industrial pollution, heavy coal burning, and automobile congestion. Recent estimates suggest that some regions of northern China are losing as many as 5.5 years of life expectancy due to the heavy air pollution! Switching from coal to low-carbon or zero-carbon energy would therefore have two huge benefits for China: climate change mitigation and improved public health.

At the COP17 in 2011 in Durban, South Africa, the Parties to the UNFCCC agreed that they would reach a more definitive agreement on climate control by 2015, at which time all countries would take binding commitments to mitigate their GHG emissions. Unlike the Framework Convention, which put the responsibility for action on the rich countries as a start, the new agreement in principle is to put responsibility everywhere. This is, at least, conceptually a breakthrough, because there is now the potential for the United States, China, and other major emitters to agree on a new approach. This was understandably hailed as a breakthrough, though it of course must be put into perspective: the decision in Durban in 2011 was taken nineteen years after the UNFCCC was signed in 1992; to be negotiated twenty-three years later in 2015; ratified twenty-six years later in 2018; and enter into force twenty-eight years later in 2020. This is not exactly a world standing on the precipice and acting with due urgency!

In practical, problem-solving terms, each region of the world needs to implement a sensible, economically efficient, deep decarbonization program built on the three pillars of energy efficiency, low-carbon electricity, and fuel switching. It can be done, if the will is there. The world should also agree to joint programs of R&D on key low-carbon challenges, such as the effective storage of renewable intermittent energy and CCS. The world should also agree to help the poorest countries take on this challenge, for example, by helping central Africa build the Grand Inga Dam. In short, the world has climate solutions. What it lacks is the time for further delay.

13

SAVING BIODIVERSITY AND PROTECTING ECOSYSTEM SERVICES

I. What Is Biodiversity?

We have examined at length what happens when a growing world economy pushes against planetary boundaries. The world now has 7.2 billion people and an output of around $90 trillion, with both the population and global output continuing to rise. The world economy is continuing to grow at 3–4 percent per year, meaning a doubling every twenty years or so. There are already huge pressures on the world's ecosystems, climate, and oceans. We have not yet found a way to reconcile that continuing growth with environmental sustainability.

This trespassing of planetary boundaries is occurring in many ways, including climate change and pollution, but one of the most dramatic ways is the loss of the planet's biodiversity, the subject of this chapter. Humanity is putting so much pressure on the Earth that it is causing a dramatic increase in the rate of species extinction, estimated to be more than a thousand times faster than before the Industrial Revolution. There are many other phenomena associated with this loss of species, such as the decline of genetic diversity within species and the abundance of particular species. *The combined effect is so large that it is causing what could be the sixth great extinction on the planet.*

There is one overriding truth to this sixth wave of extinctions and the threats to biodiversity: the threats to the survival of species are coming from many different directions. As is true of everything in sustainable development, we are dealing with a complex system, where there is not a linear effect from a single cause to a single outcome and then on to another effect. There are multiple stressors, multiple drivers of environmental change, and multiple causes of species extinction and the decline in species abundance and genetic diversity. We must understand the complexity of this system, because no single approach will be sufficient to head off this sixth great extinction that threatens millions of species, including *Homo sapiens*—human beings, ourselves!

To understand biodiversity we must start with an understanding of an *ecosystem*—the collection of plants, animals, and microbial life interacting with the abiotic (nonliving) part of the local environment. The key is that this is a set of living organisms and the nonliving environment, that all interact in a system. Ecologists study ecosystems by studying the fluxes and dynamics of the system—how does nutrient flow take place within a food web and within the processes of metabolism, oxidation, respiration, photosynthesis, and other basic processes of metabolism in the living organisms within the system? How does the diversity of the species, and the diversity of the individual organisms within a species, affect the behavior of that whole ecosystem?

Another core concept of an ecosystem is its biological diversity, or *biodiversity*. Biodiversity is the variability of life that occurs at all different levels of organization. Biodiversity includes the variability of life within a species—each of us is different from other people, with variations in our genetic makeup. Biodiversity also includes the diversity of species within an ecosystem and the various relationships of the species, such as predator and prey, mutualism and parasitism. The interactions of diverse species determine fundamental characteristics of an ecosystem, such as whether the ecosystem is *biologically productive* (e.g., in the output of photosynthesis and in the amount of living matter, or biomass, in the ecosystem) and whether it is *resilient to shocks* such as changes in climate, the introduction of new species into the system, or the overharvesting of one part of the system by human action (such as excessive fishing, logging, or hunting).

Finally, biodiversity also involves the diversity of species across ecosystems. The long-distance interactions of ecosystems, such as desert ecosystems interacting with humid ecosystems, affect the functioning of each ecosystem as well as the regulation of the Earth as a whole. If a critical biome (such as the equatorial rain forests or the Arctic region) suffers a major change, for example, as the result of human-induced climate change, the effects on other ecosystems can be profound through various long-distance interactions of Earth processes, including precipitation, winds, ocean circulation, chemical changes, and others.

Thus, to understand biodiversity we must understand the variation of life at all different levels of organization and how that variability affects the performance of ecosystems in ways that matter. One of the most important studies on the functioning of ecosystems was a major global effort that reported in 2005, called the Millennium Ecosystem Assessment, or MEA (MEA 2005). The MEA took a global view of the major ecosystems in the world and tried to give a conceptual framing of how they function, how they interact, and how they provide various functions, or *ecosystem services*, for humanity.

One of the important schematic ideas that came out of the MEA is the chart in figure 13.1. The idea of this chart is to define how ecosystems affect human wellbeing. The left side shows four categories of ecosystem services. The first, *provisioning services*, include the ways that ecosystems directly provide for human needs: providing food, fresh water, wood and fiber for building structures and clothing, and biomass for fuels. The second, *regulating services*, include various functions of ecosystems in controlling the basic patterns of climate, disease transmission, and nutrient cycling of fundamental importance to humanity, such as the fluxes of water, nitrogen, and oxygen. Ecosystems have huge effects on climate regulation. If the ecosystems of the Arctic and Antarctic regions were to be dramatically changed by human-induced climate change, there would be powerful feedbacks on the rest of the planet. For example, the melting of the great ice sheets of Antarctica and Greenland would dramatically raise ocean levels worldwide, change fundamental patterns of ocean circulation, and also fundamentally change the Earth's energy balances and overall climate. The melting of ice in the tundra could potentially release huge amounts of methane and carbon dioxide,

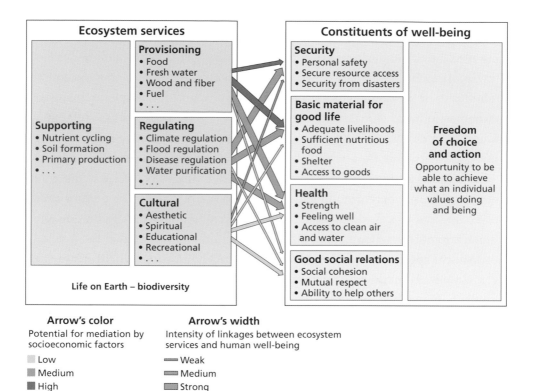

13.1 Relationships between ecosystems and wellbeing

Source: Millennium Ecosystem Assessment. 2005. Ecosystems and Human Wellbeing: Synthesis. *Washington, DC: Island Press.*

creating massive positive feedbacks that would amplify human-induced greenhouse gas emissions.

Another example of regulatory services is flood control. Topographical features of ecosystems such as mangrove swamps often protect humans living near coastlines. If these coastal features are changed by human actions, there can be terrible consequences. Regions that were naturally protected by physical and biological features of the ecosystems can become highly vulnerable to floods. This happened, for example, in the Gulf of Mexico around New Orleans, where human actions affecting the flow of the Mississippi River ended up changing the flood dynamics around New Orleans and leaving the city and its environs exposed to the devastations of Hurricane Katrina in 2005.

Ecosystems also regulate pathogens (disease-causing agents) and pests, so when ecosystems are degraded, new pathogens, pests, or invasive species can spread with devastating consequences to food production and human health. Changes in ecosystems (e.g., becoming wetter, drier, warmer, or open to new interactions of species) can lead to the emergence of new human diseases, such as *zoonotic diseases* that spread from animals to humans. HIV/AIDS is such a zoonotic disease, transmitted from chimpanzees to humans around 100 years ago somewhere in West Africa, perhaps as the result of human hunting and eating of chimpanzees as bush meat.

Invasive species are species that are introduced into a new ecosystem from the outside. Humans often bring plants and animals from one ecosystem to another, sometimes deliberately (e.g., for farming and tree cover) and sometimes by accident. The problem is that the newly introduced species may dramatically upset the regulatory function of the ecosystem; for example, if the new species has no natural predators in the new ecosystem and therefore multiplies dramatically, taking over the food chains in the ecosystem and driving out native species.

The third category of ecosystems services is called *supporting services.* These include processes like nutrient cycling and the formation of soils through the interaction of biotic and abiotic processes. Both nutrient cycling and soil formation are crucial underpinnings of agricultural productivity. Without healthy soils, nitrogen availability, and other supportive services (e.g., pollination by wild pollinators such as bumblebees), our food supplies would collapse.

The fourth and final category identified by the MEA is *cultural services,* the ways that ecosystems enhance human values, aesthetics, religion, and culture in general. One of the greatest scientists of our age, the great biologist at Harvard University, Edward O. Wilson, has argued that humanity has a deeply ingrained love of biodiversity that we inherited during the long process of human evolution. He calls this trait *biophilia,* which he defines as "the urge to affiliate with other forms of life." Wilson has given extensive, compelling evidence from the range of anthropological studies of how humanity feels at home in certain natural environments and how the degradation of those natural environments can deeply upset our cultures, our mental wellbeing, our sense of aesthetics, and thus our overall quality of life (Wilson 1984).

There is a general and important link between biodiversity and ecosystem services (Secretariat of the Convention on Biological Diversity 2010). Biodiversity promotes the health, vitality, and productivity of ecosystems, and hence enables ecosystems to deliver their provisioning, regulatory, supporting, and cultural services. When biodiversity is threatened, however, the ecosystem functions are diminished, and the services they provide are undermined. Protecting biodiversity, in other words, is key to protecting ecosystem services more generally. Scientists have verified, for example, that a reduction of biodiversity of fish species (which is currently happening all over the world) leads to a reduction in the productivity of fisheries. This is also true for farms. Crop yields in the long term are higher and more resilient in farm systems that have a higher biodiversity. Yet many farm systems around the world are experiencing a dramatic loss of biodiversity, for example, when farmers are encouraged to plant just one crop (i.e., a *monoculture*) and often with just one seed variety. Farmers may be advised, for example, that a particular seed offers them the highest yield of a staple crop. Rather than planting many varieties of rice or maize, as in their traditional practice, they therefore plant a single variety. The result in the short term might indeed be a higher yield, but the single variety within a single species leaves the farmer highly vulnerable to shocks, such as a change in weather patterns or the introduction of a new pest or pathogen. What starts out as higher farm productivity ends up as a disaster when the farm region succumbs to a devastating shock, such as drought, floods, heat waves, invasive species, or new pathogens.

In summary, ecosystem services are vital for human survival and wellbeing. Biodiversity in turn is vital for healthy ecosystem functioning. Yet biodiversity is under unprecedented threat as the result of thoughtless, unknowing human activity. We are undermining the very support structures for our biological survival and cultural vitality. Let us next turn to the ways that biodiversity and ecosystem functions are under threat. Then we will analyze what we can do to reverse these dangerous trends.

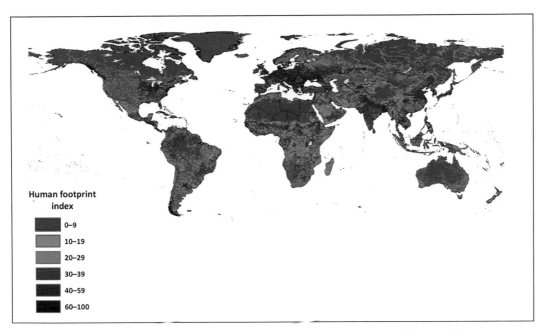

13.2 Human Footprint Index of relative human influence around the world

Copyright 2008. The Trustees of Columbia University in the City of New York. Source: Center for International Earth Science Information Network (CIESIN). "The Last of the Wild" data set. New York: Columbia University and Wildlife Conservation Society, the Bronx Zoo. http://www.sedac.ciesin.columbia.edu/wildareas.

II. Biodiversity Under Threat

In ecosystem after ecosystem, biodiversity is under massive threat. It is already being reduced, degraded, and hugely threatened across the planet. For many reasons, this will be extraordinarily difficult to bring under control. A useful starting point to understand the human impact on biodiversity is the human footprint map shown in figure 13.2, developed at the Earth Institute. My colleagues at the Earth Institute took a number of indicators, including population density, land-use change, infrastructure coverage, railroads, roads, and other human changes; aggregated and weighted these indicators; and used this information to measure the extent of the human ecological footprint in each part of the world. The map

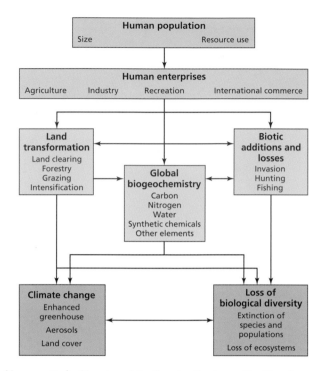

13.3 Model of humanity's direct and indirect effects on the Earth system

Source: Vitousek, Peter M., Harold A. Mooney, Jane Lubchenco, and Jerry M. Melillo. 1997. "Human Domination of Earth's Ecosystems." Science 277(5352): 494–499. Reprinted with permission from AAAS.

demonstrates the vast sweep of humanity—human activity is pervasive. The human impact is significant in all parts of the world except in the most extreme environments, notably the desert regions, some parts of the equatorial rain forests (though these are also under threat), and the poleward (high-latitude) regions that are currently too cold for agriculture. All of the rest of the planet exhibits a heavy human footprint.

The great ecologist Peter Vitousek made a similar, pioneering study more than fifteen years ago, when he and his colleagues asked the question: How much of the global ecosystems is humanity appropriating? (Vitousek et al. 1997). Their conceptual framework for that is shown in the flowchart in figure 13.3. Vitousek

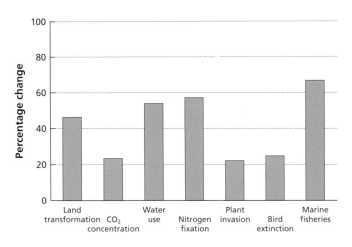

13.4 Human dominance or alteration of several major components of the Earth system

Source: Vitousek, Peter M., Harold A. Mooney, Jane Lubchenco, and Jerry M. Melillo. 1997. "Human Domination of Earth's Ecosystems." Science 277(5352): 494–499.

and his colleagues mapped the various ways that humans impact the planet and then tried to assess the human impact on ecosystems using several fascinating metrics. How much land has humanity transformed? How much has humanity changed the carbon cycle? What has humanity done to water use and the water (hydrologic) cycle?

Their conclusion (which would be even stronger today), shown in figure 13.4, revealed the extent of human impacts across all of these dimensions of the Earth's ecosystems. Humans have appropriated massive amounts of land for human use. Vitousek considered the total Net Primary Productivity (NPP) of the planet, meaning the total output of photosynthesis worldwide. He then asked how much of that NPP was taken by humans for our own species. He determined this share of NPP by adding up the human control of photosynthesis on all farms, pasturelands, and forest regions. He also added in the photosynthesis lost when humans cover the land with urban settlements and infrastructure such as roads.

The result is astounding. Humanity is now taking as much as 40–50 percent of all of the photosynthesis on the planet. We are commandeering the world's basic food supply—the output of photosynthesis—not for all species, but only for ourselves. It's like inviting 10 million guests (the roughly 10 million species on the planet) to a banquet, and then announcing that half of the food supply will go to just one of the guests, *Homo sapiens*. This is perhaps the most fundamental threat to biodiversity. Humanity is literally eating other species off the planet!

Here is another way to see this problem. A species like ours—a mammal of average size of 50–75 kilograms per adult—might normally be expected to have some tens of millions of individual members on the planet (in comparison with the numbers of other land mammals). Yet as the result of many technological and cultural revolutions, humanity no longer numbers in the millions but in the billions. As humanity has increased in number, roughly ten times since around 1750, humanity has claimed more and more land for ourselves: to grow grains, raise livestock, and provision ourselves with forest products and fibers. The human footprint is everywhere. The appropriation of NPP is astounding. The result is devastating for biodiversity.

Yet as we see in Vitousek's findings in figure 13.4, the human impact does not stop with land use. It is across the board. Humanity has fundamentally changed the carbon cycle and already raised the level of carbon dioxide in the atmosphere to 400 parts per million (ppm) compared with 280 ppm at the start of the Industrial Age. Humanity has appropriated huge amounts of water, especially to grow food, and now faces water crises in many parts of the world. Humanity has come to dominate the nitrogen cycle, turning atmospheric N_2 into reactive nitrogen (such as nitrates, nitrites, and ammonia) that can be used by plants. Humanity has introduced many invasive species into ecosystems, both intentionally and by accident, but in either case dramatically disrupting the ecosystems and food webs where those new invasive species have entered. Humanity has driven many other species to extinction (illustrated in Vitousek's study by the extinction of bird species). And last, as seen in figure 13.4, humanity has deeply undermined the abundance of fish in all parts of the world through systematic overfishing and other human-caused changes in marine ecosystems (such as ocean pollution,

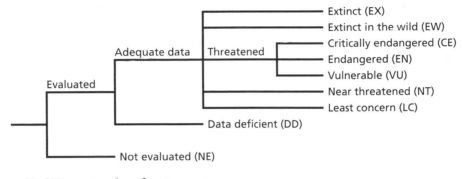

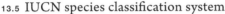

13.5 IUCN species classification system

Source: IUCN 2001.

changing ocean chemistry, and physical destruction of ocean features such as seabeds and coral reefs).

Humanity is threatening so many species that we now need a systematic scorecard to understand what we are doing. The International Union for the Conservation of Nature (IUCN) is the global scorekeeper for endangered species. The diagram in figure 13.5 explains the IUCN's classification system, ranging from species that are not threatened to those that have already been driven to extinction. We must note that the total number of species on the planet is unknown. Estimates range from around 10 million to 100 million species in total. The worldwide scientific community and the IUCN are still in the classification process for these species. Moreover, the status of most species has never been evaluated. There is no doubt that countless species are being driven to extinction before we even discover that those species exist. We are destroying the habitats of these species and appropriating their water and food supplies faster than we can even identify and name the species that are threatened by our actions.

The IUCN has a special classification called its "Red List," shown in figure 13.6, of the most endangered species. The numbers are very frightening, because even in the very short period of time covered by the Red List, the numbers of critically endangered species have soared. This is partly because there are new, additional

Critically Endangered (CR)												
Group	1996/98	2000	2002	2003	2004	2006	2007	2008	2009	2010	2011	2012	2013
Mammals	169	180	181	184	162	162	163	188	188	188	194	196	196
Birds	168	182	182	182	179	181	189	190	192	190	189	197	197
Reptiles	41	56	55	57	64	73	79	86	93	106	137	144	151
Amphibians	18	25	30	30	413	442	441	475	484	486	498	509	519
Fishes	157	156	157	162	171	253	254	289	306	376	414	415	413
Insects	44	45	46	46	47	68	69	70	89	89	91	119	120
Molluscs	257	222	222	250	265	265	268	268	291	373	487	549	548
Plants	909	1,014	1,046	1,276	1,490	1,569	1,569	1,575	1,577	1,619	1,731	1,821	1,920

Endangered (EN)												
Group	1996/98	2000	2002	2003	2004	2006	2007	2008	2009	2010	2011	2012	2013
Mammals	315	340	339	337	352	348	349	448	449	450	447	446	446
Birds	235	321	326	331	345	351	356	361	362	372	382	389	389
Reptiles	59	74	79	78	79	101	139	134	150	200	284	296	313
Amphibians	31	38	37	37	729	738	737	755	754	758	764	767	773
Fishes	134	144	143	144	160	237	254	269	298	400	477	494	530
Insects	116	118	118	118	120	129	129	132	151	166	169	207	215
Molluscs	212	237	236	243	221	222	224	224	245	328	417	480	480
Plants	1,197	1,266	1,291	1,634	2,239	2,258	2,278	2,280	2,316	2,397	2,564	2,655	2,871

Vulnerable (VU)												
Group	1996/98	2000	2002	2003	2004	2006	2007	2008	2009	2010	2011	2012	2013
Mammals	612	610	617	609	587	583	582	505	505	493	497	497	498
Birds	704	680	684	681	688	674	672	671	669	678	682	727	727
Reptiles	153	161	159	158	161	167	204	203	226	288	351	367	383
Amphibians	75	83	90	90	628	631	630	675	657	654	655	657	656
Fishes	443	452	442	444	470	681	693	717	810	1,075	1,137	1,149	1,167
Insects	377	392	393	389	392	426	425	424	471	478	481	503	500
Molluscs	451	479	481	474	488	488	486	486	500	587	769	828	843
Plants	3,222	3,331	3,377	3,864	4,592	4,591	4,600	4,602	4,607	4,708	4,861	4,914	5,038

13.6 IUCN Red List of endangered species

Source: IUCN Red List version 2014.

classifications of species, but it is also very much because human activity is driving species to critical endangerment and to extinction all over the world. Huge numbers of species are collapsing, from plants, to amphibians, to pollinators such as bees, to the great apes.

The human-induced pressures are coming from all directions: changes in land use, depletion of water supplies, nitrogen and other human-caused chemical fluxes, changes in climate patterns, overharvesting (through fishing, logging, hunting, and other extractive processes), urbanization, and more. The causes are so various and so deeply intertwined in the world economy and in the soaring numbers of the human population that reversing these adverse trends will be extremely difficult. We have yet to slow down the destruction of biodiversity more than twenty years after humanity agreed to the Convention on Biological Diversity (CBD) at the 1992 Rio Earth Summit. In other words, humanity is waking up to the problems but not yet to the solutions.

III. Oceans and Fisheries

As humanity puts pressure on terrestrial ecosystems of all kinds (e.g., polar, alpine, tropical rain forests, dryland areas), we also put tremendous pressures on our marine ecosystems and the oceans. We are changing basic ocean chemistry. We are poisoning the ocean with pollution coming from huge oil spills and other disasters. And we are degrading the biodiversity in the oceans through other forms of human activities, especially through the overfishing and overharvesting of marine life.

The oceans cover three-fourths of the Earth's surface area, so this is no small part of humanity's relation to the Earth. Our cities around the world hug the oceans and depend on the oceans for trade, for economic activity, for our food supplies, for our proteins and invaluable nutrients such as omega-3 fatty acids from our fish intake. Keeping the oceans healthy is essential for human wellbeing. Just as in other economic spheres, our technological know-how in harvesting ocean services, such as our ability to locate and capture fish, has improved enormously just in the last sixty years. We have "mastered" the oceans to the point of

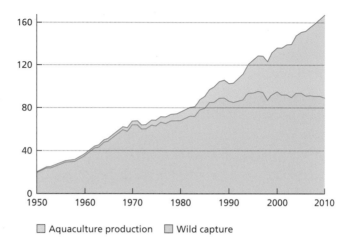

☐ Aquaculture production   ☐ Wild capture

13.7 Total fish catch (aquaculture production and wild caught) (million metric tons)

Source: Global total wild fish capture and aquaculture production in million tonnes, 1950–2010 as reported by FAO/FishStat database, Wikimedia Commons, CC BY-SA 3.0.

threatening marine life. Technological mastery, alas, does not mean intelligence, responsibility, or foresight.

The total fish catch around the world is illustrated in figure 13.7 for the years 1950–2010. This fish catch is divided into two parts: the "wild" catch, mainly from oceans (but also from rivers and lakes), and aquaculture, or fish farming. The wild catch in 1950 was about 20 million tons. By 1990, that had become about 80 million metric tons, and it then leveled off at that rate. Aquaculture rose from near zero in 1950 to around 20 million tons by 1990, and to around 75 million tons by 2010. The dramatic increase in the wild catch, up roughly four times, underestimates the true increase of fishing activity in the oceans, because such estimates do not fully include the vast biomass (fish and other marine life) that fishermen throw back in the oceans as unwanted.

The data in figure 13.7 suggest a basic lesson. The ocean catch reached a maximum extent around 1990 and further increases in fish take have come through managed aquaculture. We can say that this is bad news and good news and bad news again. The first piece of bad news is that humanity hit the limits of ocean fishing and in fact exceeded those limits. Overfishing has led many fisheries around the world into decline or complete collapse. The threats of overfishing in the oceans remain perilous today in most fishing regions of the world. The good news is that aquaculture has been able to grow to meet humanity's growing demand for fish in the diet. This is good news indeed, since fish are a key, nutritious part of the human diet, especially rich in needed oils and proteins. The second piece of bad news, however, is that aquaculture itself threatens the environment in many ways. The cultivation of fish in the managed fish farms can lead to spread of disease, excessive nutrient flows of many kinds, and threats to wild fish populations when farm fish escape into the wild. In short, aquaculture can be highly desirable if it is operated in a responsible manner, but that is a complex challenge given all the things that can go wrong.

How did the massive increase of the wild fish catch occur? It resulted from a huge increase of fishing activity, shown in figure 13.8, which compares the intensity of fish flects in different fisheries around the world in 1950 and in 2006. Fisheries back in 1950 operated along a few key coastal and river regions. By 2006, fisheries were operating throughout the oceans, including the high seas, wherever they could hunt and capture fish in large numbers. As in so many sectors of the world economy, ocean fisheries experienced many huge technical advances. These included the use of longline nets that allowed for a much greater capture of fish; the use of various kinds of remote sensing to identify where the fish are located; and the use of ocean trawling to capture bottom-dwelling fish and other marine life on the ocean floor, in a manner that often completely devastates the highly complex, biodiverse marine ecologies on the ocean bottom.

The consequence is that the technological advance of ocean fisheries, as is so often the case, has not been the friend of marine biodiversity and marine ecosystem sustainability. Technological advances in fishing have led to a huge increase in the wild catch. Yet is has also led to the depletion of ocean fisheries, a huge loss of biodiversity, and a huge threat to the productivity of marine ecosystems.

To get an indicator of the overall human impact, one can look at the amount of primary production required to feed the wild fish catch in a given region of the ocean, measured as a fraction of the total photosynthesis in that part of the

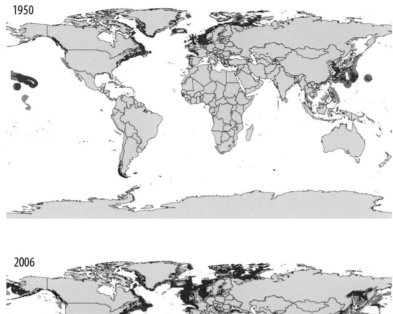

1950

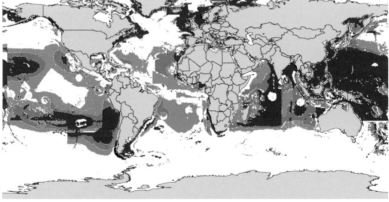

2006

13.8 Increase of global fishing fleets (1950, 2006)

ocean. For example, if the ocean feedstock of the wild catch equals one-third of all the photosynthesis in that part of the ocean, we say that the human appropriation of ocean primary production is therefore one-third. This concept, devised by marine scientist Wilf Swartz and others, is akin to Vitousek's concept of the human appropriation of NPP, which Vitousek had applied to terrestrial photosynthesis (Swartz et al. 2010). The results in figure 13.9 show fisheries where the amount of NPP associated with that amount of fish capture is already in a danger zone, of 30 percent of primary production in many fisheries around the world. Comparing 1950 and 2005, we see the massive increase in the human appropriation of marine primary production.

One implication of this finding is that not only is humanity driving down levels of fish abundance to the point of threatening their very survival, but we are also changing the *structure and functioning* of the marine ecosystems as well. One example of this is called "fishing down the trophic chain." Humanity first eats, and depletes, the large fish at the top of the food chain—the fish that eat other fish. Then, after exhausting the supplies of fish at the top of the food chain, humanity eats fish lower down the food chain, eventually exhausting their supply as well (that is, driving them to extinction or to very small populations). Step by step, we rely on smaller and smaller fish, and on fish closer to the base of the food chain (that is, fish that feed directly upon the photosynthetic output of the ocean rather than on other fish). The fish scientists (ichthyologists) measured the average trophic level of the fish that are caught; the evidence shows that over time, more of the catch is lower down the trophic chain. Humanity is very good at eating those prized fish at the top of the food chain, the predators of the predators of the predators; therefore they are being depleted rapidly, forcing humanity to go lower down on the food chain.

Ecologists speak of the "trophic level" of organisms. Plants that produce their own food are called *autotrophs* and are assigned a trophic level 1.0. All animal species must get their food by eating autotrophs or by eating other species of animals. All such species are called *heterotrophs*. Herbivores that directly eat autotrophs are assigned a trophic level 2.0. Carnivores that eat herbivores are assigned a trophic level 3.0. Carnivores that eat other carnivores are assigned a trophic level 4.0. And so on up the food chain.

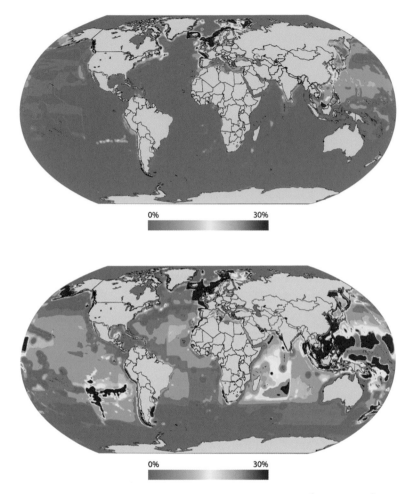

13.9 Spatial expansion and ecological footprint of fisheries (1950, 2005)

Swartz W., Sala E., Tracey S., Watson R., Pauly D. (2010) The Spatial Expansion and Ecological Footprint of Fisheries (1950 to Present). PLoS ONE 5(12): e15143. doi:10.1371/journal.pone.0015143. Sea Around Us (www.seaaroundus.org).

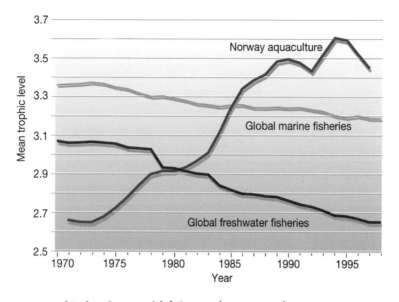

13.10 Average trophic level in world fisheries (1970–2000)

Source: Pauly, Daniel, Villy Christensen, Sylvie Guenette, Tony J. Pitcher, U. Rashid Sumaila, Carl J. Walters, R. Watson, and Dirk Zeller. 2002. "Towards Sustainability in World Fisheries." Nature 418: 689–695. Reprinted with permission from Macmillan Publishers Ltd.

When humanity fishes down the food chain, it means that our diet starts with wild fish with high trophic numbers and then shifts over time to diets of wild catch with lower trophic numbers. This phenomenon is shown in figure 13.10, which reports the average trophic level of the catch brought in from the fisheries, both wild and aquaculture. The top blue line is for the marine fisheries, and the red line is for freshwater fisheries. We see that in both cases, humanity is fishing down the trophic levels, giving further evidence that humanity is exhausting the supplies of the high-trophic species.

We also see that Norway's managed aquaculture is producing fish higher on the trophic level, such as Atlantic salmon (trophic level above 4) and Arctic char. These are highly satisfying for the human diet but complicated ecologically. High-trophic fish grown in captivity need massive quantities of fishmeal, and this in turn is provided by wild capture of lower trophic fish in the oceans. Thus, even

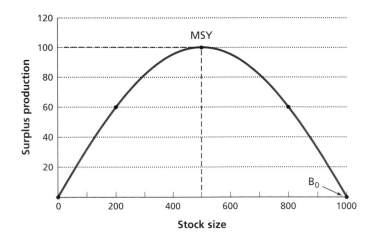

13.11 Maximum sustainable yield calculation

though fish are grown in aquaculture, it does not mean that they are not impacting the oceans. More aquaculture of high-trophic fish leads to increased demand for fishmeal, in turn putting pressure on ocean ecosystems.

Marine ecologists try to estimate the *maximum sustainable yield* (MSY) of a fish type in order to determine a safe level of wild catch. The question they ask is how much of a specific type of fish can be taken safely from a fishery (in an ocean, river, or lake) without depleting the fish stock? The typical answer is given by the upside-down U shown in figure 13.11. Consider a fishery in one part of the ocean. Suppose that if left alone, without any fishing at all, there would be 1,000 fish (or perhaps 1,000 tons) in the fishery. Since the population at 1,000 is stable (and maximized) at 1,000, any fishing at all is bound to lead to a lower fish population.

Now suppose that the fish population is at 800. If left alone, the fish population would tend to increase gradually back to 1,000. Perhaps the 800 fish would give rise to 860 fish in the following year, a net increase of 60 fish. If the fishery catches those 60 fish, then the 800 fish this year would lead to 800 fish again next year. Thus, for a fishery with a potential population of 1,000, but at 800 fish currently, the fishery could sustain annual fishing of 60 fish without raising or lowering the fish population. In figure 13.11, a fishery at 800 fish (shown on the horizontal axis) has a "surplus production" of 60 (shown on the vertical axis).

Next, suppose that the fish population is at 500. When the fish population is 500, the population tends to increase by 100 fish per year. If those 100 fish are caught each year, the fish population stays at exactly 500 fish, enabling an annual catch of 100. We see in the inverted U in figure 13.11 that a fish population of 500 has a "surplus production" of 100 fish.

At what level is the surplus production maximized? We see clearly that the MSY occurs when the fish population is exactly 500 fish, half of the potential population. At that level, the fishery can support an annual catch of 100 fish and still maintain a stable population. Yet what happens if the fishery catches 200 fish in that year? Obviously the fish population would tend to fall, becoming just 400 fish the following year. And if the overfishing continues, for example, with another 200 fish caught in the following year, the fish population would be less than 300 the third year. Eventually, the fishery would be driven to exhaustion, with no fish and no prospect of future catches!

The MSY therefore is a policy tool, telling commercial fishers how many fish it is safe to catch each year. Yet will they listen to this advice? Each individual might still try to maximize his or her own catch, while hoping the others will abide by the limits of the fishery. The result would be a "tragedy of the commons," where every commercial fisher overfishes and the combined effect is to drive the fishery to exhaustion. For this reason, the government might need to enforce a maximum level of total fishing, for example, by giving out permits that tell each fishing vessel how many fish it is allowed to catch, with the sum of the permits equaling the MSY of 100 fish per year in a fishery with 500 fish. In recent years, many fisheries around the world have successfully deployed tradable permit systems, in which the total permits are equal to the estimated MSY, and in which individual commercial fishers are allowed to buy and sell permits from other commercial fishers. In this way, the most productive buy the rights to fish from the least productive, but still do not violate the overall limit of the MSY.

The concept of the MSY has become even more complicated in recent years, as ecologists have come to understand that it is not good enough to regulate the catch one species at a time—the ecosystem as a whole must be regulated. If just

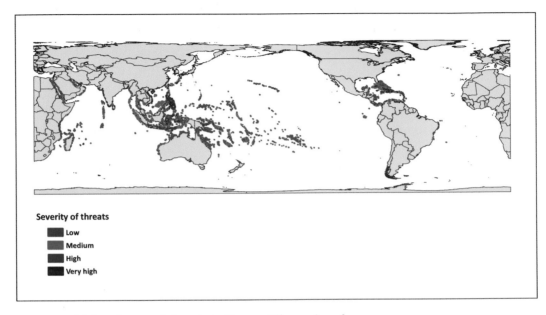

Severity of threats

- Low
- Medium
- High
- Very high

13.12 Major observed threats to the world's coral reefs

Source: Rekacewicz, Philippe. UNEP/GRID-Arendal. http://www.grida.no/graphicslib/detail/threats-to-the-worlds-coral-reefs_3601.

one species is regulated, the change in its abundance might negatively impact the abundance of other species that depend on the first species in the food chain. For this reason, marine ecologists now talk about ecosystem-wide sustainable yields, treating the ecosystem in a more holistic manner.

If the oceans were troubled only by the excessive wild catch of fish, we would have trouble enough, yet the sad fact is that humanity is assaulting marine ecosystems on many fronts. Figure 13.12 illustrates the risks facing corals around the world. Humanity is threatening coral life in various ways, including acidification of the oceans, warming of the oceans, coral destruction by tourists, overfishing, direct harvesting of the corals themselves (e.g., for home ornaments), dynamite used for fishing, pollution, and sedimentation caused by human actions (e.g., construction, mining, deforestation, and flooding, which lead to the sedimentation of coral habitats). Human activities on multiple fronts are thereby driving corals to depletion and perhaps extinction for many species. Such multiple impacts—harvesting, pollution, climate change, and so on—illustrate the more general point that the human threats to biodiversity do not arise from a single factor, but from the sum of many factors.

We have huge problems. We have some tools to address them, but the fair summary is that our oceans are at profound and still-growing risk because of the multiple pressures of human activity. We depend on the oceans in countless ways for our wellbeing and for our very survival. If we do not take care and face up to these multiple assaults, we will face growing crises in the not so distant future.

IV. Deforestation

Forests remain one of the major parts of terrestrial ecosystems on the planet, covering 31 percent of the total land area; yet the natural forest cover used to be a far higher proportion of the Earth's land area before humanity got to it. Humanity has been in the business of clearing forests for thousands of years. This is an ancient story, but we continue to lose a lot of forest area today because of the increasing human pressures on forest systems and the long-distance forces of international trade. When we lose forests, we degrade ecosystems and lose a tremendous amount of biodiversity. Our three great equatorial rain forest areas (the Amazon basin, the Congo Basin, and the Indonesian archipelago) are home to a remarkable amount of the planet's biodiversity, but this biodiversity is quickly being lost.

The map in figure 13.13 gives an indication of the extent of past deforestation and some of the challenges of current deforestation. Every shaded area of the map originally had forest cover. The very light shaded areas, say in western Europe or across China or across the Eurasian land mass, are areas that have already been deforested, with the deforestation occurring hundreds or even thousands of years ago. Only the dark shaded regions are still forested today. The main forests are in the high latitudes (e.g., northern Canada, Europe, and Russia), the eastern coast of the United States (which was converted to farmland in the nineteenth century

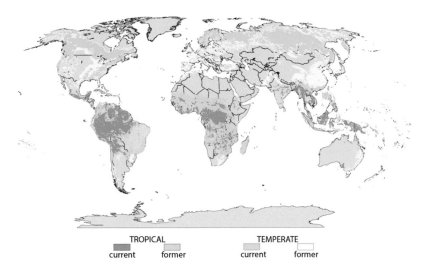

13.13 Global distribution of original and remaining forests

Source: UNEP World Conservation Monitoring Centre. 1998. "Global Generalized 'Original' Forest dataset (V 1.0) and Global Generalized 'Current' Forest dataset (V 3.0)."

but which more recently has been reconverted to forest), and along the equator, where the three great rain forest regions lie.

Today, most deforestation is taking place in the fast-growing tropical and subtropical regions, and notably in the rain forests, where population densities were traditionally low but are now rising. These rain forests regions are increasingly encroached upon for human provisioning, such as for tropical logging, for farmland and pastureland, and for provisioning of peasant smallholders who go into the forests for fuel wood or for other needs. The result is that while temperate zone areas were deforested a long time ago, it is now the tropical areas that are being deforested most quickly. The rain forests, regions of astounding biodiversity, are now facing major disturbances and human impacts. The map in figure 13.14 shows the current patterns of deforestation. The red regions are regions of rapid deforestation. We see the losses in the Amazon, the Congo Basin, and the Indonesian archipelago. The dark green regions—including the eastern coast of the

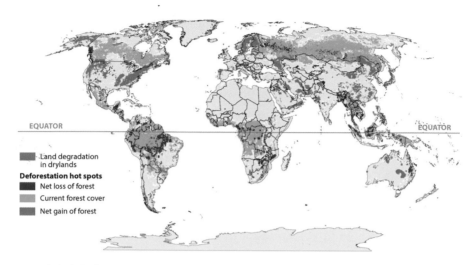

13.14 Global deforestation and land degradation

*Source: Rekacewicz, Philippe, and Emmanuelle Bournay. 2007. UNEP/GRID-Arendal. http://www.grida.no
/graphicslib/detail/locations-reported-by-various-studies-as-undergoing-high-rates-of-land-cover-change-in-the
-past-few-decades_fe3b.*

United States, Scandinavia, and parts of northern China—are undergoing refor-
estation, mostly involving the return of farmland to forestland.

James Lovelock, creator of the Gaia theory of the interconnectedness of the
world's ecosystems and the regulatory processes of those ecosystems at planetary
scale, emphasized that when we degrade one ecosystem we impede or undermine
the functioning of ecosystems in other parts of the planet. Lovelock said about
the deforestation of the tropical rain forests: "No longer do we have to justify the
existence of humid, tropical forests on the feeble grounds that they might carry
plants with drugs that cure human disease ... Their replacement by cropland could
precipitate a disaster that is global in scale" (Lovelock 1991, 14). For example, rain
forests serve to keep the planet cool by maintaining extensive cloud cover that
reflects incoming ultraviolet radiation back into space rather than allowing the
ultraviolet radiation to reach and warm the Earth. If the Amazon dries out (due
to human-induced climate change) or disappears as the result of the forest being

cleared to make way for farmland, the Amazon's cloud cover would shrink as well, thereby changing the Earth's reflectance (albedo) and causing a potentially large positive feedback to warm the planet further. Lovelock's point is that the impacts of massive deforestation can be far greater than we would recognize, beyond the direct impact of the loss of the local ecosystem services. Earth systems science teaches that the interaction of the ecosystems in their global regulation of climate, water cycle, and nutrient cycles is also of huge significance for planetary balances and for human wellbeing.

What is the cause of the mass deforestation? Some of the human impact of deforestation is internally driven, mainly by growing populations within countries. Yet a huge amount is also coming from international trade, from the demands halfway around the world for forest products. This is very difficult to control, because it means the high levels of demand, often from rich countries or rapidly growing economies like China, overwhelm local protective services, often through illegal means. One of the major drivers is the soaring demand for palm oil, which is a very versatile product. In places like Malaysia and Indonesia there has been massive deforestation, replacing the highly biodiverse existing rain forest area with a monoculture of oil palm trees. A similar driver is the rising demand for soybeans in world markets (e.g., by China), which in turn is leading to deforestation in the Amazon rain forest.

The resulting losses of biodiversity will be phenomenal in terms of the regulatory functions of these ecosystems and the threats to the survival of key endangered species such as the orangutan in Indonesia and Malaysia. The demand for tropical forest products is insatiable. If markets are not controlled, international trade will lead to continued massive deforestation. Unless we start managing tropical forests in a sustainable manner, these ecosystems will irreversibly collapse.

There are, of course, several efforts to do something about this. A notable effort is to link the conservation of the rain forests and forests in general with the climate change agenda. Perhaps 15 percent of the total carbon dioxide emissions each year come from land-use change, especially from deforestation. In recent years, an effort has been launched as part of the climate mitigation effort to reduce our carbon footprint in terms of emissions coming from deforestation in

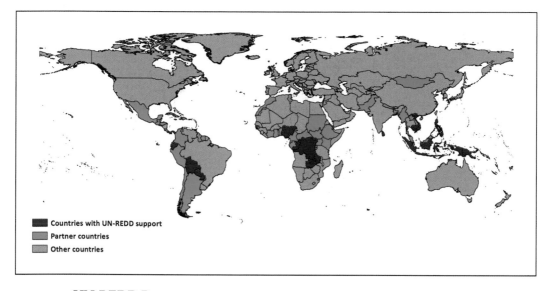

13.15 UN-REDD Programme

Source: "UN-REDD Programme Regions and Partner Countries." The United Nations REDD Programme.

addition to carbon dioxide emissions from the energy sphere. The main project to avoid deforestation is called UN-REDD+, Reduced Emissions from Deforestation and Forest Degradation. REDD+ targets both the thinning and clearing of forests. The idea, which is an excellent one, is to give financial incentives to local farmers and communities (including indigenous populations) to protect the forests.

The REDD+ programs replace part of the income that the communities would lose in the short term from their inability to overexploit forest products with a sustainable flow of income of other kinds, including a top-up of income provided by donor countries. Norway, for example, has offered $1 billion to Brazil in a REDD+ initiative for forest communities in the Amazon to play the role of protecting the Amazon rather than facilitating its loss. The map in figure 13.15 shows the countries participating in UN-REDD+. The countries in red are actively receiving United Nations support for developing REDD+ programs. Other partner countries are shown in blue. This is a very important effort, but it is still a relatively small counterpressure to the overwhelming market forces coming from global trade for the products directly from the forests themselves or for the products that are grown when the forests are cleared and replaced by other kinds of economic activity, such as farming and livestock.

V. International Dynamics

The world's nations, realizing how much danger there is in the loss of global biodiversity, have taken at least some steps to try to pull back from the sixth great extinction. Some are indirect; for example, riding on the climate control effort through REDD+. There have been treaties limiting some kinds of transboundary pollutants that threaten the oceans, and some agreements on global fisheries. In addition to these, there have been at least a couple of very important head-on attempts to focus on biological diversity.

Two of the most important attempts have been through international treaties. The single most important of these is the 1992 Convention on Biological Diversity, or CBD, the core purpose of which is to slow and reverse the loss of biodiversity (UN 1992a). The second is the 1973 Convention on International Trade in Endangered Species of Wild Fauna and Flora (CITES), which tries to restrict trade in endangered species (UN 1973). Both have experienced successes and failures. The most important point to emphasize time and again is that the pressures of the global economy are so strong that even when treaties or regulations are put in place, vested interests often give a powerful counterforce to these measures, and control mechanisms are often at the mercy of illegal activities, bribery, corruption, and other limits of enforcement. The weight, force, and momentum of the world economy are often so powerful that the world economy runs roughshod over attempts at regulation.

The CBD is a valiant attempt to try to get the human threat to biological diversity under control. It is one of the three great multilateral environmental agreements to come out of the Rio Earth Summit in 1992, along with the UNFCCC and the UN Convention to Combat Desertification (UNCCD) (UN 1994). The CBD has accomplished a bit, but it has not come near to accomplishing its core goal of heading off the massive loss of biodiversity. The convention articulates that goal as follows:

> The objectives of this Convention, to be pursued in accordance with its relevant provisions, are the conservation of biological diversity, the sustainable use of its components and the fair and equitable sharing of the benefits arising out of the utilization of genetic resources.

It is important to underscore that CBD puts great emphasis not only on conserving biological diversity but also on the fair and equitable sharing of the benefits arising out of the utilization of genetic resources. At the time, it was thought that an incredible bonanza could be expected out of what was called bioprospecting, a process in which scientists would enter the world's forest regions and identify new blockbuster drugs of both profound medical and financial value. The question was how to make sure that the host countries would benefit from these discoveries. This is not an entirely fanciful idea—nature certainly has chemical compounds of profound benefit still to be discovered. One of the greatest discoveries that I see in my everyday work is an ancient Chinese herbal remedy for fever, the wormwood plant, which became the source of the modern molecule Artemisia, used to fight malaria. But a lot of the impetus in 1992 and a lot of the mishaps with the CBD since then came from the notion that we should be focusing our efforts on the wealth from bioprospecting, rather than on limiting human activity in order to prevent a collapse of ecosystems and biodiversity for our much deeper, longer-term wellbeing.

The treaty has accomplished a certain bit, but it has fallen far short of what it should be doing. One of the main reasons for that is the disgraceful behavior of the government of the United States, my own country. Though U.S. scientists and some politicians were leading proponents of the treaty negotiations, the right-wing politicians in the United States started to lobby against the treaty even during the negotiations. By the time the treaty was finalized for the 1992 Rio Earth Summit, President George H. W. Bush decided not to sign it under pressure from members of his own party. The next year, President Bill Clinton came into office, signed the treaty, and submitted it for Senate ratification, which in the United States system requires a two-thirds vote of the Senate. The Senate committee reading this treaty gave its approval, but the Senate has never ratified the treaty.

It is quite remarkable what happened next. So-called free-market politicians in the United States rejected the idea that the world should agree to an equitable and fair sharing of biological products. Let the U.S. drug companies make a massive profit, they said. One suspects that powerful industrial lobbies made their voices heard. Then, private developers began loudly calling for the right to buy U.S. federal land to mine, to drill for oil, or for fracking shale gas. They argued that the CBD would be a menace to their profit-maximizing property rights.

This type of "free-market" sentiment is startlingly misguided, because markets should serve human purposes, not be ends in themselves or vehicles for rapacious greed that imposes huge social costs on others. When markets do not take into account the profound externalities of individual behavior such as the loss of biological diversity or species extinction, free markets become the antagonist of human wellbeing. A radical ideology that says, "Leave me alone, I have the perfect right to destroy species," can obviously create havoc. While the United States remains an observer to the CBD, its absence as a signatory has gravely weakened the implementation of the treaty. When the parties to the CBD pledged in 2002 in their strategic plan to slow and reverse the loss of biodiversity by the year 2010, there was little practical effect. By 2010, the extent of loss of biodiversity was greater than ever.

The three multilateral environmental agreements of the Rio Earth Summit were reviewed twenty years later at the Rio+20 Summit. At that time *Nature* magazine conducted an in-depth analysis of what had happened under the various treaties and created a report card for each (Tollefson and Gilbert 2012). Figure 13.16 shows the report card for the CBD. Its main assignment was to reduce the rate of biodiversity loss; it received a grade of F, total failure. This treaty has not slowed the loss of biodiversity. For its other assignments: develop targets, a D; protect ecosystems, a C; recognizing indigenous rights, a D; financing to offset the loss of biodiversity, a solid F. The one high grade that was given was creating a regulatory framework around genetically modified organisms. Whether that framework eventually serves the human purpose or inadvertently chokes off the benefits that advanced genetics might give for seed breeding remains to be seen.

The only other semidecent grade is a C in protecting ecosystems. One of the provisions in the CBD was to set aside protected zones; the graph in figure 13.17

Report Card

Convention on Biological Diversity

Main assignment

| Reduce the rate of biodiversity loss | F |

Other assignments

Develop biodiversity targets *Nations have only just started to establish focused targets for biodiversity and ways to assess it.*	D
Protect ecosystems *At least 10% of the world's ecologically valuable regions on land was protected by 2010, but only about 1% of those in the oceans.*	C
Share gene windfall *The Nagoya Protocol on the sharing of commercial benefits derived from the collection and use of genetic material has been signed by 92 countries, but is not yet in force. Only a few companies so far have shared such benefits with the source country.*	E
Recognize indigenous rights *Nations are very variable in honoring the rights of indigenous people, especially in creating protected areas within their territory.*	D
Provide funding *Countries have made many commitments but honored few of them.*	F
Regulate genetically modified organisms *The Cartagena Protocol, signed by 103 countries, is designed to help regulate the movement of genetically modified organisms between countries, and came into force in 2003.*	A

13.16 *Nature* report card: Convention on Biological Diversity

Reprinted by permission from Macmillan Publishers Ltd: Nature News, Tollefson, Jeff, and Natasha Gilbert. "Earth Summit: Rio Report Card." Copyright 2012.

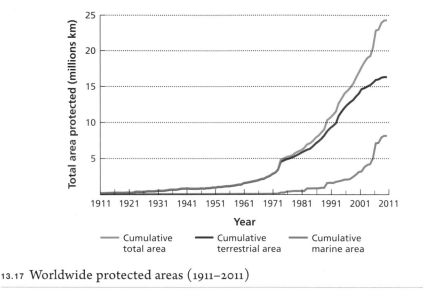

13.17 Worldwide protected areas (1911–2011)

UNEP-WCMC.

shows the cumulative protected areas in the world. National parks, national reserves, protected wildlife refuges, marine protected zones, and so forth have increased in the previous decades. This rise in protected areas, in particular of marine protected areas, is a contribution of the CBD. The treaty has had some effect, but the overall verdict of an F grade gives a fair summary of its lack of success.

Another very important treaty preceded the CBD by a couple of decades; the Convention on International Trade in Endangered Species, or CITES, which was signed in 1973 and went into effect soon thereafter. The idea of CITES is to reduce the pressures and dangers of species extinction by regulating trade specifically in endangered species. The treaty classifies endangered species; species that are not yet endangered but could become so unless trade is reduced; and species whose trade indirectly imperils species in endangerment of extinction. Within those three categories, CITES covers 35,600 plant and animal species right now.

While CITES has had an important effect, like all international law, the forces of the world economy can sweep aside what is on paper and often have absolutely devastating consequences. An example of this is the recent surge in illegal trade of rhinoceros horns and the massive kill-off of rhinoceroses because of the soaring demand. Virtually all of this demand comes from China, where rhinoceros horn is a treasured part of the pharmacopoeia of traditional Chinese medicine. It is an extraordinarily valued commodity, but the black rhinoceros is also an extraordinarily endangered species, and its numbers have fallen precipitously; in November 2013 the western black rhino subspecies was officially declared extinct. It is not surprising when you consider that the market price for a rhinoceros horn has reportedly reached $65,000 per kilo, higher than gold. There are thus bound to be tremendous pressures and a great deal of corruption along every part of the supply chain.

A very important recent study by Manfred Lenzen of the University of Sydney and his colleagues found that trade in products like rhinoceros horns and elephant tusks are a pervasive problem, not specific to a couple of headline products but involving many thousands of endangered plant and animal species (Lenzen et al. 2012). The results show that about one-third of endangered species are part of important global trading chains. This means it is not good enough to stop local pressures; essentially the full weight of the $90 trillion world economy is providing the fuel for the massive loss of biodiversity. A useful graphic from the Lenzen study, shown in figure 13.18, traces worldwide supply chains and shows both the supply and the demand sides. The main point of this graph is that the issue is in global supply chains and the many countries engaged in threats to biodiversity as both suppliers and consumers.

The conclusion is that the global efforts over many decades have not yet come to grips with the sixth great extinction wave. Humanity's power over ecosystem functions and its endangerment of biodiversity are so significant and coming from so many different directions that we still lack the public awareness, the political impulse, and the economic incentives to get this right. When the world met at Rio+20 in June 2012 and received a grade of F on implementing the Convention on Biological Diversity, the UN Framework Convention on Climate Change, and the UN Convention to Combat Desertification, it was clear to governments that something different must be done. We need a breakthrough in global policy and action under a new set of Sustainable Development Goals that at least have the potential to help us move from the very threatening path of business as usual to a true path of sustainable development.

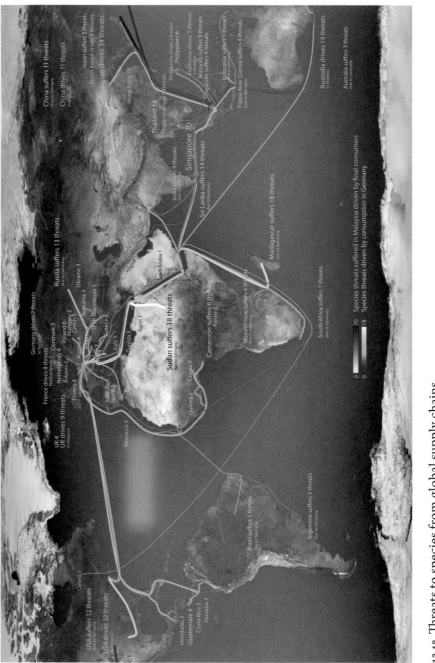

13.18 Threats to species from global supply chains

Based on the Eora global trade database (worldmrio.com). Lenzen, M., Moran, D., Kanemoto, K., Foran, B., Lobefaro, L., Geschke, A. International trade drives biodiversity threats in developing nations. Nature 486(7401).

14

SUSTAINABLE DEVELOPMENT GOALS

I. The Sustainable Development Goals (SDGs)

The world is far off course for achieving sustainable development. The issue has been on the global agenda for more than forty years at least, dating back to 1972 with the first UN Conference on the Human Environment in Stockholm and the simultaneous publication of *Limits to Growth,* which correctly pointed out that the challenge of combining economic development and environmental sustainability would pose huge threats in the twenty-first century. Twenty years later, the world met in Rio de Janeiro at the UN Conference on Environment and Development, also known as the Rio Earth Summit, and adopted two major multilateral environmental agreements, the UN Framework Convention on Climate Change and the Convention on Biological Diversity (CBD), and also laid the groundwork for a third that was adopted two years later, the UN Convention to Combat Desertification. And on the twentieth anniversary of the Rio Earth Summit, in June 2012, the world met for a third time, once again at Rio, at the UN Conference on Sustainable Development, known informally as the Rio+20 Summit (figure 14.1).

At Rio+20, leaders from all over the world took stock of forty years of international environmentalism and twenty years of three big environmental treaties. What they realized was very unsettling. All of the evidence showed that the diagnosis first made back in 1972 was fundamentally correct: the challenges of combining economic growth with social inclusion and especially environmental sustainability were still unmet, and indeed were intensifying. Back in 1972, the world's population was about 3.8 billion; now it is nearly twice that, at 7.2 billion. Back in 1972, the carbon dioxide (CO_2) concentration was around 350 parts per million (ppm) and was increasing by around 1 ppm per year. Now the CO_2 concentration stands at 400 ppm and is rising by more than 2 ppm per year. Back in 1972, the loss of biodiversity was hardly recognized; now we know that we are in the sixth great extinction.

The leaders also had to swallow a second difficult conclusion. The major environmental treaties, hailed as historic breakthroughs at the Earth Summit in 1992, had not succeeded, at least not yet. As I previously noted, *Nature* magazine gave the three treaties of the Earth Summit grades of F. Not one of them had delivered as promised by the time of the Rio+20 Summit.

With that in mind, world leaders at the Rio+20 Summit in June 2012 resolved once again to join this battle. They realized the world needs a new and dramatic approach. The first thing the world leaders said in their outcome document "The Future We Want" is that we absolutely must not give up (UN General Assembly [UNGA] 2012, 8):

> We commit ourselves to re-invigorating the global partnership for sustainable development that we launched in Rio in 1992. We recognize the need to impart new momentum to our cooperative pursuit of sustainable development and commit to work together with Major Groups and other stakeholders in addressing the implementation gaps.

They also noted something else very important: that the single most *urgent* task in all of the interconnected challenges of sustainable development is the task that the world did take on in the year 2000 with the adoption of the Millennium Development Goals (MDGs): the fight against extreme poverty. Extreme poverty is the most urgent priority, because it is a matter of life and death for at least 1 billion people, and it is a struggle for survival in the here and now. Extreme poverty can rightly be defined as a condition in which mere survival is a daily struggle. People living in extreme poverty are wondering where their next meal will come from, whether the next drink of water will carry pathogens that could

14.1 World leaders at the Rio+20 Summit

"Quarta-feira, 20 de junho," Roberto Stuckert Filho/PR, Flickr, CC BY-SA 2.0.

threaten their lives, and whether the next mosquito bite might transmit a deadly case of malaria to them or their children.

Around 6.5 million children are still dying every year before their fifth birthday (down by around half of the child deaths that occurred in the year 1990), their deaths caused mostly by diseases that are either preventable or treatable. Extreme poverty is a crisis, an opportunity, and a moral challenge. We can solve this problem. In that spirit, the world leaders at Rio+20 declared that:

> Eradicating poverty is the greatest global challenge facing the world today and an indispensable requirement for sustainable development. In this regard, we are committed to free humanity from poverty and hunger as a matter of urgency . . . We reaffirm our commitment to making every effort to accelerate the achievement of the internationally agreed development goals, including the Millennium Development Goals (MDGs) by 2015. (UNGA 2012, 1)

One of the most important steps at Rio+20 came when the world leaders said: "We recognize that the development of goals could also be useful for pursuing focused and coherent action on sustainable development" (UNGA 2012, 43).

In fact, the leaders looked at the MDGs and saw how successful they had been in scaling up the world's efforts to fight extreme poverty in the preceding dozen years since their adoption in September 2000. The leaders agreed that the world now urgently needs a similar approach of scaling up the world's efforts on sustainable development. They therefore declared:

> We further recognize the importance and utility of a set of Sustainable Development Goals (SDGs). . . . These goals should address and incorporate in a balanced way all three dimensions of sustainable development and their inter-linkages. (UNGA 2012, 43)

In essence, the world leaders agreed to make the transition from MDGs to SDGs. Just as the MDGs had inspired action by being a short set of compelling objectives, so too the new SDGs should motivate global enthusiasm, knowledge, and action. As the leaders put it in "The Future We Want":

> SDGs should be action-oriented, concise and easy to communicate, limited in number, aspirational, global in nature and universally applicable to all countries while taking into account different national realities, capacities and levels of development in respecting national policies and priorities. We also recognize that the goals should address and be focused on priority areas for the achievement of sustainable development, being guided by this outcome document ["The Future We Want"]. Governments should drive implementation with the active involvement of all relevant stakeholders, as appropriate. (UNGA 2012, 43)

The call for SDGs is a potentially historic decision, a powerful way to move to a new global agenda that engages the world community, including not only governments but also businesses, scientists, leaders of civil society, NGOs, and, of course, students everywhere. Unlike the MDGs, which apply largely to poor countries and reference the rich countries mainly as donors, the SDGs will be

14.2 UN Secretary-General Ban Ki-moon

Ban Ki-moon—World Economic Forum Annual Meeting 2011, Remy Steinegger, World Economic Forum, Wikimedia Commons, CC BY-SA 2.0.

universally applicable. The United States, just like Mali, needs to learn to live sustainably! The rich countries like the poor have to promote more social inclusion, gender equality, and of course energy systems that are low carbon and resilient.

These goals can give new impetus, new power, new social mobilization, new resources, and new political will to a challenge that has been more than forty years in public awareness and twenty years in international law, but has not successfully been addressed to date. These goals will not supplant international law. The treaties are still needed. But they can create a new global energy and atmosphere of problem solving that will help to implement the treaties.

In a follow-up to Rio+20 and the call of the world leaders to put sustainable development at the very core of the international development agenda, UN Secretary-General Ban Ki-moon (shown in figure 14.2) greatly honored me by tasking me with creation of a new a global network of sustainable development problem solving. We call this the *Sustainable Development Solutions Network* (SDSN).

The key motivation for the new SDSN is the idea that the world needs not only new goals, political motivation, and will; but also a new era of intensive problem solving in sustainable development challenges that include health, education, agriculture, cities, energy systems, conservation of biological diversity, and more.

The SDSN, guided by its Leadership Council, a distinguished group of scientific, business, civil society, and policy leaders around the world, has first taken on the challenge of advising the UN General Assembly and the UN secretary-general on what the new SDGs might be. The SDSN has proposed a concise set of ten goals that could be a framework of action. These are an illustration, because it is the governments themselves, meeting at the UN General Assembly, that will ultimately agree on the new SDGs and the post-2015 development agenda. The UN General Assembly is very hard at work on that challenge, analyzing the problems during 2013–2014 in a process known as the Open Working Group, ably co-chaired by Hungary and Kenya; the process will continue in 2015 with intergovernmental negotiations. World leaders will adopt the actual SDGs at a special summit of the UN General Assembly in September 2015. The UN Secretary-General Ban Ki-moon will make his recommendations to the world leaders in a special report by the end of 2014.

The SDSN has proposed ten SDGs (SDSN 2013a, 28-31). Each of the ten goals has three associated specific targets, and even more (usually around ten) specific numerical indicators to track progress on the goals and targets. The SDSN proposal of the ten SDGs illustrates the potential power, range, and orientation that the SDGs can offer the world for the years 2016–2030.

Here are the ten SDGs as recommended by the UN SDSN.

SDG 1: *End extreme poverty, including hunger.* The more specific goal is to end extreme poverty in all its forms; in other words, to complete the MDGs including hunger, child stunting, malnutrition, and food insecurity, and give special support to highly vulnerable countries. The World Bank leadership voted in 2013 to take on this specific objective, specifically for the Bank to contribute to ending extreme poverty by the year 2030. The overriding idea that ending extreme poverty in all its forms can actually be accomplished by our generation is becoming official policy.

SDG 2: *Achieve economic development within planetary boundaries.* This goal means all countries have a right to economic development as long as that development respects planetary boundaries, ensures sustainable production and consumption patterns, and helps to stabilize the global population by midcentury. The idea of SDG 2 (as recommended by the SDSN) is to give support to continued economic growth, especially in the developing countries, but only growth that is environmentally sustainable within the planetary boundaries. This will require huge changes in the ways we use and produce energy, grow food, design and build cities, and so forth.

SDG 3: *Ensure effective learning for all children and for youth for their lives and their livelihoods.* This education goal is stated as "effective learning," meaning children should be enabled to develop the skills they need to be productive, to be fulfilled in their lives, to be good citizens, and to be able to find decent jobs. As technology changes, the pathways to decent work also require decent skills and good education. Part of effective learning will include greater attention to early childhood development (ages 0–6), when key brain development occurs.

SDG 4: *Achieve gender equality, social inclusion, and human rights for all.* Sustainable development rests on the core dimensions of justice, fairness, social inclusion, and social mobility. Discrimination is a huge and persistent barrier to full participation in economic life and to life satisfaction. This goal will also direct the world's attention to excessive inequality of income and wealth and to the concept of "relative poverty," meaning a situation in which households are not in extreme poverty, but are still too poor to be part of the dignified life of the society.

SDG 5: *Achieve health and wellbeing at all ages.* The subtitle of this SDG is to achieve universal health coverage at every stage of life with particular emphasis on primary health services, including reproductive health, to ensure that all people receive quality health services without suffering financial hardship. All countries will also be called upon to promote policies to help individuals make healthy and sustainable decisions regarding diet, physical activity, and other individual or social dimensions of health. With proper organization, it is possible to reduce child and maternal mortality dramatically, to raise life expectancy, and to control many diseases at very low cost.

SDG 6: *Improve agricultural systems and raise rural productivity.* This goal calls on all countries to improve farming practices, rural infrastructure, and access to resources for food production to increase the productivity of agriculture, livestock, and fisheries; raise smallholder incomes; reduce environmental impacts; promote rural prosperity; and ensure resilience to climate change. Smallholder farmers face many challenges. There are the problems of freshwater depletion, the impacts of climate change, and the need to create new technology- and information-based systems that help raise the most impoverished of these families out of poverty and ensure that farm systems are more productive and resilient. At the same time, existing farm practices lead to the loss of biodiversity, groundwater depletion, excessive fluxes of nitrogen and phosphorus, chemical pollution, and other harms. Sustainable Development Goal 6 recognizes the centrality of sustainable agriculture and, as part of that, the sustainability of the food supply.

SDG 7: *Empower inclusive, productive, and resilient cities.* The goal is to make all cities socially inclusive, economically productive, environmentally sustainable, and secure and resilient to climate change and other risks. Success in SDG 7 will require new forms of participatory, accountable, and effective city governance to support rapid and equitable urban transformation.

SDG 8: *Curb human-induced climate change and ensure sustainable energy.* The aim is to curb greenhouse gas emissions from the energy industry, agriculture, the built environment, and land-use change to ensure a peak of global CO_2 emissions in the coming years and to head off the rapidly growing dangers of climate change; and to promote sustainable energy for all. The world will need to cut greenhouse gas emissions approximately by half by 2050, even as the world economy grows perhaps threefold between now and then. Success requires that the world decarbonize the energy system while also ensuring that electricity and modern energy services are available for all. Meeting this challenge will of course require a much faster transition to low-carbon energy than we have achieved to date.

SDG 9: *Secure ecosystem services and biodiversity and ensure good management of water and other natural resources.* Biodiversity and marine and terrestrial ecosystems of local, regional, and global significance should be measured, managed, and monitored to ensure the continuation of resilient and adaptive life support systems that support sustainable development. Water and other natural resources should be managed sustainably and transparently to support inclusive economic and human development.

SDG 10: *Transform governance for sustainable development.* The public sector, business, and other stakeholders should commit to good governance. Good governance for sustainable development includes transparency, accountability, access to information, participation, an end to tax havens, and efforts to stamp out corruption. The international rules governing international finance, trade, corporate reporting, technology, and intellectual property should be made consistent with achieving the SDGs. The financing of poverty reduction and global public goods, including efforts to head off climate change, should be strengthened and based on a graduated set of global rights and responsibilities.

These are the ten recommended SDGs of the UN SDSN. Taken as a whole package, these goals are meant to orient the world in clear, specific, measurable, concise, and understandable ways, to help the world to make the shift from the business as usual and increasingly dangerous course to a new trajectory of sustainable development. It is now up to the world's governments, following the mandate given at the Rio+20 summit, to choose the new SDGs by September 2015 and to set an operational agenda on how to implement the new goals.

II. Goal-Based Development

The world's governments have called for SDGs, which are currently being negotiated in the UN General Assembly. But will these new goals really make a difference? Will a new set of goals somehow help the world to do what it has not yet been able to do, to shift from a dangerous business as usual path to a path of true sustainable development? Can UN goals actually make a difference?

The evidence from the MDGs is powerful and encouraging. In September 2000, the UN General Assembly adopted the "Millennium Declaration," which included the MDGs. Those eight goals became the centerpiece of the development effort for poor countries around the world. Did they really make a difference? The answer seems to be yes. There has been a marked acceleration of poverty reduction, disease control, and increased access to schooling and infrastructure in the poorest countries of the world, and especially in Africa, as a result of the MDGs. They helped to organize a global effort.

How did they do this? Why do goals matter? There are many answers to this question. First, goals are critical for social mobilization. The world needs to be oriented in a direction to fight poverty or to help achieve sustainable development, but it is very hard in our noisy, disparate, divided, crowded, congested, distracted, and often overwhelmed world to mobilize any consistency of effort to achieve any of our common purposes. Stating goals helps individuals, organizations, and governments all over the world to agree on the direction.

A second aspect of global goals is peer pressure. After the MDGs were adopted, even if governments were not necessarily initially inclined to take on this effort, they knew that progress (or lack thereof) was going to be reported. Each country would be compared with others. Peer pressure came in when leaders were publicly and privately questioned on their progress and the steps they were taking to achieve the MDGs. That kind of dynamic has been absolutely real and effective.

A third way that goals matter is in mobilizing *epistemic communities*. Epistemic communities (or "knowledge communities") are networks of expertise, knowledge, and practice around specific challenges like growing food, fighting diseases, or designing and implementing city plans. When goals are set, those communities of knowledge and practice come together to recommend practical pathways to achieve results. I have watched how the goal to fight malaria, for example, has helped to organize and mobilize the world's leading malariologists. As a group, these experts recommended practical steps to fight the disease, and those recommendations have worked. The role of epistemic communities is extremely important, because governments by themselves do not have the expertise that exists to guide action. The expert-knowledge communities can make critical recommendations of what actually to do, such as the recommendations made by the UN Millennium Project.

And finally, goals not only mobilize knowledge networks, but they also mobilize stakeholder networks. Community leaders, politicians, government ministries, the scientific community, leading nongovernmental organizations, religious groups, international organizations, donor organizations, and foundations are all constituents that need to be pulled together. That kind of multistakeholder process is essential for the complex challenges of sustainable development and for the fight against poverty, hunger, and disease. That has happened by conscious design in area after area of the MDGs. It is one of the clearest ways that the mere statement of goals leads to improved outcomes by bringing together these multistakeholder processes.

No one has ever put the case for goal-based success better than John F. Kennedy did fifty years ago in one of the greatest speeches of the modern U.S. presidency. In his famous peace speech in June 1963, President Kennedy (figure 14.3) said: "By defining our goal more clearly, by making it seem more manageable and less remote, we can help all people to see it, to draw hope from it and to move irresistibly towards it" (Kennedy 1963c). This is the essence of the importance of goal setting.

14.3 President Kennedy giving the peace speech at American University (June 10, 1963)

Commencement Address at American University. Cecil W. Stoughton, John F. Kennedy Presidential Library and Museum.

What have been the accomplishments and weaknesses of the MDGs? Probably the biggest accomplishments have been in the area of public health. Three out of the eight MDGs are about health: reducing child mortality and maternal mortality and controlling the epidemic communicable diseases. In all of those cases, the MDGs have made a very big difference. Why were they so effective in those areas? First, the health MDGs were stated in terms of specific, quantifiable targets, so the progress and outcomes could be measured and assessed.

Second, these health MDGs seemed more manageable and less remote because the epistemic communities helped to map the pathways to achieve them. Many organizations in academia, private foundations (such as the Bill and Melinda Gates Foundation), businesses, and international agencies worked together to develop and disseminate new technologies and business models for success.

Third, there were specific funding mechanisms attached to achieve the health MDGs. Most important was the arrival of the Global Fund to Fight AIDS, Tuberculosis and Malaria (GFATM), which was established in 2001, just one year after the MDGs were adopted, and put into motion in 2002. There is no doubt that the MDGs made a very big difference in giving birth to the GFATM. Many individual countries, led by the United States, also created new national efforts motivated by the MDGs. The U.S. government adopted the President's Emergency Plan for AIDS Relief (PEPFAR) in 2003 and put billions of dollars into the fight against AIDS in poor countries. In 2005, the U.S. government adopted the President's Malaria Initiative (PMI). Both PEPFAR and PMI have played major roles in fighting these diseases. In many countries, the increase in donor funding was accompanied by an increase of domestic financial resources as well, as governments were encouraged to spend more of their own limited funds on high priorities such as disease control.

Finally, the health MDGs succeeded in those areas because of monitoring, measurement, evaluation, and feedback to program design.

The MDGs did not work as well in certain other areas. In areas like sanitation, the achievement of the MDGs is lagging. There has not been a global fund for clean water and sanitation along the same lines of the GFATM. International funding was not elaborated and increased; goals were not made to seem more manageable and less remote through detailed plans of action; the epistemic communities did not organize quite as well; and the political leaders were more neglectful. Education, surprisingly, was also not quite as high on the global agenda and not as well financed as global public health. As a result, the gains in health have outpaced the gains in education. The agriculture and hunger MDGs are also a bit further behind; while agricultural progress has been real, the changes in global policies and politics have not been as dramatic as for global public health.

What these cases show is that goals can make a galvanizing difference, but that there is nothing inevitable about achieving large-scale results after stating a goal. Stating a goal is merely the first step of implementing a plan of action. There must be good policy design to implement that program of action. There must be new financing. There must be new institutions (such as the GFATM) to help implement that goal. And when the outcomes occur, they must be measured, and strategies must be rethought and adapted in a continuing loop of policy feedback, all under the pressures and the motivations of a set of goals and a clear set of timetables.

The sustainable development agenda is of course even bigger and harder than the MDGs, which themselves are no small challenges. Sustainable Development Goals will include not only the continuation of the fight against extreme poverty but also the integration of that goal with several others, including social inclusion and environmental sustainability. The set of challenges is therefore even more complex, and there are at least two aspects of sustainable development that make the problems even more complicated. First, the natural time horizon for results is longer term. Decarbonizing the energy system is going to require a thirty- to forty-year effort, even if we move as fast as we can. Making our cities more resilient will certainly take a couple of decades and more. There is a tremendous need to look forward at least twenty to thirty years in setting the kinds of frameworks needed to actually achieve SDGs. Second, the goals this time will be universal, requiring the buy-in and action of all parts of the world, rich and poor.

There are two specific tools that will be important for translating SDGs into reality. The first of those tools is called *backcasting*. Rather than forecasting (or guessing!) what will happen in 2040 or 2050, one sets the target for a certain date and then analyzes the problem from the target to the present (backward in time) in order to chart a course between today and the future goal. Backcasting is about asking: How can we get from here to there?

The second closely related tool is *technology road-mapping*. Road-mapping asks deep questions about the pathway from today to the future goal. What does the policy terrain really look like? What are the big challenges? What are the technological barriers to overcome between now and 2030 or other future dates? On the pathway of sustainable development, we may have to cross a technological mountain range. Obstacles to overcome in the energy sector, for example, will include the intermittency of renewable power; the challenge of storing wind and solar power; the relatively high costs of certain alternative energy possibilities; and the need to create new power grids to connect major low-carbon energy sources with major population centers, such as connecting the wind power of North Dakota with the energy needs of the Eastern Seaboard of the United States. In other words, we also need a road map for passing through difficult terrain.

To some people, especially those steeped in free-market economics, the combination of goals, backcasting, and road maps might seem to be a kind of planning that is impossible and even contrary to the market process. That is not correct. In sophisticated high-technology industries, exactly this kind of road-mapping is the norm. A recent successful example of this is a great motivator: the road-mapping of Moore's Law.

Moore's Law, shown in Figure 14.4, is the information technology principle that the capacity of integrated circuits is doubling roughly every eighteen to twenty-four months. Gordon Moore first outlined this law in 1965 when he was the CEO of Intel, the leading company in integrated circuits. He noted at the time that for the past ten years or so, the number of transistors that could be packed into an integrated circuit was doubling roughly every eighteen months to two years. As these integrated circuits became more powerful, they were commensurately improving the ability to process, store, and transmit data (as bits and bytes). This tremendous increase of the information capacity of central-processing units, storage, and the ability to transmit enormous volumes of information has transformed our world. Since 1965, we have had roughly a billionfold improvement in the capacity to process, store, and transmit information. We therefore live in a world of interconnectedness of information that was unimaginable at the time that Moore first enunciated this principle. Intel's newest chip in 2013, the XEON

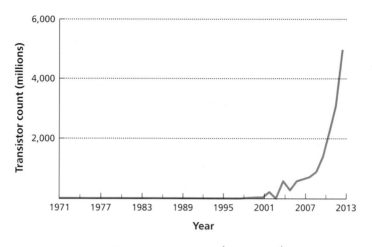

14.4 Transistor count on Intel microprocessors (1971–2012)

Source: Intel.

PHI, now has 5 billion transistors on the integrated circuit. Moore's Law has delivered a startling, indeed a world-changing, billionfold information revolution.

But how was it achieved? How did the industry succeed? It succeeded through a combination of the individual genius of engineers and scientists, industry competition, and also an industry-wide road map. One aspect of that road-mapping is a formal process called the *International Technology Roadmap for Semiconductors*, in which leaders in the industry get together and map out the steps ahead needed to ensure the continuation of Moore's Law in the upcoming decades. They have done this brilliantly and successfully. Their success has transformed our world and is also a lesson to the skeptics about looking ahead. It may not be possible to know the future, but it is certainly important to plan on how to overcome certain obstacles, to think ahead, to be prepared for the problems that will come, and to take measures to move in the target direction in the upcoming decades.

The final point that will be absolutely crucial is that like the MDGs, the SDGs will be a multistakeholder process. There will need to be hands-on effort and engagement across the public sector, the private sector, civil society, governments, individuals, academia, research centers, foundations, and so on. People from every part of society must be involved. The planetary boundaries are already so pressing that speed of action is essential, and the multistakeholder approach that mobilizes all parts of society is also vital. That is a big roundtable, a lot of discussion, a lot of harmonization of different approaches; some for profit, some not for profit, some basic science, some very applied decision making, some on-the-ground in communities all over the world, some training and educating the leaders of the future. But all of these interconnected parts of a global social network will need to be brought together for effective problem solving, decision making, and implementation of the SDGs.

III. Financing for Sustainable Development

Achieving the SDGs will require a lot of new investment: new infrastructure in water, energy, and transport; new educational systems; new health care; and other critical areas. Like anything that concerns building for the future, investing in the future—in people, technologies, infrastructure, and natural capital—is at the very heart of achieving the SDGs. Yet who will pay for the SDGs? How will they be achieved through effective financing?

In the end, we are all going to pay in some sense, because as citizens and consumers we have to pay for the goods and services that are part of our lives. We pay for them in one of two fundamentally different ways. One is that we participate in markets as consumers and suppliers. In the market sector of our economy, the interaction of supply and demand generates the economic activity and provides the motivation for financing. Businesses build factories because they anticipate they will make profits from those factories.

The other way that we buy the things that we need is by paying taxes as citizens, so governments can provide public services such as building roads; providing health care, public education, or fire and police services; and funding the scientific research that underpins technological change. In this sense, all of us will pay for sustainable development in both ways: through markets and through political institutions. But tremendous and sometimes very bitter fights ensue over the proper balancing of financing between the market-oriented, profit-seeking investments of business driven by their sales of goods to consumers, and the financing that comes through the public sector. The free-market advocates argue that markets will be more efficient than governments; whereas advocates of public leadership argue that markets are not making the necessary investments and providing the necessary services, so a public approach is needed.

In fact, these different kinds of financing are complementary mechanisms to finance sustainable development objectives. Both public and private (and philanthropic) approaches are needed. An analysis of where the right boundaries are needs to be at the core of our analytical understanding of these issues. What are the most effective ways to allocate responsibility for financing between market-based private sector financing and public financing?

There are cases in which the market has done brilliantly almost on its own. The greatest example of this is the massive expansion of mobile telephony to all parts of the world. In just roughly twenty-five years, the number of mobile subscribers has increased from a few tens of millions of subscribers around 1990 to around 7 billion subscribers today, including many of the world's poorest people. This massive scale-up was not on the basis of a government program. It was accomplished overwhelmingly by private telecommunications companies looking for profit and making investments in their base stations (for mobile transmission) and fiber-optic lines and by consumers buying the phones and the access to connectivity. This is a wonderful example in which private markets have done the job. Of course, we must keep clearly in mind that the underlying technologies that have made the global connectivity revolution possible started with the basic sciences—including solid-state physics and quantum mechanics—and then applied engineering, much of which was originally financed by the public sector and then taken up by the private sector. Indeed, some of the roots of this industry, like so many others, occurred during World War II, when the U.S. government financed science and engineering in support of the war effort.

Other kinds of critical activities in no way had the kind of dynamics of mobile phones. More than a decade ago, markets plus subsidies were the preferred mode of organization for financing malaria control, but the uptake of malaria control (e.g., bed nets) was still very small. When companies producing insecticide-treated bed nets tried to market those bed nets to the public, they found that the poorest people absolutely were not customers for the nets even though they needed those nets to stay alive. People in rural areas were so impoverished they could not afford even the very cheap nets they needed to keep themselves alive! And when it came to organizing the delivery of malaria control systems, including Community Health Workers, linkages to local clinics, malaria diagnostic services, outreach, availability, and access to the right kinds of medicines, the private sector approach simply was not delivering. The new systems did not emerge on a for-profit basis.

Then came the GFATM and the PMI. These programs mobilized increased public financing both internationally from donor countries and from the malaria-stricken countries themselves. Much more comprehensive measures could then be introduced as a provision of social services instead of as a profit-based activity. Businesses then played a key role. They manufactured high-quality, long-lasting insecticide-treated bed nets. They produced antimalaria medicines in large amounts. They devised and produced new rapid diagnostic tests that could be used in the community rather than in the laboratory. Yet these innovations depended on public financing like the GFATM and the PMI to buy the bed nets, medicines, and diagnostic kits, and to fund the Community Health Workers to utilize the new technologies.

Economics teaches us a lot about where the right boundaries are. There are a few crucial reasons why the private sector approach, which would ideally be the universal one if it actually solved problems, does not solve many critical problems in particular and important cases. The first case is when the challenge is fighting extreme poverty. Markets are basically designed to *ignore* the poor, as they are generally not good customers. But when it is a question of access to health care, for example, the poor can die as a consequence of this lack of market access. This is where the concept of "merit goods" comes in. There are areas of our economic life—health, education, and other areas—where government should provide services whether people can pay for them or not, because these are meritorious goods that should be universally accessible. Public financing is essential to ensure the poor have access to merit goods.

Public financing is also essential in areas where it is hard to recoup the returns on an investment in a direct cash sense. Consider investing in basic science. Scientific knowledge is freely available and the returns to science come via broad societal improvement, but the scientist does not patent the basic forces of nature uncovered in her or his research. Science requires public financing because the profit motive by itself will not be sufficient. Fortunately, many countries have recognized this and support the sciences. As we want to speed research and development of new low-carbon energy sources, public financing for the research and development of low-carbon energy will be absolutely essential.

Public finance is also very important for social insurance—when people are left unemployed by shifts of global markets or when they succumb to other kinds of hardships that cannot be insured effectively in private markets, government can be there as a kind of social safety net.

These are three of many reasons why public financing will be crucial for the SDGs. International help from taxpayers from the high-income countries will also play a vital role in helping poor countries that do not have an adequate tax base to meet the SDGs. There are many areas where the private sector is the natural way to finance advances. Private companies will most likely have the lead in building and operating large-scale energy systems in the future. But in order to give the right incentives for the private sector, we will need proper price signals (sometimes called "corrective pricing"). For example, a carbon tax will be advisable to shift the investments of private utility companies toward wind and solar power (shown in figure 14.5) as opposed to coal-fired power plants. Even when the financing is strictly within the private sector, a proper regulatory framework and corrective measures are very important to make sure that the private sector is investing in the right areas and is driven by market signals that are giving accurate indicators of overall social costs and social benefits.

There has been an often intense and ongoing debate about one aspect of public finance—when taxpayers in one country help to provide public services in another country through aid, also known as official development assistance (ODA). There are harsh critics of foreign aid, while others, including myself, have argued that a taxpayer-financed approach is vital, lifesaving, and crucial for organizations like the GFATM, PEPFAR, or other international development efforts aimed at helping very poor people. The debate about ODA has been running pretty strongly during the whole period of the MDGs, because MDG 8 calls for increased ODA.

14.5 Solar panels at Bear River Migratory Bird Refuge (Utah)

Solar Panels at Bear River Migratory Bird Refuge, Jason St. Sauver / USFWS, Flickr, CC BY 2.0.

To break down this debate, one line of thinking, in which I would include my own views, says aid can be useful and indeed vital in certain circumstances. Yet to be successful, such aid needs to be well targeted and well managed. The critics come in a variety of areas. One camp argues that aid is simply unnecessary and markets are always the solution. A second camp of aid skeptics believes that aid is inevitably wasted. A third argument says that aid is absolutely debilitating, not just wasted; that it leads to a kind of dependency mentality that is demeaning and diminishes motivation. And some people just want to be left alone regardless of what is happening abroad. These camps have been arguing for quite a while. I would like to give a brief explanation of why I believe aid is important.

I agree with the aid skeptics that much aid is wasted. There are cases called "aid" that are little more than shoving money in the pockets of warlords. This is not aid, at least not development aid. It may play a role in foreign policy (though I doubt it). Yet it is certainly not the kind of aid that I support to achieve development objectives.

My argument is somewhat different. My argument is that *aid can work and that it is vital in certain circumstances*. It is especially vital when people are very poor and facing life-or-death challenges, such as malaria, AIDS, safe childbirth, safe water, sanitation, or growing enough food to stay alive. Markets cannot meet the needs of the very poor. The desperately poor are not consumers who will create an immediate profit. Nor can the urgent needs be met through the very meager budgets of the poor countries. And so the poor need help through other means. Then the question is whether it is possible to help effectively through international aid, without too much corruption, theft, and debilitating bureaucratic inefficiencies. My answer is yes, if done thoughtfully (Sachs 2014).

We certainly need to think hard to design effective aid-delivery systems. My favorite examples of this are the GFATM and the Global Alliance for Vaccines and Immunizations (GAVI). Both of these new organizations (created around the time of the MDGs) pool the aid from the richer countries, have an expert process of evaluating proposals for the use of the aid by recipient countries, and then monitor very closely whether that aid is effective. To my mind, that exemplifies the ways that aid can work. Pool the resources of many donors into a common pot so the process is streamlined. Put the aid under careful scrutiny with expert review and with careful quantitative monitoring and evaluation. Have a plan of action that a certain amount of money is going to be used to procure bed nets, antimalaria medicines, diagnostics, emergency obstetrical equipment, and so forth. And then closely monitor what happens afterward.

The evidence is quite strong that this kind of aid has worked very well. While there were huge skeptics at their founding, the GFATM and GAVI have both delivered. We have seen the results on the ground. There have been cases of corruption, but because the system is closely monitored, there are quantified targets, and there is follow-up, so that even when there is money stolen (as will happen in any human system), it is possible to stop, check, correct, and move forward. That kind of feedback system is essential.

I believe we need more of these pooling mechanisms, more global funds for health, for education, for safe water and sanitation, for smallholder farming, for access for all to low-carbon energy sources, and for mechanisms to protect biodiversity. We have a Global Environment Facility, which was established under the

CBD, a major step forward. Rather than lumping together all types of policies and programs and claiming they are doomed to fail, we need to clarify and understand the directions in which success is very likely. We also need innovation in financing and more public-private partnerships, wherein private and public financing are linked in a package to the right kinds of regulatory steps. We will need public-private partnerships for clean energy and new infrastructure; we have already had successful ones for technology development, such as antiretroviral drugs to fight AIDS.

The effectiveness of ODA and public finance in general requires a serious process of planning, backcasting, road-mapping, monitoring, evaluation, and strategy updates. With this efficacious system in place, the incremental costs of meeting the SDGs are probably in a global cost range of about 1-2 percentage points of world output per year for the kinds of global transformations needed, including public financing plus ODA plus public-private partnerships. (This is a very rough estimate, in need of refinement in the coming year or two.) This level of financial effort will not break the bank. But it will not happen through market forces alone. A clear, effective strategy of official financing of national budgets for national needs, and of ODA for the poorest countries in particular, will play a vital role in the success of the SDGs.

IV. Principles of Good Governance

I see four major dimensions of sustainable development. There are the traditional three—economic development, social inclusion, and environmental sustainability. But those three require in all cases the underpinning of a fourth dimension: good governance. Good governance will play a central role in the eventual success or failure of the SDGs, so it behooves us to think clearly about what we really mean by the term. Governance is about the rules of behavior, especially in organizations. It is not only about our politics and government, but also about major organizations that are key actors in sustainable development, including our private corporations. Good governance encompasses both the public sector and the private sector, and especially the large multinational corporations in the private sector.

Clearly there are many types of governments and sets of governing principles around the world, so it would be unworkable to impose one set of political rules to dictate implementation of the SDGs. Rather than universal prescriptions, there can be certain shared principles of governance for the public and private sectors.

The first is *accountability*. Governments and businesses need to be accountable for their actions. Businesses are in part accountable to markets, but they are also accountable to the court of law, and they need to be accountable to the court of public opinion as well. Governments are accountable to their citizens in democratic elections, but they need to be accountable even in nondemocratic systems. By accountability, I do not mean a specific set of election rules, though some are better than others; but rather the idea that governments will adopt goals and be responsible for following through on outlining the measures needed to achieve those goals, to report on them, and to provide public assessments of progress toward those goals. This should be the case across all political systems.

That requires a second feature that also transcends a specific government or corporate organization: *transparency*. We as citizens, as market participants, and as fellow human beings intent on achieving sustainable development, can only hold government and business accountable for their actions if we know those actions and behaviors. This means that we must press our powerful institutions to resist secrecy, including the institutionalization of secrecy in the form of tax havens and "secrecy havens" around the world that allow people to hide their money and behavior, even when this behavior hugely impacts the global goals of ending poverty and saving the planet. Governments of all different political systems have a responsibility of transparency.

A third key tenet is *participation*: the ability of citizens and of stakeholders vis-à-vis business to participate in decision making. There are, of course, many different views about this and many different ways of participating. Elections are a kind of participation, but they must not be the only kind. The ability to participate through public discourse, through public deliberations, and through hearings on regulation are all extremely important. Businesses similarly need to engage their stakeholders through institutional means and clear processes not only with the shareholders but also with the workforce, the suppliers, and the consumers. Good businesses always have a multistakeholder approach.

A fourth aspect of good governance that falls under accountability is the polluter pays principle, which says we all need to clean up after ourselves. Whenever we as individual consumers or as parts of companies are imposing costs on others not reflected in market prices, such as when companies pollute the waterways or air, we need to bear that cost. Economists call this "internalizing the externality," meaning that companies and consumers need to bear the full social costs of their actions.

This raises the question of corporate responsibility. For example, is it "right" for a business operating in a poor country with weak environmental standards to pollute in that country, even if it is technically not against the law? Some extremist views would say that it is actually the *responsibility* of the company to pollute if it is not illegal, in order to maximize shareholder profit. To my mind, this is a clearly mistaken view. We should instead insist that companies desist from creating external damages (such as pollution) if those actions are technically legal. My view is closely related to an ancient and very important doctrine in Latin known as *primum non nocere*, which means, "First, do no harm." A principle of good governance in my view is first, do no harm. Even if the law for whatever reason allows an enterprise to impose costs on others, it is the company's responsibility not to do so, because our higher responsibility is an ethical responsibility to do no harm.

Finally, I would say good governance includes a clear affirmative commitment to sustainable development. Governments have a responsibility to the planetary needs. It is not feasible or good enough in our interconnected world for politicians to deny a responsibility beyond their narrow constituency. Good governance is also a responsibility toward a sense of universal commitment and universal participation in sustainable development.

We are not there yet. The process of elaborating SDGs until 2015 and then their implementation beyond will be a vital opportunity for improving global governance. If these basic principles of accountability, transparency, participation, the polluter pays principle, and commitment to sustainable development are universally adopted in some form, I believe that we can make important headway. Governments can work far more effectively together and businesses can play a responsible role. They can pursue sustainable development rather than spend profits doing more harm, such as in anti–climate change propaganda or corporate lobbying. As we move to the good governance needed to underpin the SDGs, we will need that kind of good governance and responsible leadership in both our public and our private sectors.

V. Is Sustainable Development Feasible?

Is sustainable development feasible? Can we elaborate SDGs and carry them out in time? In our confusing, confused, and distracted world, we are running powerfully off course in many ways—climate change, the sixth great extinction, cities in danger, food supplies under threat, massive dislocations, widening inequalities of income, high youth unemployment, broken politics. Is it even conceivable that we can get back on course? This is a very deep and real concern.

Some of the most important thinkers in the world have expressed some very serious doubts. Three authors whom I tremendously admire have all recently made me shudder with their pessimism. Jane Jacobs, who was one of the world's greatest urbanists and champion of vibrant sustainable urban areas, in her last years of life wrote a book called *Dark Age Ahead*. It was a troubling book to read from such a wonderful thinker. She argued that not only are we on the wrong track but that the tendencies will continue to worsen. Communities are fraying; public spirit is disappearing; there is dysfunctional higher education; governments are responsive to vested interests rather than real needs; and we have a culture that distracts us from the central challenges. Also, who could rival the pessimism of the title of the great astronomer Lord Martin Rees's book *Our Final Century: Will the Human Race Survive the Twenty-First Century?* The title says it all—Lord Rees argues that there is a way out, but that our circumstances are extraordinarily dangerous. And the great ecologist, the pioneer of the interconnectedness of global ecosystems as well as inventor of the Gaia theory, James Lovelock, declared in his recent book *The Revenge of Gaia* that we have already passed the planetary safety margins and major parts of the world are doomed to disaster. He said afterward that perhaps he was too pessimistic, but we still are making a serious mistake if we are glib about the road ahead.

Simply speaking, sustainable development is the greatest, most complicated challenge humanity has ever faced. Climate change alone is extraordinarily difficult, but then add in these other challenges of a rapidly urbanizing world, a great extinction process underway due to human domination of ecosystems, increasing population, overextraction from oceans and land resources, massive illegal trade, and all the other issues already discussed. These are complex problems, and are science-based issues without the necessary worldwide public literacy in the scientific underpinnings. These are issues of tremendous uncertainty in chaotic, nonlinear, complex systems. This is a multigenerational problem that we are unequipped by tradition to think about. It goes to core areas of our economic life like energy, transport, infrastructure, and food supply, all of which need major technological overhauls. There are powerful vested interests like Big Oil that have hindered clarity and progress on implementation. There are long lead times in rebuilding our infrastructure because infrastructure has such a long life expectancy, 50–100 years or more. And we have very limited space left, partly because we have in a way frittered away the last twenty-two years since the Rio Earth Summit, even though we had been on notice decades earlier.

We must not give up hope. We have identified very specific ways through our backcasting and our road-mapping of how we can get from here to where we need to be. We have identified technologies that can decarbonize the energy system and lead to tremendous energy efficiency. We have identified technologies that can economize tremendously on land, raise agricultural productivity, and reduce the fluxes of nitrogen and phosphorous and their poisoning of the estuaries. We have shown how cities can plan ahead and design smart infrastructure. These are opportunities within our grasp, not fanciful science fiction, but things that we know how to do where the costs are absolutely within reach. In many cases, as with wind power and solar power, the costs are already close to the traditional technologies, at least in some favored regions of the world.

We can see how we could succeed with the SDGs, just as the world has made tremendous progress with the MDGs. I believe that despite the cynicism, the darkness, the confusion, and the miserable politics on many of these issues, we can make a breakthrough. Even though it looks as if the political systems are unresponsive, things can change. The most important message I would send is that ideas count. They can have an effect on public policy far beyond anything that can be imagined by the hard-bitten cynics.

Ideas have been transformative throughout history and have sparked some of the greatest transformational movements of the last two centuries (the time of our modern economic growth). First consider the end of slavery. The outlawing of slavery in the British Empire was the result of a massive social movement, the first of its kind in modern history. In the late eighteenth century and the early nineteenth century, English leaders such as William Wilberforce, Thomas Clarkson, Granville Sharp, Charles James Fox, and William Pitt the Younger took on the deeply economically embedded institution of slavery. It took a few decades in the face of much cynicism and dirty dealings, but in 1807 the British Empire abolished the slave trade and in 1833 abolished slavery in the British holdings entirely. This flew against powerful and entrenched British economic interests. In the end, the ideas and morality were the underlying forces of change.

The struggle against European colonial rule, led by Mahatma Gandhi (figure 14.6) and by many of his contemporaries in Africa and in Asia, also at first seemed impossible. One would have bet in 1910 or 1930 that Gandhi would have been long

14.6 Mahatma Gandhi

Gandhi during prayer at Mumbai, September 1944.

forgotten by now and the British Empire would have continued to rule over India and Africa. But of course it is Gandhi's leadership in helping to end colonialism that we regard as the correct moral answer for our age, and it is one that inspired many in the civil rights movement, the human rights movement, and beyond. Ideas played a role so powerfully that the interests and entrenched power structures were in the end completely overwhelmed.

The human rights movement followed, partly led by Eleanor Roosevelt, who championed the UN Universal Declaration of Human Rights. This moral charter is sadly violated massively every day, but the Universal Declaration of Human Rights has changed the world. It has expanded the recognition and reach of human rights, empowering major initiatives like the MDGs that have turned into real results on the ground.

These ideas of course all inspired the civil rights movement. As the great civil rights leader Martin Luther King Jr. said, "The arc of the moral universe is long, but it bends toward justice." Ideas and morality have repeatedly paved the way for great breakthroughs. The women's rights movement, which is playing such a magnificent and crucial role in enabling the world to get on a path of sustainable development, is another idea of our time. It has been literally hundreds of year in the making, but it has had great advances in recent decades, often in the least likely places in the world and largely due to brave crusaders.

This brings us to the key ideas of our own time. The idea that we can end extreme poverty is now an official doctrine of major institutions like the World Bank, and it will likely soon be at the core of a new set of SDGs. The idea of sustainable development is now a worldwide commitment to a safer, more prosperous, and more just planet. There is an underpinning of ethics in all these ideas. When we talk about moving to global SDGs, we are also talking about the need for and possibility of a shared global ethics. It is heartening that many of the world's religious leaders have come together and declared that the world's religions share a common ethical underpinning that could underpin a shared commitment like SDGs, including the Golden Rule; the commitment to "first, do no harm;" and the standards of good governance, including human rights, accountability, transparency, and participation.

It has been a half-century since two great episodes in U.S. history where values changed history: the U.S. civil rights movement and President John F. Kennedy's quest to make peace with the Soviet Union. They both give us inspiration for our challenges today. In 1963 Kennedy succeeded in negotiating the Partial Nuclear Test Ban Treaty with the Soviets, which was a major step back from the nuclear arms race and an easing of the tensions that had brought the two parties to the brink of nuclear war in the 1962 Cuban Missile Crisis. It is astounding and inspiring that Kennedy used ideas and words, not force, to bring about this advance of peace. His words can teach us today about how sustainable development can be achieved.

President Kennedy gave what is called his "Peace Speech" on June 10, 1963 (Kennedy 1963c). It is a speech about values, human rights, and ideas; and the most important idea is that humanity can solve its problems peacefully and can live together, because what we have in common is so much more important than what divides us. Kennedy said:

No problem of human destiny is beyond human beings. Man's reason and spirit have often solved the seemingly unsolvable, and we believe they can do it again. I am not referring to the absolute, infinite concept of universal peace and good will of which some fantasies and fanatics dream. I do not deny the value of hopes and dreams but we merely invite discouragement and incredulity by making that our only and immediate goal. Let us focus instead on a more practical, more attainable peace, based not on a sudden revolution in human nature but on a gradual evolution in human institutions—on a series of concrete actions and effective agreements which are in the interest of all concerned. There is no single, simple key to this peace; no grand or magic formula to be adopted by one or two powers. Genuine peace must be the product of many nations, the sum of many acts. It must be dynamic, not static, changing to meet the challenge of each new generation. For peace is a process—a way of solving problems.

Sustainable development also is a process, a way of solving problems peacefully and globally, using our science and technology, our know-how, and our shared global ethics to address our deep common needs. Kennedy was grappling with the divide between the United States and the Soviet Union, the divide of deep values, political systems, and nuclear arms pointed at each other. But his message was that we have common interests, and can resolve our problems peacefully. He had an absolutely magnificent way of describing those common interests that resonates today:

> So let us not be blind to our differences, but let us also direct attention to our common interests and the means by which those differences can be resolved. And if we cannot end now our differences, at least we can help make the world safe for diversity. For in the final analysis, our most basic common link is that we all inhabit this small planet. We all breathe the same air. We all cherish our children's futures. And we are all mortal.

Yes, we are all today breathing the same air now with 400 parts per million of CO_2; it is a threat to our wellbeing and future survival. We all cherish our children's futures. And we know what needs to be done.

After Kennedy made the Peace Speech, he visited Europe and made a stop in his own family's ancestral country of Ireland. He spoke magnificently in the Irish Parliament (Kennedy 1963b):

> This is an extraordinary country. George Bernard Shaw, speaking as an Irishman, summed up an approach to life: Other people, he said "see things and . . . say 'Why?' . . . But I dream things that never were—and I say: 'Why not?'" It is that quality of the Irish—that remarkable combination of hope, confidence and imagination—that is needed more than ever today. The problems of the world cannot possibly be solved by skeptics or cynics, whose horizons are limited by the obvious realities. We need men who can dream of things that never were, and ask why not.

Kennedy got his peace treaty signed and helped the world step back from the nuclear brink as part of a practical and step-by-step process. We need that kind of practical process again. We will need, as Kennedy said, to look beyond the skeptics and the cynics. They have every reason to point out our difficulties. But we will need to look forward to what needs to be done and to find the pathways to achieve it.

In Kennedy's last speech to world leaders, in the United Nations in the fall of 1963, just after the Partial Nuclear Test Ban Treaty was signed, he ended his speech with absolutely remarkable words that must guide us today: "Archimedes," said Kennedy, "in explaining the principles of the lever was said to have declared to his friends: 'Give me a place where I can stand and I shall move the world.' My fellow inhabitants of this planet: Let us take our stand here in this Assembly of nations. And let us see if we, in our own time, can move the world to a just and lasting peace." (Kennedy 1963a)

Now it's our turn to see if we can move the world toward sustainable development.

BIBLIOGRAPHY

Arrhenius, Svante. 1896. "On the Influence of Carbonic Acid in the Air Upon the Temperature of the Ground." *Philosophical Magazine and Journal of Science* 5(41): 237–276.

Arrow, Kenneth. 1963. "Uncertainty and the Welfare Economics of Medical Care." *American Economic Review* 53(5): 941–973.

Bairoch, Paul, and Gary Goertz. 1985. "Factors of Urbanisation in the Nineteenth Century Developed Countries: A Descriptive and Econometric Analysis." *Urban Studies* 23: 285–305.

Brundtland, Gro Harlem, and World Commission on Environment and Development. 1987. *Our Common Future: Report of the World Commission on Environment and Development*. Oxford: Oxford University.

Cook, Benjamin I., Jason E. Smerdon, Richard Seager, and Sloan Coats. 2014. "Global Warming and 21st Century Drying." *Climate Dynamics*. doi: 10.1007/s00382–014–2075-y.

Corak, Miles. 2009. *Chasing the Same Dream, Climbing Different Ladders: Economic Mobility in the United States and Canada*. Washington, DC: Pew Charitable Trusts Economic Mobility Project.

——. 2013. "Income Inequality, Equality of Opportunity, and Intergenerational Mobility." *Journal of Economic Perspectives* 27(3): 79–102.

Dobermann, Achim, and Rebecca Nelson. 2013. "Opportunities and Solutions for Sustainable Food Production." Background Paper for the High-Level Panel of Eminent Persons on the Post-2015 Development Agenda. New York: Sustainable Development Solutions Network.

Eidlin, Eric. 2010. "What Density Doesn't Tell Us About Sprawl." *ACCESS* 37: 2–9.

Gallup, John, Andrew Mellinger, and Jeffrey D. Sachs. 2000. "Climate, Coastal Proximity, and Development." In *The Oxford Handbook of Economic Geography*, ed. Gordon L. Clark, Maryann P. Feldman, and Meric S. Gertler, 169–194. Oxford: Oxford University Press.

Hansen, James, and Makiko Sato. 2012. "Paleoclimate Implications for Human-Made Climate Change." In *Climate Change: Inferences from Paleoclimate and Regional Aspects*, ed. André Berger, Fedor Mesinger, and Djordjie Šijački, 21–48. Heidelberg: Springer.

Hansen, James, Makiko Sato, and Reto Ruedy. 2012. "Perception of Climate Change." *Proceedings of the National Academy of Sciences* 109(37): E2415–E2423.

Heckman, James J. 2006. "Skill Formation and the Economics of Investing in Disadvantaged Children." *Science* 31(5782): 1900–1902. doi:10.1126/science.1128898.

——. 2008. "Schools, Skills, and Synapses." IZA Discussion Paper No. 3515. http://ftp.iza.org/dp3515.pdf.

Helliwell, John, Richard Layard, and Jeffrey D. Sachs. 2013. *World Happiness Report.* New York: Sustainable Development Solutions Network.

High-Level Panel of Eminent Persons on the Post-2015 Development Agenda. 2013. *A New Global Partnership: Eradicate Poverty and Transform Economies Through Sustainable Development.* New York: United Nations Publications.

International Conference on Primary Health Care, Alma-Ata, USSR, 6-12 September. 1978. "Declaration of Alma-Ata."

Institute of Medicine. 2013. *Best Care at Lower Cost: The Path to Continuously Learning Health Care in America.* Washington, DC: National Academies Press.

International Monetary Fund. 2014. "World Economic Outlook Database." Last Modified April 2014. www.imf.org/external/pubs/ft/weo/2014/01/weodata/index.aspx.

International Union for Conservation of Nature. 1980. *World Conservation Strategy: Living Resource Conservation for Sustainable Development.* Gland, Switzerland: IUCN.

IPCC. 2013. "Summary for Policymakers." In *Climate Change 2013: The Physical Science Basis. Contribution of Working Group I to the Fifth Assessment Report of the Intergovernmental Panel on Climate Change,* ed. T.F. Stocker, D. Qin, G.-K. Plattner, M. Tignor, S.K. Allen, J. Boschung et al. Cambridge: Cambridge University Press.

Kennedy, John F. 1963a. "Address to the UN General Assembly." Speech to UN General Assembly, New York, September 20.

——. 1963b. "Address Before the Irish Parliament." Speech to Irish Parliament, Dublin, June 28.

——. 1963c. "A Strategy of Peace." Commencement address to American University, Washington, DC, June 10.

Keynes, John Maynard. 1920. *The Economic Consequences of the Peace.* Library of Economics and Liberty. Accessed June 26, 2014. http://www.econlib.org/library/YPDBooks/Keynes/kynsCP2.html.

——. 1930. "Economic Possibilities for Our Grandchildren." www.econ.yale.edu/smith/econ116a/keynes1.pdf.

Kiszewski, Anthony, Andrew Mellinger, Andrew Spielman, Pia Malaney, Sonia Ehrlich Sachs, and Jeffrey Sachs. 2004. "A Global Index Representing the Stability of Malaria Transmission." *American Journal of Tropical Medicine and Hygiene* 70(5): 486–498.

Lackner, Klaus S., Sarah Brennan, Jürg M. Matter, A.-H. Alissa Park, Allen Wright, and Bob van der Zwaan. 2012. "The Urgency of the Development of CO_2 Capture from Ambient Air." *Proceedings of the National Academy of Sciences* 109(33): 13156–13162.

Lenzen, Manfred, Dan Moran, Keiichiro Kanemoto, Barney Foran, Leonarda Lobefaro, and Arne Geschke. 2012. "International Trade Drives Biodiversity Threats in Developing Nations." *Nature* 486: 109–112.

Lovelock, James E. 1991. "The Earth as a Living Organism." In *Learning to Listen to the Land*, ed. Bill Willers, 11–16. Washington, DC: Island Press.

Maddison, Angus. 2006. *The World Economy*. Paris: Organization for Economic Co-operation and Development.

Marx, Karl, and Frederick Engels. 1848. *Manifesto of the Communist Party*. https://www.marxists.org /archive/marx/works/download/pdf/Manifesto.pdf.

McCord, Gordon, and Jeffrey Sachs. 2013. "Development, Structure, and Transformation: Some Evidence on Comparative Economic Growth." NBER Working Paper 19512. Washington, DC: National Bureau of Economic Research.

Millennium Ecosystem Assessment. 2005. *Ecosystems and Human Wellbeing: Synthesis*. Washington, DC: Island Press.

Millennium Villages Project. 2011. *The Millennium Villages Project: The Next Five Years: 2011–2015*. New York: Millennium Villages Project.

Muthayya, Sumithra, Jee Hyun Rah, Jonathan D. Sugimoto, Franz F. Roos, Klaus Kraemer, and Robert E. Black. 2013. "The Global Hidden Hunger Indices and Maps: An Advocacy Tool for Action." *PLoS One* 8(6): e67860.

National Center for Education Statistics. 2000. "Postsecondary Attainment." http://nces.ed.gov/ssbr /pages/attainment.asp.

National Scientific Council on the Developing Child and the National Forum on Early Childhood Policy and Programs. 2010. *The Foundations of Lifelong Health Are Built in Early Childhood*. Cambridge, Mass.: Harvard University Center on the Developing Child.

Nellemann, Christian, Monika MacDevette, Ton Manders, Bas Eickhout, Birger Svihus, Anne Gerdien Prins, and Bjørn P. Kaltenborn, eds. 2009. *The Environmental Food Crisis: The Environment's Role in Averting Future Food Crises*. A UNEP Rapid Response Assessment. Arendal, Norway: UNEP/GRID-Arendal.

One Million Community Health Workers Campaign. 2013. *Fact Sheet*. New York: Sustainable Development Solutions Network.

Organization for Economic Co-operation and Development. 2010. "History of the 0.7% ODA Target." www.oecd.org/dac/stats/45539274.pdf.

——. 2011. "Health at a Glance 2011: OECD Indicators." www.oecd-ilibrary.org/sites/health_glance -2011-en/07/01/index.html?itemId=/content/chapter/health_glance-2011-60-en.

——. 2014. "Official Development Assistance—Definition and Coverage." www.oecd.org/dac/stats /officialdevelopmentassistancedefinitionandcoverage.htm#Coverage.

Peters, Glen P., Robbie M. Andrew, Tom Boden, Josep G. Canadell, Philippe Ciais, Corinne Le Quéré, Gregg Marland et al. 2014. "The Challenge to Keep Global Warming Below 2 °C." *Nature Climate Change* 3: 4–6.

PlaNYC. 2009. *Inventory of New York City Greenhouse Gas Emissions: September 2009*. New York: Mayor's Office of Long-Term Planning & Sustainability.

——. 2011. *Update April 2011: A Greener, Greater New York*. New York: Mayor's Office of Long-Term Planning & Sustainability.

———. 2013. *Progress Report 2013: A Greener, Greater New York*. New York: Mayor's Office of Long-Term Planning & Sustainability.

Rockström, Johan, Will Steffen, Kevin Noone, Åsa Persson, F. Stuart Chapin, Eric F. Lambin, Timothy M. Lenton et al. 2009. "A Safe Operating Space for Humanity." *Nature* 461(24): 472–475.

Rockström, Johan, Jeffrey D. Sachs, Marcus C. Öhman, and Guido Schmidt-Traub. 2013. "Sustainable Development and Planetary Boundaries." Background Paper for the High-Level Panel of Eminent Persons on the Post-2015 Development Agenda. New York: Sustainable Development Solutions Network.

Rosegrant, Mark W., Simla Tokgoz, Prapti Bhandary, and Siwa Msangi. 2012. "Scenarios for the Future of Food." In *2012 Global Food Policy Report*, 89–101. Washington, DC: International Food Policy Research Institute.

Rubel, Franz, and Markus Kottek. 2010. "Observed and Projected Climate Shifts 1901–2100 Depicted by World Maps of the Köppen-Geiger Climate Classification." *Meteorologische Zeitschrift* 19: 135–141.

Sachs, Jeffrey D. 2005. *The End of Poverty: Economic Possibilities for Our Time*. New York: Penguin.

———. 2014. "The Case for Aid." *Foreign Policy*, January 21.

Sachs, Jeffrey D., and Guido Schmidt-Traub. 2013. "Financing for Development and Climate Change Post-2015." Background Paper for the High-Level Panel of Eminent Persons on the Post-2015 Development Agenda. New York: Sustainable Development Solutions Network.

Sachs, Jeffrey D., John W. McArthur, Guido Schmidt-Traub, Margaret Kruk, Chandrika Bahadur, Michael Faye, and Gordon McCord. 2004. "Ending Africa's Poverty Trap." *Brookings Papers on Economic Activity* 35(1): 117–240.

Schwartz, Marc, Donna Heimiller, Steve Haymes, and Walt Musial. 2010. *Assessment of Offshore Energy Resources for the United States*. Technical Report NREL/TP-500–45889. Golden, CO: National Renewable Energy Laboratory.

Schweinhart, Lawrence J., Jeanne Montie, Zongping Xiang, W. Steven Barnett, Clive R. Belfield, and Milagros Nores. 2005. *The High/Scope Perry Preschool Study Through Age 40: Summary, Conclusions, and Frequently Asked Questions*. Ypsilanti, Mich.: HighScope.

Scripps Institution of Oceanography. 2014. "The Keeling Curve." Accessed June 4, 2014. http://keeling-curve.ucsd.edu.

Secretariat of the Convention on Biological Diversity. 2010. *Global Biodiversity Outlook 3*. Montreal: Secretariat of the Convention on Biological Diversity.

Shell. 2014. "Prelude FLNG." Accessed June 4, 2014. www.shell.com/global/aboutshell/major-projects-2/prelude-flng.html.

Smith, Adam. 1776. *An Inquiry into the Nature and Causes of the Wealth of Nations*. http://www2.hn.psu.edu/faculty/jmanis/adam-smith/wealth-nations.pdf.

Stern, Nicholas. 2006. *The Stern Review Report: The Economics of Climate Change*. London: HM Treasury.

Squires, David A. 2012. "Explaining High Health Care Spending In the United States: An International Comparison of Supply, Utilization, Prices, and Quality." *Issues in International Health Policy* 10: 1–14.

Sustainable Development Solutions Network. 2012a. *Draft Framework for Sustainable Development*. New York: Sustainable Development Solutions Network.

———. 2012b. "Global Profile of Extreme Poverty." Background Paper for the High-Level Panel of Eminent Persons on the Post-2015 Development Agenda. New York: Sustainable Development Solutions Network.

———. 2013. *An Action Agenda for Sustainable Development.* New York: Sustainable Development Solutions Network.

Sustainable Development Solutions Network Thematic Group on Challenges of Social Inclusion. 2013. *Achieving Gender Equality, Social Inclusion, and Human Rights for All: Challenges and Priorities for the Sustainable Development Agenda.* New York: Sustainable Development Solutions Network.

Sustainable Development Solutions Network Thematic Group on Early Childhood Development, Education and Transition to Work. 2014. *The Future of Our Children: Lifelong, Multi-Generational Learning for Sustainable Development.* New York: Sustainable Development Solutions Network.

Sustainable Development Solutions Network Thematic Group on Health for All. 2014. *Health in the Framework of Sustainable Development: Technical Report for the Post-2015 Development Agenda.* New York: Sustainable Development Solutions Network.

Sustainable Development Solutions Network Thematic Group on Sustainable Agriculture and Food Systems. 2013. *Solutions for Sustainable Agriculture and Food Systems: Technical Report for the Post-2015 Development Agenda.* New York: Sustainable Development Solutions Network.

Sustainable Development Solutions Network Thematic Group on Sustainable Cities. 2013. "The Urban Opportunity: Enabling Transformative and Sustainable Development." Background Paper for the High-Level Panel of Eminent Persons on the Post-2015 Development Agenda. New York: Sustainable Development Solutions Network.

Sutton, Mark A., Clare M. Howard, Jan Willem Erisman, Gilles Billen, Albert Bleeker, Peringe Grennfelt, Hans van Grinsven et al., eds. 2011. *The European Nitrogen Assessment: Sources, Effects and Policy Perspectives.* Cambridge: Cambridge University Press.

Swartz, Wilf, Enric Sala, Sean Tracey, Reg Watson, and Daniel Pauly. 2010. "The Spatial Expansion and Ecological Footprint of Fisheries (1950 to Present)." *PLoS One* 5(12): e15143.

Tollefson, Jeff, and Natasha Gilbert. 2012. "Rio Report Card." *Nature* 468: 20–23.

Transparency International. 2013. "Corruption Perceptions Index 2013." Accessed June 4, 2014. http://cpi.transparency.org/cpi2013/results.

United Nations. 1973. Convention on International Trade in Endangered Species of Wild Fauna and Flora (993 UNTS 243).

———. 1992a. Convention on Biological Diversity (1760 UNTS 79).

———. 1992b. United Nations Framework Convention on Climate Change (1771 UNTS 107).

———. 1994. United Nations Convention to Combat Desertification in Countries Experiencing Serious Drought and/or Desertification, Particularly in Africa (1954 UNTS 3).

———. 1998. Kyoto Protocol to the United Nations Framework Convention on Climate Change. (FCCC/CP/1997/7/Add.1).

United Nations Convention to Combat Desertification. 2013. *The Economics of Desertification, Land Degradation and Drought: Methodologies and Analysis for Decision-Making.* Background Document for UNCCD 2nd Scientific Conference, Bonn, Germany, 9–12 April 2013.

United Nations Department of Economic and Social Affairs Division for Social Policy and Development, Secretariat of the Permanent Forum on Indigenous Issues. 2009. *State of the World's Indigenous Peoples*. New York: United Nations.

United Nations Department of Economic and Social Affairs Population Division (DESA Population Division). 2012. "World Urbanization Prospects: The 2011 Revision." http://esa.un.org/unup /CD-ROM/Urban-Agglomerations.htm.

——. 2013. "World Population Prospects: The 2012 Revision." http://esa.un.org/wpp/Excel-Data /population.htm.

United Nations Development Programme. 2013. *Human Development Report 2013. The Rise of the South: Human Progress in a Diverse World*. New York: United Nations Development Programme.

——. 2013a. "Gender Inequality Index (GII)." Accessed June 4, 2014. http://hdr.undp.org/en/statistics /gii.

——. 2013b. "Human Development Index (HDI)." Accessed June 4, 2014. http://hdr.undp.org/en /statistics/hdi.

United Nations Environment Programme. 2012. *Global Environment Outlook 5: Summary for Policy Makers*. Nairobi: United Nations Environment Programme.

——. 2014. *Assessing Global Land Use: Balancing Consumption with Sustainable Supply. A Report of the Working Group on Land and Soils of the International Resource Panel*. Nairobi: United Nations Environment Programme.

United Nations General Assembly, 3rd Session. 1948. "Universal Declaration of Human Rights." (217 A (III)). New York: United Nations.

——, 21st Session. 1966a. "International Covenant on Civil and Political Rights" (A/RES/21/2200). New York: United Nations.

——, 21st Session. 1966b. "International Covenant on Economic, Social and Cultural Rights" (A/RES /21/2200). New York: United Nations.

——, 55th Session. 2000. "United Nations Millennium Declaration" (A/RES/55/2). New York: United Nations.

——, 66th Session. 2012. "The Future We Want" (A/RES/66/288). New York: United Nations.

United Nations Millennium Project. 2005a. *Halving Hunger: It Can Be Done. Summary Version of the Report of the Task Force on Hunger*. New York: Earth Institute.

——. 2005b. *Investing in Development: A Practical Plan to Achieve the Millennium Development Goals. Overview*. New York: United Nations Development Programme.

United Nations Secretary-General's High-Level Group on Sustainable Energy for All. 2012. *Sustainable Energy for All: A Framework for Action*. New York: United Nations.

United Nations Statistics Division. 2008. *Official List of MDG Indicators*. http://unstats.un.org/unsd /mdg/Host.aspx?Content=Indicators/OfficialList.htm.

Vitousek, Peter M., Harold A. Mooney, Jane Lubchenco, and Jerry M. Melillo. 1997. "Human Domination of Earth's Ecosystems." *Science* 277(5352): 494–499.

Walker, Susan P., Theodore D. Wachs, Sally Grantham-McGregor, Maureen M. Black, Charles A. Nelson, Sandra L. Huffman, Helen Baker-Henningham et al. 2011. "Inequality in Early Childhood: Risk and Protective Factors for Early Child Development." *Lancet* 378(9799): 1325–1338.

Wildlife Conservation Society and Center for International Earth Science Information Network, Columbia University. 2005. *Last of the Wild Project, Version 2, 2005 (LWP-2): Global Human Footprint Dataset (IGHP)*. Palisades, NY: NASA Socioeconomic Data and Applications Center. http://dx.doi.org/10.7927/H4GF0RFQ.

Williams, James H., Andrew DeBenedictis, Rebecca Ghanadan, Amber Mahone, Jack Moore, William R. Morrow III, Snuller Price et al. 2012. "The Technology Path to Deep Greenhouse Gas Emissions Cuts by 2050: The Pivotal Role of Electricity." *Science* 335(6064): 53–59.

Wilson, Edward O. 1984. *Biophilia*. Cambridge, Mass.: Harvard University Press.

World Bank. 2014a. "EdStats: Education Statistics." Accessed June 4, 2014. http://datatopics.worldbank.org/education.

——. 2014b. "World Development Indicators: Maternal Mortality Ratio (Modeled Estimate, per 100,000 Live Births)." Accessed June 4, 2014. http://data.worldbank.org/indicator/SH.DYN.MORT.

——. 2014c. "World Development Indicators: Mortality Rate, Under-5 (per 1,000 Live Births)." Accessed June 4, 2014. http://data.worldbank.org/indicator/SH.DYN.MORT.

——. 2014d. "World Development Indicators: Poverty Headcount Ratio at $1.25 a Day (PPP) (% of Population)." Accessed June 4, 2014. http://data.worldbank.org/topic/poverty.

World Health Organization. 2001. *Macroeconomics and Health: Investing in Health for Economic Development: Report of the Commission on Macroeconomics and Health*. Geneva: World Health Organization.

World Summit on Sustainable Development. 2002. *Plan of Implementation of the World Summit on Sustainable Development*. http://www.un.org/esa/sustdev/documents/WSSD_POI_PD/English/WSSD_PlanImpl.pdf.

INDEX